ASYMPTOTIC METHODS IN THE THEORY OF PLATES WITH MIXED BOUNDARY CONDITIONS

ASYMPTOTIC METHODS IN THE THEORY OF PLATES WITH MIXED BOUNDARY CONDITIONS

Igor V. Andrianov
RWTH Aachen University, Germany

Jan Awrejcewicz
Lodz University of Technology, Poland

Vladyslav V. Danishevs'kyy
Prydniprovska State Academy of Civil Engineering and Architecture, Ukraine

Andrey O. Ivankov
Southtrans Ltd, Russia

WILEY

This edition first published 2014

Registered office
John Wiley & Sons Ltd, The Atrium, Southern Gate, Chichester, West Sussex, PO19 8SQ, United Kingdom

For details of our global editorial offices, for customer services and for information about how to apply for permission to reuse the copyright material in this book please see our website at www.wiley.com.

Library of Congress Cataloging-in-Publication Data
J. Awrejcewicz, I.V. Andrianov.
Asymptotic methods in the theory of plates with mixed boundary conditions / I.V. Andrianov, J. Awrejcewicz, V.V. Danishevs'kyy, A.O. Ivankov.
pages cm
Includes bibliographical references and index.
ISBN 978-1-118-72519-1 (cloth)
1. Plates (Engineering) – Mathematical models. 2. Asymptotic expansions. A.O. Ivankov, V.V. Danishevs'kyy. II. Title.
TA660.P6A97 2014
624.1′7765015114 – dc23

2013034287

A catalogue record for this book is available from the British Library.

ISBN: 978-1-118-72519-1

Typeset in 10/12pt TimesLTStd by Laserwords Private Limited, Chennai, India

1 2014

Contents

Preface

It is evident that plate structural elements are widely used in various branches of engineering. In industrial and civil engineering they serve as covers, working elements and parts of the various foundations; in the machine building they are elements of technological design. The above-mentioned construction members are intended to accommodate various static and dynamic excitations, and their strength, resistance and technical stability require increasing engineering expectations. In real constructions the boundary conditions are usually of a complicated character: free edge, clamping, elastic clamping, as well as various types of mixed boundary conditions. Similar conditions may occur in constructing various supports of different and mixed types. On the other hand, mixed boundary conditions may appear during the linkage of design structural members with the a use of various laps as well as intermittent welding. Furthermore, mixed boundary conditions may appear in supporting a plate beam on a nonsmooth surface. Finally, computation of plates with slits and cracks in many cases may be reduced to the computation of constructions with mixed boundary conditions. It should be emphasized that the computational scheme of a construction can be changed in the exploitation time due to the action of external loads (occurrence of corrosion and cracks, damage of part of a resistance support, etc.). In this case one may also expect a mixed boundary support, which was not predicted by the previous engineering analysis and design.

Nowadays, a wide spectrum of applications devoted to computations of the above-mentioned engineering objects can be solved by FEM (Finite Element Method). In practice, any problem can be solved via application of the appropriately chosen finite elements. However, it should be emphasized that FEM also suffers from a few drawbacks: it is rather difficult to estimate the validity of the FEM obtained results; in many cases instability in the vicinity of points occurs, where boundary conditions undergo changes, etc. This is why from the point of view of theory of plates and shells, as well as engineering practice, analytical approximate methods still play an important role in the study of a wide class of constructions with mixed boundary conditions. It seems that among analytical approaches, the asymptotic ones are most appropriate and successful in solving the problems discussed above.

It has recently been observed that asymptotic approaches again attract a big attention of many scientists in spite of the big development of numerical techniques [1]. The reason is mainly motivated by the intuition development of a researcher/engineer through asymptotic analysis. Even in a case where we are interested only in numerical solutions, a priori asymptotical analysis allows us to choose the most suitable numerical method and sheds light on usually disordered and largely numerically obtained material.

Moreover asymptotic analysis is extremely useful in providing the external value of parameters, where direct numerical computation meets serious difficulties in obtaining reliable results. This aspect of asymptotic methods has been well illustrated by the English scientist D.G. Crighton [2]: "*Design of computational or experimental schemes without the guidance of asymptotic information is wasteful at best, dangerous at worst, because of possible failure to identify crucial (stiff) features of the process and their localization in coordinate and parameter space. Moreover, all experience suggests that asymptotic solutions are useful numerically far beyond their nominal range of validity, and can often be used directly, at least at a preliminary product designs stage, for example, saving the need for accurate computation until the final design stage where many variables have been restricted to narrow ranges.*"

Since asymptotic methods play a key role in our book, the first part (Chapter 1) has been devoted to their description. We mainly rely on examples and avoid unnecessary generalizations. We have aimed to keep the book self-organized and discrete. In other words, the material in this book should be sufficient for the reader without need for supplementary material. In particular, we have focused on asymptotic approaches, which are either not well known or not well reported, such as the method of summation and construction of asymptotically equivalent functions, methods of small and large delta, homotopy perturbations method, etc.

Let us look briefly at the latter mentioned approach, which has recently been very popular. Its main idea is as follows. We introduce the parameter ε into either differential equations or boundary conditions in such a way that for $\varepsilon = 0$ we obtain the boundary value problem allowing us to find a simple solution, whereas for $\varepsilon = 1$ it gives the input boundary value problem. In the next step we apply the splitting method regarding ε, and in the finally obtained solution we put $\varepsilon = 1$. In other words, we apply a certain homotopic transformation. It is clear that this approach is not new, since it has already been successively applied by H. Poincaré [3] and A.M. Liapunov [4]. However, it has rarely been applied for many years because the obtained series are divergent in the majority of cases. This is why the homotopy perturbation method is supplemented by the effective summation method of the yielded series.

In particular, in order to solve this problem the application of the Padé approximation has been proposed in reference [5], which has been further developed in [6], [7], [8]. The method of boundary conditions perturbation also stands in the forefront of novel asymptotic development trends.

The second part of this book is devoted to application of the latter method to solve various problems of the theory of plates with mixed boundary conditions. Both free and forced vibrations of plates are studied, as well as their stress states and stability problems. One of the important benefits is that the results obtained are presented in simple analytical forms, and they can be directly used in engineering practice.

Furthermore, as we show, our analytical results possess high accuracy, since they have been compared either with known analytical or with numerical solutions.

Many of the results included this book have been obtained with the help of our colleagues, R.G. Barantsev, W.T. van Horssen, L.V. Kurpa, L.I. Manevitch, Yu.V. Mikhlin, V.O. Olevs'kyy, A.V. Pichugin, V.N. Pilipchuk, G.A. Starushenko, S. Tokarzewski, H. Topol, A. Vakakis, D. Weichert and we warmly acknowledge their input through numerous discussions and ideas exchanged at many conferences, meetings, congresses, symposia, etc.

J. Awrejcewicz acknowledges a financial support by the National Science Centre of Poland under the grant MAESTRO 2, No. 2012/04/A/ST8/00738, for years 2013-2016.

Aachen, Dnipropetrovs'k, Lodz, Moscow, 2013 *Authors*

References

[1] Bauer S.M., Filippov S.B., Smirnov A.L., Tovstik P.E.: Asymptotic methods in mechanics with applications to thin shells and plates. Asymptotic Methods in Mechanics. *CRM Proc. Lect. Notes*, 3, 3–140 (1994).
[2] Crighton D.G.: Asymptotics –an indispensable complement to thought, computation and experiment in applied mathematical modelling. *Seventh European Conf. on Mathematics in Industry*. Eds. Fasano A., Primicerio M.B. Stuttgart: G.Teubner, 3–19 (1994).
[3] Poincaré H.: *New Methods in Celestial Mechanics*. Ed. and introd. D.L. Goroff. Woodbury, AIP, 1993.
[4] Liapunov A.M.: *Stability of Motion*. Academic Press, New York, 1966.
[5] Andrianov I.V., Ivankov A.O.: Application of the Padé approximant in the method of introducing a parameter when investigating biharmonic equations with complex boundary conditions. *USSR Comp. Math. Math. Phys.*, 27(1), 193–7 (1987).
[6] Andrianov I.V., Gristchak V.Z., Ivankov A.O.: New asymptotic method for the natural, free and forced oscillations of rectangular plates with mixed boundary conditions. *Technische Mechanik*, 14(3/4), 185–93 (1994).
[7] Andrianov I.V., Ivankov A.O.: New asymptotic method for solving of mixed boundary value problem. *International Series of Numerical Mathematics*, 106, 39–45 (1992).
[8] Andrianov I.V., Ivankov A.O.: On the solution of plate bending mixed problems using modified technique of boundary conditions perturbation. *ZAMM*, 73(2), 120–2 (1993).

List of Abbreviations

ADM	Adomian decomposition method
AEF	asymptotically equivalent function
BC(s)	boundary condition(s)
BVP(s)	boundary value problem(s)
DE(s)	differential equation(s)
FEM	finite element method
HPM	homotopy perturbation method
l.h.s.	left hand side
LAE(s)	linear algebraic equation(s)
ODE(s)	ordinary differential equation(s)
PA	Padé approximants
PDE(s)	partial differential equation(s)
PS	perturbation series
r.h.s.	right hand side
SSS	stress-strain state
TPPA	two-point Padé approximants

1

Asymptotic Approaches

Asymptotic analysis is a constantly growing branch of mathematics which influences the development of various pure and applied sciences. The famous mathematicians Friedrichs [109] and Segel [217] said that an asymptotic description is not only a suitable instrument for the mathematical analysis of nature but that it also has an additional deeper intrinsic meaning, and that the asymptotic approach is more than just a mathematical technique; it plays a rather fundamental role in science. And here it appears that the many existing asymptotic methods comprise a set of approaches that in some way belong rather to art than to science. Kruskal [151] even introduced the special term "asymptotology" and defined it as the art of handling problems of mathematics in extreme or limiting cases. Here it should be noted that he called for a formalization of the accumulated experience to convert the art of asymptotology into a science of asymptotology.

Asymptotic methods for solving mechanical and physical problems have been developed by many authors. We can mentioned excellent monographs by Eckhaus [96], [97], Hinch [133], Holms [134], Kevorkian and Cole [147], Lin and Segel [162], Miller [188], Nayfeh [62], [63], Olver [197], O'Malley [198], Van Dyke [244], [246], Verhulst [248], Wasov [90] and many others [15], [20], [34], [71], [72], [110], [119], [161], [169], [173]-[175], [216], [222], [223], [250], [251]. The main feature of the present book can be formulated as follows: it deals with new trends and applications of asymptotic approaches in the fields of Nonlinear Mechanics and Mechanics of Solids. It illuminates developments in the field of asymptotic mathematics from different viewpoints, reflecting the field's multidisciplinaiy nature. The choice of topics reflects the authors' own research experience and participation in applications. The authors have paid special attention to examples and discussions of results, and have tried to avoid burying the central ideas in formalism, notations, and technical details.

1.1 Asymptotic Series and Approximations

1.1.1 Asymptotic Series

As has been mentioned by Dingle [92], theory of asymptotic series has just recently made remarkable progress. It was achieved through the seminal observation that application of

Asymptotic Methods in the Theory of Plates with Mixed Boundary Conditions, First Edition.
Igor V. Andrianov, Jan Awrejcewicz, Vladislav V. Danishevs'kyy, Andrey O. Ivankov.

asymptotic series is tightly linked with the choice of a summation procedure. A second natural question regarding the method of series summation emerges. It is widely known that only in rare cases does a simple summation of the series terms lead to satisfactory and reliable results. Even in the case of convergent series, many problems occur, which increase essentially in the case of a study of divergent series [64]. In order to clarify the problems mentioned so far, let us consider the general form of an asymptotic series widely used in physics and mechanics [65]:

$$\sum_{n=1}^{\infty} M_n \left(\frac{\varepsilon}{\varepsilon_0}\right)^n \Gamma(n+a), \tag{1.1}$$

where a denotes an integer, and Γ is a Gamma function (see [2], Chapter 6).

The quantity ε_0 is often referred to as a singulant, and M_n denotes a modifying factor. The sequence M_n tends to a constant for $n \to \infty$ and yields information on the slowly changed series part, whereas the constant ε_0 is associated with the first singular point of the initially studied either integral or differential equation linked to the series (1.1).

In what follows we recall the classical definition: a power type series is the asymptotic series regarding the function $f(\varepsilon)$, if for a fixed N and essentially small $\varepsilon > 0$, the following relation holds

$$\left| f(x) - \sum_{j=0}^{N} a_j \varepsilon^j \right| \sim O(\varepsilon^{N+1}),$$

where the symbol $O(\varepsilon^{N+1})$ denotes the accuracy order of ε^{N+1} (see Section 1.2).

In other words we study the interval for $\varepsilon \to 0$, $N = N_0$.

Although series (1.1) is divergent for $\varepsilon \neq 0$, its first terms vanish exponentially fast for $\varepsilon \ll \varepsilon_0$. This underscores an important property of asymptotic series, related to a game between decaying terms and factorial increase of coefficients. An optimal accuracy is achieved if one takes a smallest term of the series, and then the corresponding error achieves $\exp(-\alpha/\varepsilon)$, where $\alpha > 0$ is the constant, and ε is the small/perturbation parameter. Therefore, a truncation of the series up to its smallest term yields the exponentially small error with respect to the initial value problem. On the other hand, sometimes it is important to include the above-mentioned exponentially small terms from a computational point of view, since it leads to improvement of the real accuracy of an asymptotic solution [52], [53], [64], [65], [226], [230].

Let us consider the following Stieltjes function (see [65]):

$$S(\varepsilon) = \int_0^{\infty} \frac{\exp(-t)}{1+\varepsilon t} dt. \tag{1.2}$$

Postulating the approximation

$$\frac{1}{1+\varepsilon t} = \sum_{j=0}^{N} (-\varepsilon t)^j + \frac{(-\varepsilon t)^{N+1}}{1+\varepsilon t}, \tag{1.3}$$

and putting series (1.3) into integral (1.2) we get

$$S(\varepsilon) = \sum_{j=0}^{N} (-\varepsilon^j) \int_0^{\infty} t^j \exp(-t) dt + E_N(\varepsilon), \tag{1.4}$$

where

$$E_N(\varepsilon) = \int_0^{\infty} \frac{\exp(-t)(-\varepsilon t)^{N+1}}{1+\varepsilon t} dt. \tag{1.5}$$

Computation of integrals in Equation (1.4) using integration by parts yields

$$S(\varepsilon) = \sum_{j=0}^{N} (-1)^j j! \varepsilon^j + E_N(\varepsilon).$$

If N tends to infinity, then we get a divergent series. It is clear, since the under integral functions have a simple pole in the point $t = -1/\varepsilon$, therefore series (1.3) is valid only for $|t| < 1/\varepsilon$. The obtained results cannot be applied in the whole interval $0 \le t < \infty$.

Let us estimate an order of divergence by splitting the function $S(\varepsilon)$ into two parts, i.e.

$$S(\varepsilon) = S_1(\varepsilon) + S_2(\varepsilon) = \int_0^{1/\varepsilon} \frac{\exp(-t)}{1+\varepsilon t} dt + \int_{1/\varepsilon}^{\infty} \frac{\exp(-t)}{1+\varepsilon t} dt.$$

Since $1/(1+\varepsilon t) \le 1/2$ for $t > 1/\varepsilon$, the following estimation is obtained: $S_2(\varepsilon) < 0.5 \exp(-1/\varepsilon)$.

Therefore, the exponential decay of the error is observed for decreasing ε, which is a typical property of an asymptotic series.

Let us now estimate an optimal number of series terms. This corresponds to the situation in which the term $t^{N+1} \exp(-t)$ in Equation (1.4) is a minimal one, which holds for $t = 1/(N+1)$. For $t \ge 1/\varepsilon$ we observe the divergence, and this yields the following estimation: $N = [1/\varepsilon]$, where $[\ldots]$ denotes an integer part of the number. The optimally truncated series is called the super-asymptotic one [65], whereas the hyperasymptotic series [52], [53] refers to the series with the accuracy barrier overcome. It means that after the truncation procedure one needs novel ideas to increase accuracy of the obtained results. Problems regarding a summation of divergent series are discussed in Chapters 1.3–1.5.

One may, for instance, transform the series part

$$S(\varepsilon) \approx \sum_{j=0}^{2N} (-1)^j j! \varepsilon^j \tag{1.6}$$

into the PA, i.e. into a rational function of the form

$$S(\varepsilon) \approx \frac{1 + \sum_{j=1}^{N} \alpha_j \varepsilon^j}{1 + \sum_{i=1}^{N} \beta_i \varepsilon^i}, \tag{1.7}$$

where constants α_j, β_i are chosen in a such a way that first $2N+1$ terms of the MacLaurin series (1.7) coincide with the coefficients of series (1.6). It has been proved that a sequence of PA (1.7) is convergent into a Stieltjes integral, and the error related to estimation of $S(\varepsilon)$ decreases proportionally to $\exp(-4\sqrt{N/\varepsilon})$.

The definition of an asymptotic series indicates a way of numerical validation of an asymptotic series [62]. Let us for instance assume that the solution $U_a(\varepsilon)$ is the asymptotic of the exact solution $U_T(\varepsilon)$, i.e.

$$E = U_T(\varepsilon) - U_a(\varepsilon) = K\varepsilon^{\alpha}.$$

One may take as U_T a numerical solution. In order to define α, usually graphs of the dependence $\ln E$ versus $\ln \varepsilon$ for different values of ε are constructed. The associated relations should be closed to linear ones, whereas the constant α can be defined using the method of least squares. However, for large ε the asymptotic property of the solution is not clearly exhibited, whereas for small ε values it is difficult to get a reliable numerical solution. Let us study an example of the following integral

$$I(\varepsilon) = \varepsilon e^{\varepsilon} \int_{\varepsilon}^{\infty} \frac{e^{-t}}{t} dt$$

for large values of ε. Although the infinite series

$$I(\varepsilon) = \sum_{n=0}^{\infty} \frac{(-1)^n n!}{\varepsilon^n}$$

is divergent for all values of ε, series parts

$$I_M(\varepsilon) = \sum_{n=0}^{M} \frac{(-1)^n n!}{\varepsilon^n} \tag{1.8}$$

are asymptotically equivalent up to the order of $O(\varepsilon^{-M})$ with the error of $O(\varepsilon^{-M-1})$ for $x \to \infty$. In Figure 1.1 the dependence $\log E_M(\varepsilon)$ vs. $\log \varepsilon$, where $E_M(\varepsilon) = I(\varepsilon) - I_M(\varepsilon)$, is reported (curves going down correspond to decreasing values of $M = 1, \ldots, 5$).

It is clear that curve slopes are different. However, results reported in Table 1.1 of the least square method fully prove the high accuracy of the method applied.

Let us briefly recall the method devoted to finding asymptotic series, where the function values are known in a few points. Let a numerical solution be known for some values of the parameter ε: $f(\varepsilon_1), f(\varepsilon_2), f(\varepsilon_3)$. If we know a priori that the solution is of an asymptotic-type,

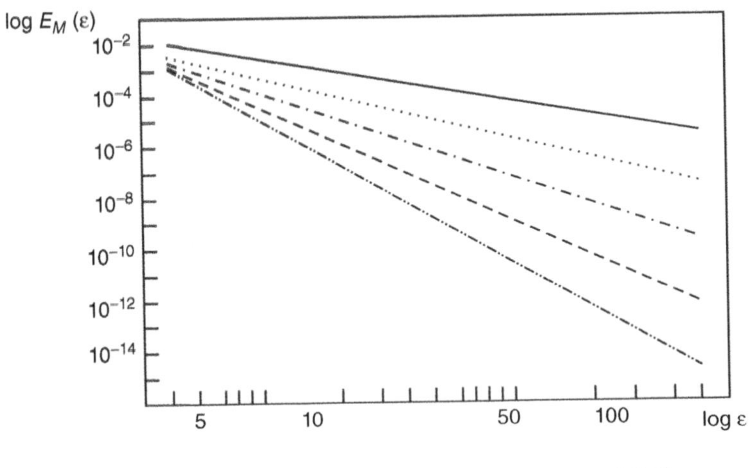

Figure 1.1 Asymptotic properties of partial sums of (1.8)

Table 1.1 Slope coefficient log $E_M(\varepsilon)$ as the function of log ε defined via the least square method

$E_M(\varepsilon)$	$\varepsilon \in [5, 50]$	$\varepsilon \in [50, 200]$	$\varepsilon \in [200, 500]$	slope
1	−1.861	−1.972	−1.991	−2.0
2	−2.823	−2.963	−2.988	−3.0
3	−3.789	−3.954	−3.985	−4.0
4	−4.758	−4.945	−4.981	−5.0
5	−5.729	−5.937	−5.999	−6.0

and its general properties are known (for instance it is known that the series corresponds only to integer values of ε), then the following approximation holds

$$f(\varepsilon_i) = \sum_{i=0}^{3} \varepsilon^i a_i,$$

and the coefficients a_i can be easily identified. The latter approach can be applied in the following briefly addressed case. In many cases it is difficult to obtain a solution regarding small values of ε, whereas it is easy to find it for ε of order 1. Furthermore, assume that we know a priori the solution asymptotic for $\varepsilon \to 0$, but it is difficult or unnecessary to define it analytically. In this case the earlier presented method can be applied directly.

1.1.2 Asymptotic Symbols and Nomenclatures

In this section we introduce basic symbols and a nomenclature of the asymptotic analysis considering the function $f(x)$ for $x \to x_0$. In the asymptotic approach we focus on monitoring the function $f(x)$ behavior for $x = x_0$. Namely, we are interested in finding another arbitrary function $\varphi(x)$ being simpler than the original (exact) one, which follows $f(x)$ for $x \to x_0$ with increasing accuracy. In order to compare both functions, a notion of the order of a variable quantity is introduced accompanied by the corresponding relations and symbols.

We say that the function $f(x)$ is of order $\varphi(x)$ for $x \to x_0$, or equivalently

$$f(x) = O(\varphi(0)) \qquad \text{for} \qquad x \to x_0,$$

if there is a number A, such that in a certain neighborhood Δ of the point x_0 we have $|f(x)| \leq A|\varphi(x)|$.

Besides, we say that $f(x)$ is the quantity of an order less than $\varphi(x)$ for $x \to x_0$, or equivalently

$$f(x) = o(\varphi(0)) \qquad \text{for} \qquad x \to x_0,$$

if for an arbitrary $\varepsilon > 0$ we find a certain neighborhood Δ of the point x_0, where $|f(x)| \leq \varepsilon|\varphi(x)|$.

In the first case the ratio $|f(x)|/|\varphi(x)|$ is bounded in Δ, whereas in the second case it tends to zero for $x \to x_0$. For example, $\sin x = O(1)$ for $x \to \infty$; $\ln x = o(x^\alpha)$ for an arbitrary $\alpha > 0$ for $x \to \infty$.

Symbols $O(\dots)$ and $o(\dots)$ are often called Landau's symbols (see [62], [63]). It should be emphasized that Edmund Landau introduced these symbols in 1909, whereas Paul Gustav Heinrich Bachman had already done so in 1894. Sometimes it worthwhile to apply additional symbols introducing other ordering relations. Namely, if $f(x) = O(\varphi(x))$, but $f(x) \neq o(\varphi(x))$ for $x \to x_0$, then the following notation holds $f(x) = \tilde{O}(\varphi(x))$ for $x \to x_0$, where the symbol $\tilde{O}(\varphi(x))$ is called the symbol of the exact order (note that in some cases also the following symbol is applied $Oe(\varphi(x))$. If $f(x) = O(\varphi(x))$, $\varphi(x) = O(f(x))$ for $x \to x_0$, (it means that $f(x)$ asymptotically equals to $\varphi(x)$ for $x \to x_0$), which is abbreviated by the notation $f(x) \asymp \varphi(x)$ for $x \to x_0$. Recall that in some cases the symbol $\asymp$ is used. Asymptotic relations give rights for the existence of the numbers $a > 0$ and $A > 0$, where in the vicinity of the point x_0 the following approximation holds: $a|\varphi(x)| \leq |f(x)| \leq A|\varphi(x)|$.

Symbols $\tilde{O}$ and $\asymp$ might be expressed by O, o and are used only for a brief notation. One may distinguish the following steps while constructing an asymptotic approximation. In the beginning high (low) order estimations are constructed of the type $f(x) = O(\varphi(x))$. Usually this first approximation is overestimated, i.e. we have $f(x) = O(\varphi(x))$.

In order to improve this first approximation the following exact order is applied $f(x) = \tilde{O}(\varphi_0(x))$, and the following asymptotic approximation is achieved $f(x) \sim a_0\varphi_0(x)$. Carrying out this kind of a cycle, we may get the asymptotic chain $f(x) - a_0\varphi_0(x) \sim a_1\varphi_1(x)$, and go further with the introduced analysis. We say that the sequence $\{\varphi_n(x)\}$, $n = 0, 1, \dots$ for $x \to x_0$ is an asymptotic one, if $\varphi_{n+1}(x) = o(\varphi_n(x))$. For instance, the following sequence $\{x^n\}$ is an asymptotic one for $x \to 0$.

A series $\sum_{n=0}^{\infty} a_n\varphi_n(x)$ with constant coefficients is called an asymptotic one, if $\{\varphi_n(x)\}$ is an asymptotic sequence. We say that $f(x)$ has an asymptotic series with respect to the sequence $\{\varphi_n(x)\}$, or equivalently

$$f(x) \sim \sum_{n=0}^{N} a_n\varphi_n(x), \qquad N = 0, 1, 2, \dots, \tag{1.9}$$

if

$$f(x) = \sum_{n=0}^{m} a_n\varphi_n(x) + o(\varphi_m(x)), \qquad m = 0, 1, 2, \dots, N. \tag{1.10}$$

Let us investigate the uniqueness of the asymptotic series. Let the function $f(x)$ for $x \to x_0$ be developed into a series with respect to the asymptotic sequence $\{\varphi_n(x)\}, f(x) \sim \sum_{n=0}^{\infty} a_n\varphi_n(x)$. Then the coefficients a_n are defined uniquely via the following formula

$$a_n = \lim_{x \to x_0} \left[f(x) - \sum_{k=0}^{n-1} a_k\varphi_k(x) \right] \varphi_n^{-1}(x).$$

Observe that the same function $f(x)$ can be developed with respect to another sequence $\chi_n(x)$, for instance

$$\frac{1}{1-x} \sim \sum_{n=0}^{\infty} x^n \quad \text{for} \quad x \to 0; \qquad \frac{1}{1-x} \sim \sum_{n=0}^{\infty} (1+x)x^{2n} \quad \text{for} \quad x \to 0.$$

On the other hand, one asymptotic series may correspond to a few functions, for instance

$$\frac{1+e^{-1/x}}{1-x} \sim \sum_{n=0}^{\infty} x^n \qquad \text{for} \quad x \to 0.$$

In other words an asymptotic series represents a class of asymptotically equivalent functions. The latter property can be applied directly in many cases (see Chapter 1.5).

Asymptotic expansion of functions $f(x)$ and $g(x)$ for $x \to x_0$ regarding the sequence $\{\varphi_n(x)\}$ follows

$$f(x) \sim \sum_{n=0}^{\infty} a_n \varphi_n(x), \quad g(x) \sim \sum_{n=0}^{\infty} b_n \varphi_n(x),$$

and the following property holds

$$\alpha f(x) + \beta g(x) \sim \sum_{n=0}^{\infty} (\alpha a_n + \beta b_n)\varphi_n(x).$$

In general, a direct multiplication of the series $\{\varphi_n(x) \cdot \varphi_m(x)\}$ $(m, n = 0, 1, \dots)$ is not allowed, since they sometimes cannot be ordered into an asymptotic sequence. However, it can be done, for instance, in the case $\varphi_n(x) = x^n$. Power series allow division if $b_0 \neq 0$.

Finding logarithms is generally allowed. For instance, let us consider the function $f(x) = (\sqrt{x}\ln x + 2x)e^x$, for which the following relation holds

$$f(x) = [2x + o(x)]e^x \qquad \text{for} \qquad x \to \infty. \tag{1.11}$$

Let $g(x) = \ln[f(x)]$, then according to (1.11), we have

$$g(x) = x + \ln[2x + o(x)] = x + \ln x + \ln 2 + o(1) \sim x + o(x) \quad \text{for} \quad x \to \infty.$$

Raising $g(x)$ to a power we find $f(x) \sim e^x$ for $x \to \infty$. Note that the multiplier $2x$ is lost. The reason is that the carried out involution in series approximation of $g(x)$ does not include terms $\ln x$ and $\ln 2$ acting on the main term of the asymptotic of $f(x)$, and only the quantities of order $o(1)$ do not change the coefficient, since $\exp\{o(1)\} \sim 1$.

The power form asymptotic series

$$f(x) \sim \sum_{n=2}^{\infty} a_n x^{-n} \qquad \text{for} \qquad x \to \infty,$$

may be integrated step by step. Differentiation of asymptotic series are not allowed in general. For example, the function

$$f(x) = e^{-1/x}\sin(e^{-1/x})$$

possesses the following singular power form series

$$f(x) \sim 0 \cdot 1 + 0 \cdot x + 0 \cdot x^2 + \dots,$$

whereas the associated derivative of the function $f(x)$ does not allow a power type series development. If the function $f(x)$ and its continuous derivative $f'(x)$ for $x \geq d > 0$ possess a power type asymptotic series for $x \to \infty$, then this derivative can be obtained via step by step differentiation of the series terms of the function $f(x)$.

Let us emphasize that the majority of errors regarding the application of asymptotic methods occur through incorrect change of orders of limiting transitions and differentiations (see [244]). This remark is followed by an example. Let the method of Bubnov-Galerkin be applied for a thin-walled problem. The following natural question arises: How many terms N should remain in order to keep a reliable solution? N parameter should be linked with α parameter characterizing thinness of the studied construction ($L/\sqrt{F}$ for a beam, R/h for a shell, etc.). However, in general

$$\lim_{N\to\infty}\lim_{\alpha\to 0}(\dots) \neq \lim_{\alpha\to 0}\lim_{N\to\infty}(\dots).$$

Additional information regarding the state-of-art of the asymptotic series can be found in [23], [25], [39], [96], [97], [133], [62], [63], [244], [246].

1.2 Some Nonstandard Perturbation Procedures

1.2.1 Choice of Small Parameters

The choice of an asymptotic method and the introduction of small dimensionless parameters to an investigated system is very often the most significant and informal part of the analytical study of physical problems. This should be carried out with the help of experience and intuition, analysis of the physical nature of the problem, as well as with the use of experimental and numerical results. It is often dictated by physical considerations, which are evidently shown through dimensionless and scaling procedures. However, it seems to be sometimes advantageous to use an initial approximation guess although this is not obvious, and may perhaps seem even strange at first glance. To illustrate this, consider a simple example [42], i.e. an algebraic equation of the form

$$x^5 + x = 1. \tag{1.12}$$

We seek a real root of Equation (1.12), the exact value of which can be determined numerically: $x = 0.75487767\dots$. A small parameter ε is not included explicitly in Equation (1.12). Consider various possibilities of introducing a parameter ε into Equation (1.12).

1. We introduce a small parameter ε as the multiplier to a nonlinear term in Equation (1.12)

$$\varepsilon x^5 + x = 1, \tag{1.13}$$

and present x as a series of ε, i.e.

$$x = a_0 + a_1\varepsilon + a_2\varepsilon^2 + \dots. \tag{1.14}$$

Substituting series (1.14) into Equation (1.13), and equating terms of equal powers, we obtain

$$a_0 = 1, \quad a_1 = -1, \quad a_2 = 5, \quad a_3 = -35, \quad a_4 = 285, \quad a_5 = -2530, \quad a_6 = 23751.$$

These values can be predicted by a closed expression for the coefficients a_n:

$$a_n = \frac{(-1)^n (5n)!}{n!(4n+1)!}.$$

The radius R of convergence of series (1.14) is $R = \frac{4^4}{5^5} = 0.08192$. Consequently, for $\varepsilon = 1$ series (1.14) diverges very fast, so the sum of the first six terms is 21476. The situation can be corrected by the method of PA. Constructing a PA (see Chapter 1.4) with three terms in the numerator and denominator and calculating it with $\varepsilon = 1$, we obtain the value of the root $x = 0.76369$ (the error in comparison to the exact value is 1.2%).

2. We now introduce a small parameter ε as multiplier to the linear term in Equation (1.12)

$$x^5 + \varepsilon x = 1. \tag{1.15}$$

Presenting the solution of Equation (1.15) in the form

$$x(\varepsilon) = b_0 + b_1\varepsilon + b_2\varepsilon^2 + \ldots, \tag{1.16}$$

and after applying the standard procedure of perturbation method, we get

$$b_0 = 1, \quad b_1 = -1, \quad b_2 = -1\frac{1}{25}, \quad a_3 = -1\frac{1}{125}, \quad b_4 = 0,$$

$$b_5 = \frac{21}{15625}, \quad b_6 = \frac{78}{78125}.$$

In this case we can also construct a general expression for the coefficients

$$b_n = -\frac{\Gamma[(4n-1)/5]}{5\Gamma[(4-n)/5]n!},$$

and determine the radius of convergence of the series (1.15): $R = \frac{5}{4^{(4/5)}} = 1.64938\ldots$. The value of $x(1)$, taking into account the first six terms of the series (1.16), deviates from the exact by 0.07%.

3. Now, let us introduce a "small parameter" δ in the exponent

$$x^{1+\delta} + x = 1, \tag{1.17}$$

and let us present x in the form

$$x = c_0 + c_1\delta + c_2\delta^2 + \ldots. \tag{1.18}$$

In addition, we use the expansion:

$$x^{1+\delta} = x(1 + \delta \ln|x| + \ldots).$$

Coefficients of series (1.18) are determined easily, i.e. they read:

$$c_0 = 0.5, \quad c_1 = 0.25 \ln 2, \quad c_2 = -0.125 \ln 2, \quad \ldots.$$

The radius of convergence is equal to 1 in this case. Using PA with three terms in the numerator and denominator, if $\varepsilon = 1$, we find $x = 0.75448$, which only deviates from the exact result by 0.05%. Calculating c_i for $i = 0, 1, \ldots, 12$ and constructing PA with six terms in the numerator and denominator, we find $x = 0.75487654$ (0.00015% error). The method is called "the method of small delta" (see Section 1.2.3) [42], [43].

4. We now assume the exponent to be a large parameter. Consider the equation

$$x^n + x = 1. \tag{1.19}$$

Assuming $n \to \infty$ (the method of large δ, see Section 1.2.4), we present the desired solution in the form

$$x = \left[\frac{1}{n}(1 + x_1 + x_2 + \dots)\right]^{1/n}, \tag{1.20}$$

where $1 > x_1 > x_2 > \dots$.
Substituting the Ansatz (1.20) in Equation (1.19), and taking into account that

$$n^{1/n} = 1 + \frac{1}{n}\ln n + \dots, \quad x^{1/n} = 1 + \frac{1}{n}\ln(1 + x_1 + x_2 + \dots) + \dots,$$

one obtains the following hierarchy with increasing accuracy

$$x \approx \left(\frac{\ln n}{n}\right)^{1/n}, \tag{1.21}$$

$$x \approx \left(\frac{\ln n - \ln \ln n}{n}\right)^{1/n}, \tag{1.22}$$

$$\dots$$

For $n = 2$ formula (1.21) gives $x = 0.58871$; the error compared to the exact solution $(0.5(\sqrt{5} - 1) \approx 0.618034)$ is 4.7%. When $n = 5$ from (1.21) we obtain $x = 0.79715$ (from numerical solution one obtains $x = 0.75488$; error of (1.21) 5.6%). Equation (1.22) for $n = 5$ gives $x = 0.74318$ (error 1.5%). Thus, even the first terms of the large δ asymptotics give excellent results.

Hence, in this case the method of large delta already provides good accuracy even for low orders of the perturbation method. Approximations (1.21), (1.22) illustrate an example of nonpower type asymptotics.

In particular, thanks to A.V. Pichugin, the obtained solution can be improved using the Lambert functions $W(z)$, which is governed by the following equation [87]

$$z = W(z)e^{W(z)}.$$

Then, the solution to our problem has the form

$$x \approx \left[\frac{1+C}{n} W\left(\frac{n}{1+C}\right)\right]^{1/n}, \quad \text{where} \quad C = \frac{1}{2n}\ln\left(\frac{W(n)}{n}\right).$$

Note that for $n = 5$ the above formula yields $x = 0.75443$ (error 0.06%).

1.2.2 *Homotopy Perturbation Method*

In recent years the so-called homotopy perturbation method (HPM) has received much attention [1], [43], [130], [44], [132], [157], [158] (the term "method of artificial small parameters" is also used). Its essence is as follows. In the equations or BCs the parameter ε is introduced so that for $\varepsilon = 0$ one obtains a BVP which admits a simple solution, and for $\varepsilon = 1$

one obtains the governing BVP. Then the perturbation method regarding ε is applied and we put $\varepsilon = 1$ in the final formula. Apparently, this approach is not new and has already been used in references [115], [159] and [207]. However, the above term, emphasizing the continuous transition from the initial value $\varepsilon = 0$ to the value of $\varepsilon = 1$ (homotopy deformation), seems to be most adequate. Let us analyze an example of the homotopy perturbation parameter method using an approach taken from reference [8], [9]. The occurence of internal resonance between modes belongs to a special feature of nonlinear systems with distributed parameters. This is why in many cases the neglect of higher modes can lead to significant errors. The following approach describes the asymptotic method of solving problems of nonlinear vibrations of systems with distributed parameters, allowing us to broadly take into account all modes. The vibrations of a square membrane lying on a nonlinear elastic foundation can be governed by the following PDE:

$$\frac{\partial^2 w}{\partial x^2} + \frac{\partial^2 w}{\partial y^2} - \frac{\partial^2 w}{\partial t^2} - cw - \varepsilon w^3 = 0, \tag{1.23}$$

where ε is the dimensionless small parameter ($\varepsilon \ll 1$).

The BCs are as follows

$$w|_{x=0,L} = w|_{y=0,L} = 0. \tag{1.24}$$

The desired periodic solution must satisfy the periodicity conditions of the form

$$w(t) = w(t+T), \tag{1.25}$$

where $T = \frac{2\pi}{\omega}$ is the period, and Ω is the natural frequency of vibrations. We seek the natural frequencies corresponding to these forms of natural vibration frequencies at which the linear case ($\varepsilon = 0$) is realized by one half-wave in each direction x and y. We introduce the transformation of time

$$\tau = \omega t. \tag{1.26}$$

The solution is sought in the form of power series

$$w = w_0 + \varepsilon w_1 + \varepsilon^2 w_2 + \dots, \tag{1.27}$$

$$\omega = \omega_0 + \varepsilon \omega_1 + \varepsilon^2 \omega_2 + \dots. \tag{1.28}$$

Substituting Ansatzes (1.27), (1.28) to Equations (1.23)–(1.25), and equating terms of equal powers, we obtain the following recurrent sequence of linear BVPs:

$$\frac{\partial^2 w_0}{\partial x^2} + \frac{\partial^2 w_0}{\partial y^2} - \omega_0^2 \frac{\partial^2 w_0}{\partial \tau^2} - cw_0 = 0, \tag{1.29}$$

$$\frac{\partial^2 w_1}{\partial x^2} + \frac{\partial^2 w_1}{\partial y^2} - \omega_1^2 \frac{\partial^2 w_1}{\partial \tau^2} - cw_1 = 2\omega_0\omega_1 \frac{\partial^2 w_0}{\partial \tau^2} + w_0^3, \tag{1.30}$$

$$\dots$$

The BCs (1.24) and periodicity conditions (1.25) take the following form for $i = 1, 2, \dots$:

$$w_i|_{x=0,L} = w_i|_{y=0,L} = 0, \tag{1.31}$$

$$w_i(\tau) = w_i(\tau + 2\pi). \tag{1.32}$$

The solution to Equation (1.29) is as follows:

$$w_{0,0} = \sum_{m=1}^{\infty} \sum_{m=1}^{\infty} A_{m,n} \sin\left(\frac{\omega_{m,n}}{\omega_0}\tau\right) \sin\left(\frac{\pi m}{L}x\right) \sin\left(\frac{\pi n}{L}y\right), \tag{1.33}$$

where $\omega_{m,n} = \sqrt{\pi^2 \frac{(m^2+n^2)}{L} + c}$, $m, n = 1, 2, 3, \ldots$, and $A_{1,1}$ is the amplitude of the fundamental tone of vibrations; $A_{m,n}$, $m, n = 1, 2, 3 \ldots$, $(m, n) \neq (1, 1)$ is the amplitude of the subsequent modes; $\omega_{m,n}$ are the natural frequencies of the counterpart linear system, $\omega_0 = \omega_{1,1}$.

Next approximation results in solving the BVP (1.30)–(1.32). To prevent the appearance of secular terms in the right hand side of Equation (1.30), the coefficients standing by the terms of the form

$$\sin\left(\frac{\omega_{m,n}}{\omega_0}\tau\right) \sin\left(\frac{\pi m}{L}x\right) \sin\left(\frac{\pi n}{L}y\right), \quad m, n = 1, 2, 3, \ldots$$

should be compared with zero.

These conditions lead to the following infinite system of nonlinear algebraic equations:

$$\frac{2A_{m,n}\omega_1}{\beta_2\omega_0}(\omega_{m,n})^2 = \sum_{i=1}^{\infty}\sum_{j=1}^{\infty}\sum_{k=1}^{\infty}\sum_{l=1}^{\infty}\sum_{p=1}^{\infty}\sum_{s=1}^{\infty} C_{m,n}^{(ijklps)} A_{i,j}A_{k,l}A_{p,s}, \tag{1.34}$$

where $m, n = 1, 2, 3, \ldots$.

Coefficients are found by substituting Ansatz (1.33) into the right hand side of Equation (1.30) and carrying out the relevant simplifications. System (1.34) can be solved by reduction. However, a sufficiently large number of equations produces significant computational difficulties. In addition, this approach does not take into account the influence of higher modes of vibrations. Therefore, in order to omit the above-mentioned difficulties we use further the HPM.

On the right side of each (m, n)-th equation of system (1.34) we introduce the parameter μ associated with those members of $A_{i,j}A_{k,l}A_{p,s}$, for which the following condition is valid: $(i > m) \cup (k > m) \cup (p > m) \cup (j > n) \cup (l > n) \cup (s > n)$. Thus, for $\mu = 0$ system (1.34) takes the "triangular" form, and for $\mu = 1$ it returns to its original form. Next, we seek a solution in the form of the following series:

$$\omega_1 = \omega^{(0)} + \mu\omega^{(1)} + \mu^2\omega^{(2)} + \ldots, \tag{1.35}$$

$$A_{m,n} = A_{m,n}^{(0)} + \mu A_{m,n}^{(1)} + \mu^2 A_{m,n}^{(2)} + \ldots, \tag{1.36}$$

where $m, n = 1, 2, 3, \ldots, \quad (m, n) \neq (1, 1)$.

In the so-far obtained solution we put $\mu = 1$.

This approach allows us to keep any number of equations in system (1.34). Below we limit ourselves to the first two terms in expansions (1.35), (1.36). We analyze the solutions and note that in this problem the parameter c plays the role of a bifurcation parameter. In general, for $c \neq 0$, $c \sim 1$, the system (1.34) admits the following solution:

$$A_{i,j}, \quad i,j = 1, 2, 3, \ldots, \quad (i,j) \neq (m, n),$$

$$\omega_1 = \frac{27}{128}\frac{A_{m,n}^2\omega_0}{\omega_{m,n^2}}, \quad m, n = 1, 2, 3, \ldots.$$

Amplitude-frequency response is given by the following formula

$$\Omega_{m,n} = \omega_{m,n} + 0.2109375\frac{A_{m,n}^2}{\omega_{m,n}}\varepsilon + \dots .$$

It is of particular interest to the case when the linear component of the restoring force is zero ($c = 0$), and the phenomenon of internal resonance between modes of vibrations occurs. Solving system (1.34) by the method described so far we find

$$A_{m,\,n} = 0, \quad m, n = 1, 2, 3, \dots, \quad (m, n) \neq (1, 1), \quad (m, n) \neq (2i-1, 2i-1), \quad i = 1, 2, 3 \dots,$$

$$A_{3,3} = -4.5662 \cdot 10^{-3} A_{1,1}, \quad A_{5,5} = 2.1139 \cdot 10^{-5} A_{1,1}, \dots, \omega_1 = 0.211048 A_{1,1}^2/\omega_0.$$

If vibrations are excited by the mode $(1, 1)$ all odd modes $(3, 3)$, $(5, 5)$ etc. are also realized. However, if the vibrations are excited by one of the higher modes, the result of energy redistribution of modes appear at lower orders until the fundamental mode $(1, 1)$.

1.2.3 Method of Small Delta

In references [42], [43] the effective method of small δ has been proposed, which we are going to explain through a few examples. Let us construct a periodic solution to the following Cauchy problem

$$x_{tt} + x^3 = 0, \tag{1.37}$$

$$x(0) = 1, \quad x_t(0) = 0. \tag{1.38}$$

We introduce a homotopy parameter δ in Equation (1.37), and hence

$$x_{tt} + x^{1+2\delta} = 0. \tag{1.39}$$

At the final expression one should put $\delta = 1$, but in the process of solving we assume $\delta \ll 1$. Then

$$x^{2\delta} = 1 + \delta \ln x^2 + 0.5\delta^2(\ln x^2)^2 + \dots . \tag{1.40}$$

We assume a solution to Equation (1.37) in the form

$$x = \sum_{k=0}^{\infty} \delta^k x_k, \tag{1.41}$$

and carry out the change of independent variable

$$t = \frac{\tau}{\omega}, \tag{1.42}$$

where $\omega^2 = 1 + \alpha_1\delta + \alpha_2\delta^2 + \dots$.

The constants α_i ($i = 1, 2, \dots$) are determined during solution process. After substituting Ansatzes (1.40)–(1.42) in Equation (1.39), and splitting with respect to δ, the following recurrent sequence of Cauchy problems is obtained

$$x_{0\tau\tau} + x_0 = 0, \tag{1.43}$$

$$x_0(0) = 1, \quad x_{0\tau}(0) = 0; \tag{1.44}$$

$$x_{1\tau\tau} + x_1 = -x_0 \ln(x_0^2) - \alpha_1 x_{0\tau\tau}, \tag{1.45}$$

$$x_1(0) = x_{1_\tau} = 0; \tag{1.46}$$

$$x_{2\tau\tau} + x_2 = -x_1 \ln(x_1^2) - 2x_1 - x_0(\ln(x_0^2))^2 - \alpha_2 x_{0\tau\tau} - \alpha_1 x_{1\tau\tau}; \tag{1.47}$$

$$x_1(0) = x_{1_\tau} = 0; \tag{1.48}$$

$$\ldots .$$

A Cauchy problem regarding zero order approximation (1.43), (1.44) has the following solution.

$$x_0 = \cos\tau .$$

In the first approximation, one obtains

$$x_{1\tau\tau} + x_1 = -\cos\tau \ln(\cos^2\tau) + \alpha_1 \cos\tau \equiv L_0 .$$

The condition of absence of secular terms in the solution of this equation can be written as follows

$$\int_0^{\pi/2} L_0 \cos t \, dt = 0,$$

and it allows us to determine the constant $\alpha_1 = 1 - 2\ln 2$.

The period of vibration can be written as

$$T = 2\pi[1 + \delta(\ln 2 - 0.5)].$$

For $\delta = 1$, we have $T = 6.8070$, while the exact value is $T = 7.4164$ (the error introduced by the approximate solution is 8.2%). A solution to the Cauchy problem of the next approximation (1.47), (1.48) gives the period value practically coinciding with the exact one ($T = 7.4111$).

We now consider the wave equation

$$u_{tt} = u_{xx}, \tag{1.49}$$

with nonlinear BCs of the form

$$u(0,t) = 0, \tag{1.50}$$

$$u_x(1,t) + u(1,t) + u^3(1,t) = 0. \tag{1.51}$$

We introduce the parameter δ into Equation (1.51) as follows

$$u_x(1,t) + u(1,t) + u^{1+2\delta}(1,t) = 0. \tag{1.52}$$

In the final expression we put $\delta = 1$, but in the asymptotical process we assume $\delta \ll 1$. We have

$$u^3 \equiv u^{1+2\delta} = u\left[1 + \delta \ln u^2 + \frac{\delta^2}{2}\left(\ln u^2\right)^2 + \ldots\right]. \tag{1.53}$$

We assume the solution to Equation (1.49) to be in the form

$$u = \sum_{k=0}^{\infty} \delta^k u_k. \tag{1.54}$$

After substituting Ansatzes (1.54), (1.52) into Equations (1.49), (1.50), (1.52), and after splitting regarding the parameter δ, we obtain the following recurrent sequence of BVPs:

$$u_{0\tau\tau} = u_{0xx}; \tag{1.55}$$

$$\text{at } x = 0, \quad u_0 = 0; \tag{1.56}$$

$$\text{at } x = 1, \quad u_{0x} + 2u_0 = 0; \tag{1.57}$$

$$u_{0\tau\tau} = u_{0xx} - \sum_{p=0}^{1} \alpha_{i-p} u_{p\tau\tau}; \tag{1.58}$$

$$\text{at } x = 0, \quad u_1 = 0; \tag{1.59}$$

$$\text{at } x = 1, \quad u_{1x} + 2u_1 = -u_0 \ln u_0^2; \tag{1.60}$$

$$u_{0\tau\tau} = u_{0xx} - \sum_{p=0}^{2} \alpha_{i-p} u_{p\tau\tau}; \tag{1.61}$$

$$\text{at } x = 0, \quad u_2 = 0; \tag{1.62}$$

$$\text{at } x = 1, \quad u_{2x} + 2u_2 = -u_1 \ln u_0^2 - 2u_1 - 0.5u_0 (\ln u_0^2)^2; \tag{1.63}$$

$$\ldots,$$

where $\alpha_0 = 0$.

The solution of the BVP of the zero order approximation (1.55)–(1.57) can be written as

$$u_0 = A \sin(\omega_0 x) \sin(\omega_0 \tau), \tag{1.64}$$

where the frequency ω_0 is determined from the transcendental equation

$$\omega_0 = 2 \tan \omega_0. \tag{1.65}$$

The first few nonzero values of ω are given in Table 1.2.

When $k \to \infty$, we have the asymptotics: $\omega^{(k)} \to 0.5\pi(2k+1)$.

Table 1.2 First few roots of transcendental equation (1.65)

$\omega_0^{(1)}$	$\omega_0^{(2)}$	$\omega_0^{(3)}$	$\omega_0^{(4)}$	$\omega_0^{(5)}$	$\omega_0^{(6)}$	$\omega_0^{(7)}$	$\omega_0^{(8)}$	$\omega_0^{(9)}$	$\omega_0^{(10)}$
2.289	5.087	8.096	11.173	14.276	17.393	20.518	23.646	26.778	29.912

BVP problem of the first approximation is as follows:

$$u_{1xx} - u_{1\tau\tau} = \alpha_1 A\omega_0^2 \sin(\omega_0 x)\sin(\omega_0 \tau), \tag{1.66}$$

$$\text{at } x = 0, \quad u_1 = 0, \tag{1.67}$$

$$\text{at } x = 1, \quad u_{1x} + 2u_1 = A_1 \sin(\omega_0\tau)[\ln(A^2\sin^2\omega_0) + \ln\sin^2(\omega_0\tau)], \tag{1.68}$$

where $A_1 = -A\sin\omega_0$.

The particular solution to Equation (1.66) satisfying the BC (1.67) has the form

$$u_1^{(1)} = -0.5\alpha_1 A\omega_0 x\cos(\omega_0 x)\sin(\omega_0\tau). \tag{1.69}$$

We choose the constant α_1 in such a way that it compensates the secular term on the r.h.s. of Equation (1.68)

$$\alpha_1 = \frac{2R_1}{\omega_0(6+\omega_0^2)},$$

where $R_1 = \ln(0.25eA^2\sin^2\omega_0)$.

Nonsecular harmonics on the r. h. s. of Equation (1.68) yield the solution

$$u_1^{(2)} = 4A_1 \sum_{k=2}^{\infty} T_k \sin(\omega_0 kx)\sin(\omega_0 k\tau)\frac{1}{k^2-1}, \tag{1.70}$$

where $T_k = 1/[k\omega_0\cos(k\omega_0) + 2\sin(k\omega_0)]$.

The complete solution of the first approximation has the form

$$u_1 = u_1^{(1)} + u_1^{(2)}.$$

Assuming $\delta = 1$, we obtain the solution of Equations (1.49)–(1.51).

Let us now consider the Schrödinger equation

$$\Psi_{xx} - x^{2N}\Psi + E\Psi = 0, \tag{1.71}$$

$$\Psi(\pm\infty) = 0. \tag{1.72}$$

Here Ψ is the wave function; E is the energy and plays the role of an eigenvalue.

It is shown that the eigenvalue problem (1.71)–(1.72) has a discrete countable spectrum E_n, $n = 0, 1, 2, \ldots$ [228]. For $N = 2$ the eigenvalue problem (1.71), (1.72) has an exact solution. Now let N differ slightly from 2, i.e.

$$\Psi_{xx} - x^{2+2\delta}\Psi + E\Psi = 0. \tag{1.73}$$

We assume the expansion

$$x^{2\delta} = 1 + \delta\ln(x^2) + \ldots,$$

and we will search for the eigenfunction Ψ and the eigenvalue E in the form of the following series

$$\Psi = \Psi_0 + \delta\Psi + \delta^2\Psi^2 + \ldots, \tag{1.74}$$

$$E = E_0 + \delta E + \delta^2 E^2 + \ldots. \tag{1.75}$$

As a result, after the asymptotic splitting, we obtain the following hierarchy of recursive sequence of eigenvalue problems

$$\Psi_{0xx} - x^2\Psi_0 + E_0\Psi_0 = 0, \tag{1.76}$$

$$\Psi_{1xx} - x^2\Psi_1 + E_0\Psi_1 + E_1\Psi_0 = x^2\Psi_0 \ln(x^2), \tag{1.77}$$

$$\dots,$$

$$|\Psi_i| \to 0 \quad \text{at} \quad |x| \to \infty, \quad i = 1, 2, 3, \dots. \tag{1.78}$$

The solution to the eigenvalue problem (1.76), (1.78) has the form

$$E_0^{(n)} = 2n + 1, \quad \Psi_0^{(n)} = e^{-x^2/2}H_n(x), \quad n = 1, 2, 3, \dots,$$

where $H_n(x)$ is the Struve function ([2], Chapter 12).

From the eigenvalue problem (1.77), (1.78) we find

$$E_1^{(n)} = \frac{\int\limits_{-\infty}^{\infty} x^2 e^{-x^2} H_n^2(x) \ln(x^2)\, dx}{\sqrt{\pi}\, 2^n n!}.$$

For $n = 0$ one obtains $H_0(x) = 1$, and

$$\int\limits_{-\infty}^{\infty} x^2 \ln x e^{-x^2} dx = \frac{\sqrt{\pi}}{8}(2 - 2\ln 2 - C),$$

where $C = 0.577215\dots$ is the Euler constant. Hence

$$E_0^{(0)} = 1 + \frac{1}{16}(2 - 2\ln 2 - C)\delta + \dots. \tag{1.79}$$

1.2.4 *Method of Large Delta*

An alternative method of small delta is the method of large delta, which we demonstrate using as an example the following nonlinear equation

$$x_{tt} + x^n = 0, \quad n = 3, 5, 7, \dots. \tag{1.80}$$

This equation can be integrated with the functions Cs and Sn, introduced by Liapunov in [159] (inversions of incomplete beta functions, see also [219]). Note that much later the same (up to normalization) function have been proposed by Rosenberg, who called them Ateb-functions [213], [214]. However, working with these objects is inconvenient, and therefore the problem arises of finding the approximate analytical solution to Equation (1.80) in expressed through elementary functions. We construct asymptotics of periodic solutions of Equation (1.80) at $n \to \infty$. Let the initial conditions for Equation (1.80) be

$$x(0) = 0, \quad \dot{x}(0) = 1. \tag{1.81}$$

The first integral of the Cauchy problem (1.80), (1.81) can be written as follows

$$\left(\frac{dx}{dt}\right)^2 = 1 - \frac{2x^{n+1}}{n+1}. \tag{1.82}$$

The replacement of $x = \lambda^{-\lambda/2}$, $\lambda = 2/(n+1)$, and integration gives us a solution in the following implicit form

$$\lambda^{\lambda/2}t = \int_0^{0\le\xi\le 1} \frac{d\xi}{\sqrt{1-\xi^{2/\lambda}}}.$$

After replacing $\xi = \sin^{\lambda}\theta$ this implicit solution is transformed into an expression that contains a small parameter in the exponent of the integrand, namely we have

$$\lambda^{\lambda/2}t = \lambda \int_0^{0\le\theta\le\pi/2} \sin^{-1+\lambda}\theta d\theta.$$

We now consider the integrand separately:

$$\sin^{-1+\lambda}\theta = \theta^{-1+\lambda}\left(\frac{\theta}{\sin\theta}\right) = \theta^{-1+\lambda}\left[\frac{\theta}{\sin\theta} - \lambda\ln\frac{\theta}{\sin\theta} + \dots\right].$$

Expanding this function into a Maclaurin series, one obtains

$$\sin^{-1+\lambda}\theta = \theta^{-1+\lambda} + \frac{\theta^{-1+\lambda}}{3} + \dots + O(\lambda).$$

The first term of this expression makes the main contribution, so in the first approximation we can suppose

$$\lambda^{\lambda/2}t \approx \theta^{\lambda}, \quad \text{i.e.} \quad \theta \approx \lambda^{1/2}t^{1/\lambda}.$$

In the original variables one obtains

$$x \approx \lambda^{-\lambda/2}\sin^{\lambda}\left(\lambda^{1/2}t^{1/\lambda}\right). \tag{1.83}$$

The solution (1.83) should be used on a quarter-period, which yields

$$T = 4\left(\frac{\pi}{2\lambda^{1/2}}\right)^{\lambda}. \tag{1.84}$$

Let us analyze the solution (1.83), (1.84). At $n = 1$ one obtains the exact values $x = \sin t$, $T = 2\pi$, whereas for $n \to \infty$ one obtains $T \to 4$. Expanding the r.h.s. of Equation (1.83) into a series of t, and restricting our considerations to the first term only, we obtain a nonsmooth solution [67]. We estimate the error of the solution (1.84). For this purpose we use the expression

$$\lambda\int_0^{\pi/2} \sin^{-1+\lambda} d\theta = 0.5\lambda B(0.5\lambda, 0.5) \equiv A_1, \tag{1.85}$$

where $B(\dots,\dots)$ is the beta function ([2], Chapter 6).

Table 1.3 Comparison of exact and approximate solutions

n	1	3	5	...	∞
A_1	π^2	1.30	1.20	...	1
A_2	π^2	1.25	1.16	...	1
$\Delta, \%$	0	5	3	...	~ 0

The approximate value of the integral on the l.h.s. of Equation (1.83) is calculated as follows: $A_2 = (\pi/2)^{\lambda}$. Numerical comparison of the values A_1, A_2 and error estimation Δ is given in Table 1.3.

Thus, the first approximation of the asymptotics for $n \to \infty$ already gives quite acceptable accuracy for practical purposes, even for not very large values of n. Note that expression (1.83) gives an approximation of incomplete beta function ([2], Chapter 6) from $n = 1$ (sinus function) to $n = \infty$ (linear function).

1.2.5 Application of Distributions

Asymptotic methods are based, generally speaking, on the use of Taylor series. In this connection the question arises: what to do with functions of the form $\exp(-\varepsilon^{-1}x)$, which cannot be expanded into a Taylor series for $\varepsilon \to 0$ via smooth functions [202]. The way out lies in the transition to the following distribution [100]:

$$H(x)\exp(-\varepsilon^{-1}x) = \sum_{n=0}^{\infty} (-1)^n \varepsilon^{n+1} \delta^{(n)}(x), \tag{1.86}$$

where $\delta(x)$ is the Dirac delta function, representing the derivative of the Heaviside function $H(x)$; $\delta^{(n)}(x)$, $n = 1, 2, \ldots$ are the derivatives of the delta function.

We show how formally the formula (1.86) can be obtained. Applying the Laplace transform to function $\exp(-\varepsilon^{-1}x)$, one obtains:

$$\int_0^{\infty} \exp(-\varepsilon^{-1}x)\,\mathrm{d}x = \frac{\varepsilon}{\varepsilon p + 1}.$$

Expanding the r.h.s. of this equation into a Maclaurin series of ε, and then calculating inverse transform term by term, one obtains expansion (1.86). Thus, we again use the Taylor series, but now in the dual space.

Here is another interesting feature of the approach using distributions: a singular perturbated problem can be regarded as a regular perturbated one [100]. Suppose, for example, we deal with the ODE:

$$\varepsilon y' + y = 0 \quad x > 0, \quad y = 1 \quad \text{at} \quad x = 0.$$

This is a singularly perturbated problem: for $\varepsilon = 0$ one obtains a smooth solution $y = 0$, which does not satisfy the given initial condition. However, one can seek a solution in the form of a nonsmooth function. Namely, assuming $z(x) = H(x)y(x)$, one obtains

$$\varepsilon z' = -z + \varepsilon\delta(x). \tag{1.87}$$

A solution of Equation (1.87) is sought in the following series form

$$z = \sum_{n=0}^{\infty} z_n \varepsilon^n.$$

As a result one obtains

$$z_0 = 0, \quad z_1 = \delta(x), \quad z_{n+1} = (-1)^n \delta^{(n)}, \quad n = 1, 2, \ldots. \tag{1.88}$$

Note that expressions (1.88) allow us to reach smooth functions. To do this, it is possible to apply the Laplace transform, then the PA in the dual space, and then one may calculate inverse Laplace transforms.

We show other application of the asymptotic method using distribution [21]. Consider the equation of the membrane, reinforced with fibers of small but finite width ε. The governing PDE is

$$[1 + 2\varepsilon\Phi_0(y)]u_{xx} + u_{yy} = 0, \tag{1.89}$$

where

$$\Phi_0(y) = \sum_{k=-\infty}^{\infty} [H(y + kb - \varepsilon) + H(y - kb + \varepsilon)].$$

Let us expand the function $\Phi_0(y)$ in a series of ε. Applying the two-sided Laplace transform [241], one obtains

$$\overline{\Phi}(p, \varepsilon) = \int_{-\infty}^{\infty} e^{-p|y|}\Phi(y, \varepsilon)\, dy.$$

Expanding the function $\overline{\Phi}(p, \varepsilon)$ in a series of ε, and performing the inverse Laplace transform, we obtain

$$\Phi_0(y) = 2\varepsilon\Phi(y) + 2\varepsilon \sum_{k=1,3,5,\ldots} \varepsilon^n \Phi^{(n)}(y), \tag{1.90}$$

where $\Phi(y) = \sum_{k=-\infty}^{\infty} \delta(y - kb)$.

Now, let us consider a solution to Equation (1.89) in the form

$$u = u_0 + \varepsilon u_1 + \varepsilon^2 u_2 + \ldots. \tag{1.91}$$

Substituting Ansatzes (1.90), (1.91) into Equation (1.89), and splitting the resulting equation with respect to ε, we arrive at the recursive sequence of BVPs:

$$[1 + 2\varepsilon\Phi(y)]u_{0xx} + u_{0yy} = 0, \tag{1.92}$$

$$[1 + 2\varepsilon\Phi(y)]u_{1xx} + u_{1yy} = -\varepsilon u_{0xx}\Phi_y(y), \tag{1.93}$$

$$\ldots$$

Thus, in the zero approximation, we obtain the problem with one-dimensional fibers (1.92), and the influence of the width of fibers is taken into account in the first approximation (1.93).

1.3 Summation of Asymptotic Series

1.3.1 Analysis of Power Series

Here we follow [133], [242], [243], [245].

We assume that one obtains the following series as the result of an asymptotic study:

$$f(\varepsilon) \sim \sum_{n=0}^{\infty} C_n \varepsilon^n \quad \text{for} \quad \varepsilon \to 0. \tag{1.94}$$

As it is known, the radius of convergence ε_0 of series (1.94) is determined by the distance to the nearest singularity of the function $f(\varepsilon)$ on the complex plane, and can be found using the following Cauchy-Hadamard formula:

$$\frac{1}{\varepsilon_0} = \overline{\lim_{n\to\infty}} |C_n|^{1/n}.$$

If the nearest singularity lies on the positive real axis, then the coefficients C_n usually have one and the same algebraic sign, for example

$$\frac{1}{1-\varepsilon} \sim 1 + \varepsilon + \varepsilon^2 + \varepsilon^3 + \dots.$$

If the nearest singularity is located on the negative axis, the algebraic signs of the coefficients C_n are usually alternated, for example

$$\frac{1}{1+\varepsilon} \sim 1 - \varepsilon + \varepsilon^2 - \varepsilon^3 + \dots.$$

The pattern of signs is usually set pretty quickly. If there are several features of the same radius, which could happen to a real function with complex singularities necessarily occurring in complex conjugate pairs, then the rule of alternation of signs may be more complex, such as

$$\frac{1+\varepsilon}{1+\varepsilon^2} \sim 1 + \varepsilon - \varepsilon^2 - \varepsilon^3 + \varepsilon^4 + \varepsilon^5 - \varepsilon^6 - \varepsilon^7 \dots.$$

Here we have a pattern of signs $++--$. To define ε_0 it may be useful to apply the so-called Domb-Sykes plot [133], [242], [243], [245]. Let the function f have one of the nearest singularities at a point $\varepsilon = \pm\varepsilon_0$ with an index of α, i.e.

$$f(\varepsilon) \sim \begin{cases} (\varepsilon_0 \pm \varepsilon)^\alpha & \text{for} \quad \alpha \neq 0, 1, 2, \dots, \\ (\varepsilon_0 \pm \varepsilon)^\alpha \ln(\varepsilon_0 \pm \varepsilon) & \text{for} \quad \alpha = 0, 1, 2, \dots, \end{cases}$$

then we get

$$\frac{C_n}{C_{n-1}} \sim \pm\frac{1}{\varepsilon_0}\left(1 - \frac{1+\alpha}{n}\right)n.$$

Constructing a graph of C_n/C_{n-1} on the vertical axis and $1/n$ on the horizontal axis, one obtains the radius of convergence (as the reciprocal of the intercepts on the axis C_n/C_{n-1}), and then, knowing the slope, the required singularity. Figure 1.2 shows the numerical results

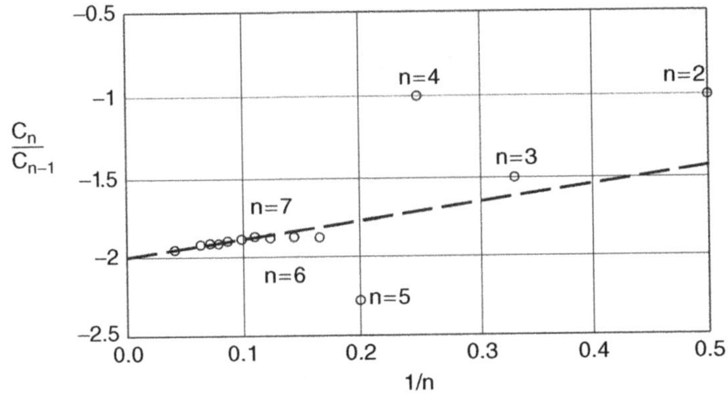

Figure 1.2 The Domb-Sykes plot for $f(\varepsilon) = \varepsilon(1+\varepsilon)(1+2\varepsilon)^{-1/2}$

for the function

$$f(\varepsilon) = \varepsilon(1+\varepsilon)(1+2\varepsilon)^{-1/2} \sim \varepsilon - \varepsilon^2 + \frac{3}{2}\varepsilon^3 - \frac{3}{2}\varepsilon^4 + \frac{27}{8}\varepsilon^5 - \frac{51}{8}\varepsilon^6 + \frac{191}{16}\varepsilon^7 - \frac{359}{16}\varepsilon^8 + \ldots, \quad (1.95)$$

starting with $n = 7$ points arranged in a linear relationship.

If ε_0 or α are known from physical considerations, they can be used for the construction of the Domb-Sykes plot. If several singularities have the same convergence radius, so that the signs of the coefficients oscillate, one way is to try to construct a dependence on the value $(C_n/C_{n-1})^{1/2}$. If the radius of convergence tends to infinity and $C_n/C_{n-1} \sim k/n$, then the analyzed function has a factor $\exp(k\varepsilon)$, where $C_n/C_{n-1} \sim k/n^{1/p}$ has a factor $\exp(\varepsilon^p)$. If the radius of convergence tends to zero, then the analyzed function has an essential singularity and asymptotic expansion diverges. If the coefficients behave like $C_{n-1}/C_n \sim 1/(kn)$, then we can write $C_n \sim Ck^n n!$, where C is a constant.

Knowledge of singular solutions can eliminate them from the perturbation series and thus its convergence can be significantly improved. We describe some techniques for removing singularities. If the singularity lies on the positive real axis, then it often means that the function $f(\varepsilon)$ is multivalued, and that there is a maximum attainable point $\varepsilon = \varepsilon_0$. Then, the inverse of the original function $\varepsilon = \varepsilon(f)$ can be single valued. For example, consider the function

$$f(\varepsilon) = \arcsin \varepsilon = \varepsilon + \frac{1}{6}\varepsilon^3 + \frac{3}{40}\varepsilon^5 + \frac{5}{112}\varepsilon^7 + \ldots, \quad (1.96)$$

and the inverse of this function is

$$\varepsilon \sim f - \frac{1}{6}f^3 + \frac{1}{1}20f^5 - \frac{1}{5040}f^7 + \ldots. \quad (1.97)$$

Numerical results are shown in Figure 1.3, where the solid curve denotes the function $\arcsin \varepsilon$, the dotted and dashed curve shows the n-term expansions (1.96) and k-terms expansions (1.97) for different numbers of terms. It is evident that the expansion (1.97) achieves a good approximation of the second branch of the original function.

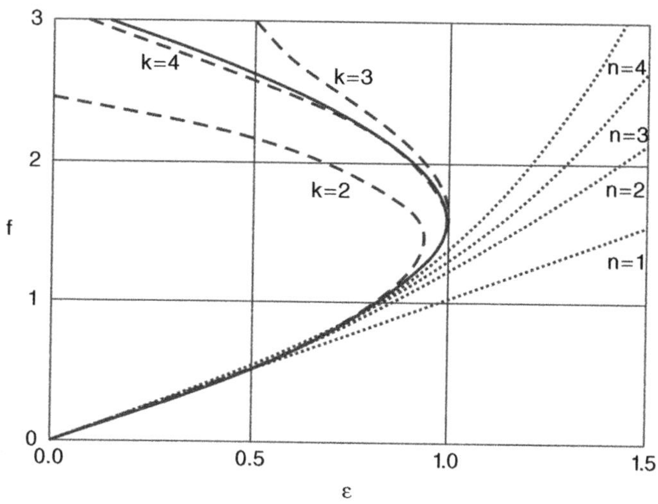

Figure 1.3 Application of the inversion method of a power series

If

$$f \sim A(\varepsilon_0 - \varepsilon)^{\alpha} \quad \text{for} \quad \varepsilon \to \varepsilon_0, \quad 0 < \alpha < 1,$$

the transition to the function $f^{1/\alpha}$ removes the singularity.

Consider the following function

$$f(\varepsilon) = e^{-\varepsilon/2}\sqrt{1+2\varepsilon} \sim 1 + \frac{1}{2}\varepsilon - \frac{7}{8}\varepsilon^2 + \frac{41}{48}\varepsilon^3 - \frac{367}{384}\varepsilon^4 + \frac{4849}{3840}\varepsilon^5 + \dots. \tag{1.98}$$

The radius of convergence of this expansion is equal to $1/2$, while the radius of convergence of functions

$$f^2 \sim 1 + \varepsilon - \frac{3}{2}\varepsilon^2 + \frac{5}{6}\varepsilon^3 - \frac{7}{24}\varepsilon^4 + \frac{3}{40}\varepsilon^5 + \dots. \tag{1.99}$$

is infinite.

Numerical results are shown in Figure 1.4, where the solid curve denotes the function $f(\varepsilon) = e^{-\varepsilon/2}\sqrt{1+2\varepsilon}$, the dotted and dashed curve show the n-term expansions (1.98) and the square roots of k-term expansions (1.99), respectively.

In addition, knowing the singularity, one can construct a new function $f_M(\varepsilon)$ (multiplicative extraction rule)

$$f(\varepsilon) = (\varepsilon_0 - \varepsilon)^{\alpha} f_M(\varepsilon),$$

or $f_A(\varepsilon)$ (additive extraction rule)

$$f(\varepsilon) = A(\varepsilon_0 - \varepsilon)^{\alpha} f_A(\varepsilon).$$

The functions $f_M(\varepsilon)$ and $f_A(\varepsilon)$ should not contain singularities at ε_0. In many cases, one can effectively use the conformal transformation of the PS [145], a fairly complete catalog of which is given in [52]. In particular, it sometimes turns out to be a successful Euler transformation

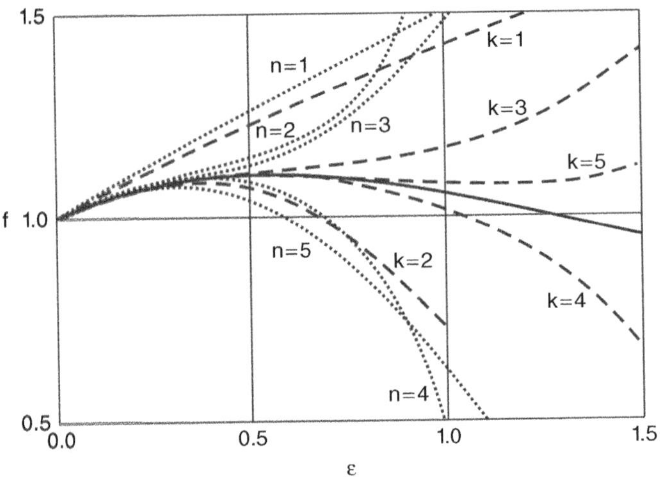

Figure 1.4 Functions, expansions and square roots (see text)

[38], [39], [242], [243], [244], [245] based on the introduction of a new variable

$$\tilde{\varepsilon} = \frac{\varepsilon}{1 - \varepsilon/\varepsilon_0}. \tag{1.100}$$

Recasting the function f in terms of $\tilde{\varepsilon}$, $f \sim \sum d_n \tilde{\varepsilon}^n$ has the singularity pushed out at the point $\tilde{\varepsilon} = \infty$. For example, the function (1.95) is singular at the $\varepsilon = -1/2$, which can be eliminated with the Euler transformation $\tilde{\varepsilon} = \varepsilon/(1 + 2\varepsilon)$. The expansion of the function (1.95) in terms of $\tilde{\varepsilon}$ is

$$f(\tilde{\varepsilon}) \sim 1 + \frac{1}{2}\tilde{\varepsilon} + \frac{1}{8}\tilde{\varepsilon}^2 - \frac{31}{48}\tilde{\varepsilon}^3 - \frac{895}{384}\tilde{\varepsilon}^4 - \frac{22591}{3840}\tilde{\varepsilon}^5 + \ldots. \tag{1.101}$$

Some numerical results are shown in Figure 1.5, where dotted and dashed curves show the n-term expansions (1.95) and k-terms in the expansion (1.101).

A natural generalization of Euler transformation is

$$\tilde{\varepsilon} = \frac{\varepsilon}{(1 - \varepsilon/\varepsilon_0)^{\alpha}},$$

where α is the real number.

1.3.2 *Padé Approximants and Continued Fractions*

The coefficients of the Taylor series in the aggregate have a lot more information about the values of features than its partial sums. It is only necessary to be able to retrieve it, and some of the ways to do this is to construct a Padé approximation [244]. Padé approximation (PA) allow us to implement among the most salient natural transformation of power series in a fractional rational function. Let us define a PA following [29], [18]. Suppose we have power series

$$f(\varepsilon) = \sum_{i=1}^{\infty} c_i \varepsilon^i. \tag{1.102}$$

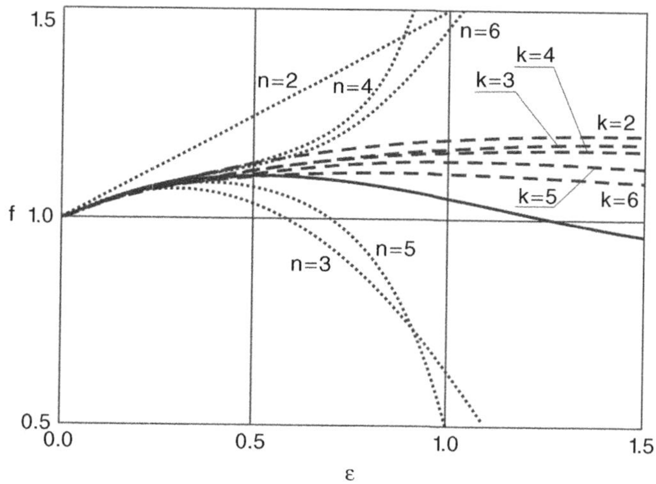

Figure 1.5 Illustration of Euler transformation

Its PA can be written as the following expression

$$f_{[n/m]}(\varepsilon) = \frac{a_0 + a_1\varepsilon + \cdots + a_n\varepsilon^n}{1 + b_1\varepsilon + \cdots + b_m\varepsilon^m}, \tag{1.103}$$

whose coefficients are determined from the condition

$$\begin{aligned}&(1 + b_1\varepsilon + \cdots + b_m\varepsilon^m)(c_0 + c_1\varepsilon + c_2\varepsilon^2 + \ldots) = \\ &a_0 + a_1\varepsilon + \cdots + a_n\varepsilon^n + O(\varepsilon^{m+n+1}).\end{aligned} \tag{1.104}$$

Equating coefficients of the same powers ε, one obtains a system of LAEs

$$\begin{array}{llll} b_m c_{n-m+1} & + b_{m-1}c_{n-m+2} & + c_{n+1} & = 0; \\ b_m c_{n-m+2} & + b_{m-1}c_{n-m+3} & + c_{n+2} & = 0; \\ \vdots & \quad \vdots & \quad \vdots & = \vdots \\ b_m c_n & + b_{m-1}c_{n+1} & + c_{n+m} & = 0, \end{array} \tag{1.105}$$

where $c_j = 0$ for $j < 0$.

The coefficients a_i can now be obtained from Equation (1.104) by comparing coefficients standing by the same powers ε:

$$\begin{aligned} a_0 &= c_0; \\ a_1 &= c_1 + b_1 c_0; \\ \vdots & \quad \vdots \\ a_n &= c_n + \sum_{i=1}^{p} b_i c_{n-i}, \end{aligned} \tag{1.106}$$

where $p = \min(n, m)$.

Equations (1.104), (1.105) are called Padé equations. In the case where the system (1.105) is solvable, one can obtain the Padé coefficients of the numerator and denominator of the PA.

Table 1.4 Padé table

m \ n	0	1	2	...
0	$f_{[0/0]}(\varepsilon)$	$f_{[1/0]}(\varepsilon)$	$f_{[2/0]}(\varepsilon)$	...
1	$f_{[0/1]}(\varepsilon)$	$f_{[1/1]}(\varepsilon)$	$f_{[2/1]}(\varepsilon)$	...
2	$f_{[0/2]}(\varepsilon)$	$f_{[1/2]}(\varepsilon)$	$f_{[2/2]}(\varepsilon)$	...
...	...	...	...	...

Functions $f_{[n/m]}(\varepsilon)$ at different values of n and m form a set which is usually written in the form of a table, called the Padé table (Table 1.4). The terms of the first row of the Padé table correspond to the finite sums of the Maclaurin series. In case of $n=m$ one obtains the diagonal PA, the most common in practice. Note that the Padé table can have gaps for those indices $n,\ m$, for which the PA does not exist.

We note some properties of the PA (see [5], [7], [29], [18], [231] for more details).

1. If the PA at the chosen m and n exists, then it is unique.
2. If the PA sequence converges to a function, the roots of its denominator tend to the poles of the function. This allows for a sufficiently large number of terms to determine the pole, and then to perform an analytical continuation.
3. The PA has meromorphic continuation regarding a given power series functions.
4. The PA on the inverse function is treated as the PA function inverse itself. This property is called duality and more exactly formulated as follows. Let
$$q(\varepsilon)=f^{-1}(\varepsilon) \quad \text{and} \quad f(0)\neq 0, \quad \text{then} \quad q_{[n/m]}(\varepsilon)=f^{-1}_{[n/m]}(\varepsilon), \tag{1.107}$$
provided that one of these approximations exists.
5. Diagonal PA are invariant under fractional linear transformations of the argument. Suppose that the function is given by their expansion (1.102). Consider the linear fractional transformation that preserves the origin $W=(a\varepsilon)/(1+b\varepsilon)$, and the function $q(W)=f(\varepsilon)$. Then $q_{[n/n]}$, provided that one of these approximations exist. In particular, the diagonal PA is invariant concerning the Euler transformation (1.100).
6. Diagonal PA are invariant under fractional linear transformations of functions. Let us analyse a function (1.102). Let
$$q(\varepsilon)=\frac{a+bf(\varepsilon)}{c+df(\varepsilon)}.$$
If $c+df(\varepsilon)\neq 0$, then
$$q_{[n/n]}(\varepsilon)=\frac{a+bf_{[n/n]}(\varepsilon)}{c+df_{[n/n]}(\varepsilon)},$$
provided that there is $f_{[n/n]}(\varepsilon)$. Because of this property infinite values of PA can be considered on a par with the end.
7. The PA can get the upper and lower bounds for $f_{[n/n]}(\varepsilon)$. For the diagonal PA one has the estimate
$$f_{[n/n-1]}(\varepsilon)\leq f_{[n/n]}(\varepsilon)\leq f_{[n/n+1]}(\varepsilon). \tag{1.108}$$
Typically, this estimate is valid for the function itself, i.e. $f_{[n/n]}(\varepsilon)$ in Equation (1.108) can be replaced by $f(\varepsilon)$.

8. Diagonal and close to it a sequence of PA often possess the property of autocorrection [163]–[166]. It consists of the following. To determine the coefficients of the numerator and denominator of PA one has to solve systems of LAE. This is an ill-posed procedure, so the coefficients of PA can be determined with large errors. However, these errors are in a certain sense of self-consistency, the PA can approximate the searching function with a higher accuracy. This is a radical difference between the PA and the Taylor series.

Autocorrection property is verified for a number of special functions. At the same time, even for elliptic functions the so-called Froissart doublets phenomenon arises, consisting of closely spaced zeros and poles to each other (but different and obviously irreducible) in the PA. This phenomenon is not of a numerical nature, but due to the nature of the elliptic function [232]. Thus, in general, having no information about the location of the poles of the PA, but relying solely on the PA (computed exactly), one cannot say that a good approximation for the approximated function is found.

To overcome these defects several methods are suggested, and in particular the smoothing method [35]. Its essence is that instead of the usual-term diagonal PA for complex functions $f_{[n/n]}(\varepsilon) = p_n(\varepsilon)/q_n(\varepsilon)$ the following expression is used

$$f_{[n/n]}(\varepsilon) = \frac{\overline{q_n(\varepsilon)}p_n(\varepsilon) + \overline{q_{n-1}(\varepsilon)}p_{n-1}(\varepsilon)}{\overline{q_n(\varepsilon)}q_n(\varepsilon) + \overline{q_{n-1}(\varepsilon)}q_{n-1}(\varepsilon)},$$

where $\overline{f}$ denotes complex conjugation of f.

Now consider the question: in what sense can the available mathematical results on the convergence of the PA facilitate the solution of practical problems? Gonchar's theorem [120] states: if none of the diagonal PA $f_{[n/n]}(\varepsilon)$ has poles in the circle of radius R, then the sequence $f_{[n/n]}(\varepsilon)$ is uniformly convergent in the circle to the original function $f(\varepsilon)$. Moreover, the absence of poles of the sequence of the $f_{[n/n]}(\varepsilon)$ in a circle of radius R must be original and confirm convergence of the Taylor series in the circle. Since for the diagonal PA invariant under fractional linear maps we have $\varepsilon \to (\varepsilon)/(a\varepsilon + b)$, the theorem is true for any open circle containing the point of splitting, and for any area, which is the union of these circles. The following theorem holds [87]: suppose the sequence of diagonal PA of the function $w(\varepsilon)$, which is holomorphic in the unit disc and has no poles outside this circle. Then this sequence converges uniformly to $w(\varepsilon)$ in the disc $|z| < r_0$, where $0.583R < r_0 < 0.584R$. A significant drawback in practice is the need to check all diagonal PA.

How can we use these results? Suppose that there are a few terms of the perturbation series and someone wants to estimate its radius of convergence R. Consider the interval $[0, \varepsilon_0]$, where the truncated PS and the diagonal PA of the maximal possible order differ by no more than 5%. If none of the previous diagonal PA has poles in a circle of radius ε_0, then it is a high level of confidence to assert that $R \geq \varepsilon_0$ [9].

The procedure of constructing the PA is much less labor-intensive than the construction of higher approximations of the PS. The PA is not limited to power series, but to the series of orthogonal polynomials. PA is locally the best rational approximation of a given power series. They are constructed directly on its odds and allow the efficient analytic continuation of the series outside its circle of convergence, and their poles in a certain sense localize the singular points (including the poles and their multiplicities) of the continuation function at the corresponding region of convergence and on its boundary. This PA is fundamentally different from rational approximations to (fully or partially) fixed poles, including those from the polynomial approximation, in which case all the poles are fixed in one, infinity, the point. Currently, the PA method is one of the most promising nonlinear methods of summation of power series and the localization of its singular points. This includes the reason why the theory of the PA turned into a completely independent section of approximation theory, and these approximations have found a variety of applications both directly in the theory of rational approximations, and in perturbation theory. Thus, the main advantages of PA compared with the Taylor series are as follows:

1. Typically, the rate of convergence of rational approximations greatly exceeds the rate of convergence of polynomial approximation. For example, the function e^{ε} in the circle of convergence approximated by rational polynomials $P_n(\varepsilon)/Q_n(\varepsilon)$ is 4^n times better than an algebraic polynomial of degree $2n$. More tangible it is a property for functions of limited smoothness. Thus, the function $|\varepsilon|$ on the interval $[-1, 1]$ cannot be approximated by algebraic polynomials, so that the order of approximation was better than $1/n$, where n is the degree of polynomial. PA gives the rate of convergence $\sim \exp(-\sqrt{2n})$.
2. Typically, the radius of convergence of rational approximation is large compared with power series. Thus, for the function arctan(x) Taylor polynomials converge only if $|\varepsilon| \leq 1$, and PA - everywhere in $C\backslash((-i\infty, -i] \cup [i, i\infty))$.
3. PA can establish the position of singularities of the function.

Similarly, the PA method is a method of continued fractions [136]. There are several types of continued fractions. The regular C-fraction has the form of an infinite sequence, in which the N-th term can be written as follows

$$f_N(\varepsilon) = a + \cfrac{c_0}{1 + \cfrac{c_1\varepsilon}{1 + \cfrac{c_2\varepsilon}{1 + \cfrac{\vdots}{\cfrac{c_{N+1}\varepsilon}{1 + c_N\varepsilon}}}}}. \tag{1.109}$$

The coefficients c_i are obtained after the decomposition of expression (1.109) into a Maclaurin series, and then equating the coefficients of equal powers of ε. When $a = 0$ one obtains the fraction of Stieltjes or S-fraction. For the function of Stieltjes

$$S(\varepsilon) = \int_0^{\infty} \frac{\exp(-t)}{1 + \varepsilon t}\, dt,$$

the coefficients of expansion (1.109) have the form: $a = 0$, $c_0 = 1$, $c_{2n-1} = c_{2n} = n$, $n \geq 1$.

Description of the so-called J-, T-, P-, R-, g-fractions, algorithms for their construction and the range of applicability are described in detail in [136].

Continued fractions are a special case of continuous functional approximation [44]. This is the sequence in which the $(n + 1)$-th term $c_n(\varepsilon)$ has the form n-th iteration of a function $F(\varepsilon)$. For the Taylor series one obtains $F(\varepsilon) = 1 + \varepsilon$, for the continuous fraction $F(\varepsilon) = 1/(1 + \varepsilon)$.

If $F(\varepsilon) = \exp(\varepsilon)$ one obtains a continuous exponential approximation

$$c_n(\varepsilon) = a_0 \exp\{a_1\varepsilon \exp[a_2\varepsilon \ldots \exp(a_n\varepsilon)]\},$$

for $F(\varepsilon) = \sqrt{1+\varepsilon}$

$$c_n(\varepsilon) = a_0\sqrt{1 + a_1\varepsilon\sqrt{1 + a_2\varepsilon\sqrt{1 + \ldots a_{n-1}\varepsilon\sqrt{1 + a_n\varepsilon}}}},$$

for $F(\varepsilon) = \ln 1 + \varepsilon$

$$c_n(\varepsilon) = a_0 \ln\{a_1\varepsilon \ln[a_2\varepsilon \ldots \ln(a_n\varepsilon)]\}.$$

In some cases, such approximations can converge significantly faster than power series. As an example, we take a solution of the transcendental equation

$$x = \varepsilon \ln x$$

for large values of ε (see [23])

$$x_0 = \varepsilon \ln \varepsilon; \quad x_1 = \varepsilon \ln(\varepsilon \ln \varepsilon); \quad x_2 = \varepsilon \ln[\varepsilon \ln(\varepsilon \ln \varepsilon)]; \ldots.$$

1.4 Some Applications of PA

1.4.1 Accelerating Convergence of Iterative Processes

The efficiency of PA or other methods of summation depends largely on the availability of higher approximations of the asymptotic process. Sometimes they can be obtained by using computer algorithms [188], but in general this problem remains an open one. Iterative methods are significantly easier to be implemented. As a result of an iterative procedure a sequence of S_n is obtained. Suppose that it converges and has a limit value. We introduce the parameter a defined by the ratio

$$a = \lim_{n\to\infty} \frac{S_{n+1} - S_n}{S_n - S}.$$

It is called superlinear convergence, if $a = 0$, a linear for $a < 1$ and logarithmic at $a = 1$. The biggest issues are, of course, logarithmically convergent sequences. Very often linearly convergent sequences may also cause a problem. Therefore, it is often necessary to improve the convergence. One method of improving the convergence is to move to a new sequence T_n with the aid of a transformation so that

$$\lim_{n\to\infty} \frac{T_n - S_n}{S_n - S} = 0.$$

In such cases we say that the sequence T_n converges faster than sequence S_n. There are linear and nonlinear methods to improve convergence. Linear methods are described by formulas

$$T_n = \sum a_{ni}S_i, \; n = 0, 1, 2, \ldots,$$

where the coefficients a_{ni} do not depend on the terms of the sequence S_n.

Since linear methods improve the convergence of a restricted class of sequences, currently nonlinear methods belong to the most popular ones. Among them the Aitken method [18] stands out for its easiness, which is described by the formula

$$T_n = S_n - \frac{(S_{n+1} - S_n)(S_n - S_{n-1})}{S_{n+1} - 2S_n + S_{n-1}}, \qquad n = 0, 1, 2, \ldots . \tag{1.110}$$

The Aitken method accelerates the convergence of all linear and many of logarithmically convergent sequences. It is very easy to calculate, and in some cases it can be applied iteratively. A natural generalization of the Aitken transformation is the Shanks transformation [224] of the form

$$T_p^{sh} = \frac{D_{kp}^{(1)}}{D_{kp}^{(1)}}, \tag{1.111}$$

where

$$D_{kp}^{(1)} = \begin{vmatrix} S_{p-k} & S_{p-k+1} & \cdots & S_p \\ \Delta S_{p-k} & \Delta S_{p-k+1} & \cdots & \Delta S_p \\ \cdots & \cdots & \cdots & \cdots \\ \Delta S_{p-1} & \Delta S_p & \cdots & \Delta S_{p+k-1} \end{vmatrix},$$

$$D_{kp}^{(2)} = \begin{vmatrix} 1 & 1 & \cdots & 1 \\ \Delta S_{p-k} & \Delta S_{p-k+1} & \cdots & \Delta S_p \\ \cdots & \cdots & \cdots & \cdots \\ \Delta S_{p-1} & \Delta S_p & \cdots & \Delta S_{p+k-1} \end{vmatrix},$$

$$\Delta S_k = S_{k+1} - S_k.$$

Equation (1.111) is called the Shanks transformation of the order k of the sequences S_k to the sequence S_k. For $k = 1$ one obtains the Aitken transform (1.110). Shanks method requires the calculation of determinants, which is not always easy. One can also use the Wynn algorithm, which is described by the formulas

$$T_{-1}^{(n)} = 0, \quad T_0^{(n)} = S_n, \quad T_{k+1}^{(n)} = T_{k+1}^{(n+1)} + \frac{1}{T_k^{(n+1)} - T_k^{(n)}}. \tag{1.112}$$

The Wynn algorithm is related to the transformation of Shanks (1.111) in the following way

$$T_{2k}^{(n)} = T_k^{(sh)}(S_n), \quad T_{2k+1}^{(n)} = \frac{1}{T_k^{(sh)}}(\Delta S_n).$$

The Wynn algorithm is a quadratic convergent method for solving systems of nonlinear equations [68], [74], [114]. There are many other techniques for accelerating sequences' convergence. One can use them consistently, for example, to convert the original sequence into a linearly convergent one, and then apply the method of Aitken. One can also use different methods to improve convergence, at each stage by comparing the obtained results [67]. All the described methods have a close relationship with the PA. The Aitken method corresponds to the PA $[n/1]$, the Shanks method to the PA $[p/k]$, and for the method of Wynn one obtains $T_{2k}^{(n)} = [n + k/k]$.

1.4.2 Removing Singularities and Reducing the Gibbs-Wilbraham Effect

Consider the problem of uniform plane flow of an incompressible inviscid fluid streamlines a thin elliptic airfoil ($|x| \le 1,\ |y| \le \varepsilon,\ \varepsilon \ll 1$). The expression for the relative velocity q^* of the flow follows [244]:

$$q^* = \frac{q}{V} = \frac{(1+\varepsilon)\sqrt{1-x^2}}{\sqrt{1-x^2(1+\varepsilon^2)}}, \tag{1.113}$$

where V is the free-stream speed.

The splitting of the r.h.s. of Equation (1.113) in a series of ε can be expressed as

$$q^*(x,\varepsilon) = 1+\varepsilon-\frac{1}{2}\varepsilon^2\frac{x^2}{1-x^2}-\frac{1}{2}\varepsilon^3\frac{x^2}{1-x^2}+\dots . \tag{1.114}$$

This expression diverges at $x = 1$. No wonder it is not: the expansion (1.114) is obtained as a result of the limiting process $\lim\limits_{\varepsilon\to 0} q(x,\varepsilon),\ x > 1$, and to get the value of $q(1,\varepsilon)$, it is necessary to perform the limit as $\lim\limits_{x\to 1} q(x,\varepsilon)$ for $\varepsilon > 0$. Divergence of series (1.114), when $x = 1$ indicates that the limit processes cannot be interchanged. Now, let us apply PA to the r.h.s. of Equation (1.114) and then pass to the limit $x \to 1$. After trying various options, we conclude that the best result is given by the PA

$$q^*(x,\varepsilon) = \frac{(1-x^2)(1+\varepsilon)}{1-x^2}. \tag{1.115}$$

Numerical results for $\varepsilon = 0.5$ are shown in Figure 1.6, where the dashed curve denotes the solution (1.114), curves 1 and 2 the exact solution (1.115) and the PA (1.115), respectively. It is seen that the use of PA significantly improves the accuracy of the approximate solution.

PA can be also successfully applied for suppression of the Gibbs-Wilbraham effect (see discussion in references [36], [69], [95], [196], [218]). Consider, for example, the function sign$x$ of the form

$$\operatorname{sign} x = \begin{cases} -1, & -\pi < x < 0, \\ 1, & 0 < x < \pi. \end{cases}$$

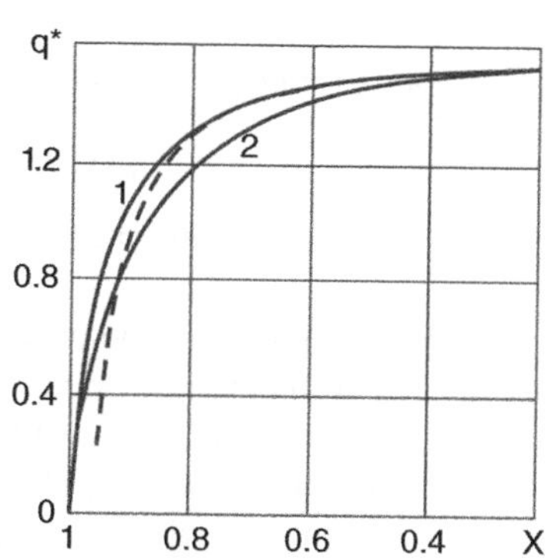

Figure 1.6 Removing singularities by the PA

Its Fourier series expansion has the form

$$\operatorname{sign} x = \frac{4}{\pi}\sum_{j=0}^{\infty}\frac{\sin(2j+1)x}{2j+1}. \tag{1.116}$$

Direct summation of series (1.116) leads to the Gibbs-Wilbraham effect in the neighborhood of $x = 0$, while the defect of convergence reaches 18%, i.e. instead of 1 one obtains the value of 1.1789797 …. Diagonal PA for series (1.116) can be written as follows

$$\operatorname{sign} x_{[N/N]} = \frac{\sum\limits_{j=0}^{[(N-1/2)]} q_{2j+1}\sin((2j+1)x)}{1+\sum\limits_{j=0}^{[(N/2)]} s_{2j}\cos(2jx)]}, \tag{1.117}$$

where

$$q_{2j+1} = \tfrac{4}{\pi}(2j+1)\left[\frac{1}{(2j+1)^2}+\sum_{i=1}^{[N/2]}\frac{s_{2i}}{(2j+1)^2-(2i)^2}\right],$$

$$s_{2i} = 2(-1)^i\frac{(N!)^4(2N+2i)!(2N-2i)!}{(N-1)!(N+1)!(N-2i)!(N+2i)![(2N)!]^2}.$$

Numerical studies show that the Gibbs-Wilbraham effect for PA (1.117) does not exceed 2% [196].

1.4.3 Localized Solutions

We consider the stationary Schrödinger equation

$$\nabla^2(x,y) - u(x,y) + u^3(x,y) = 0. \tag{1.118}$$

We seek the real, localized axisymmetric solutions of the Equation (1.118). In polar coordinates (ξ,θ) we construct a solution $\varphi(\xi)$, which does not depend on θ. As a result, we obtain the following BVP

$$\varphi''(\xi) + \frac{1}{\xi}\varphi'(\xi) - \varphi(\xi) + \varphi^3(\xi) = 0, \tag{1.119}$$

$$\varphi(x) = 0, \tag{1.120}$$

$$\lim_{\xi\to\infty}\varphi(\xi) = 0. \tag{1.121}$$

BVP (1.119)–(1.121) can be regarded as an eigenvalue problem, and the role of an eigenvalue is unknown quantity $A = \varphi(0)$. This problem plays an important role in nonlinear optics, quantum field theory, and theory of magnetic media. As shown in [191], BVP (1.119)–(1.121) has a countable set of "eigenvalues" A_n, the solution $\varphi(\xi, A_n)$ has exactly n zeros, and the solution $\varphi(\xi, A_0)$ has no zeros and decreases monotonically on ξ. That is the last solution, which is the most interesting from the standpoint of physical applications, and we will focus on obtaining it.

The problem of computing the decaying solutions of BVP (1.119)–(1.121) is identical to the problem of computing homoclinic orbits in the 3D phase space for the nonlinear oscillator, or equivalently, for computing the initial conditions for these orbits (see [99]).

Since the solutions sought are expected to be analytical functions of ξ, they can be expressed in the Maclaurin series about $\xi = 0$:

$$\varphi(\xi) = A_0 + \sum_{j=1}^{\infty} C_{2j}\xi^{2j}. \tag{1.122}$$

Substituting Ansatz (1.122) into Equation (1.119), producing a splitting of the powers of the ξ and solving the relevant equations, one obtains [99]

$$C_2 = 0.25A_0(1 - A_0^2); \quad C_4 = 0.25C_2\tilde{C}; \quad C_6 = \frac{C_4}{6} - \frac{3A_0^2\tilde{C}^2}{16};$$
$$C_8 = \frac{1}{64}(\tilde{C}C_6 - 6A_0C_2C_4 - C_2^3);$$
$$C_{10} = -0.01(\tilde{C}C_8 + 6A_0C_2C_6 + 3A_0C_4^2 + 3C_2^2C_4);$$
$$C_{12} = -\frac{1}{144}(-\tilde{C}C_{10} + 6A_0C_2C_8 + 6A_0C_4C_6 + 3C_2c_4^2),$$

where $\tilde{C} = 1 - 3A_0^2$.

Then we construct PA for the truncated series (1.122) of the form

$$\varphi(\xi) = \frac{A_0 + \sum_{j=0}^{N} a_{2j}\xi^{2j}}{1 + \sum_{k=1}^{N} b_{2j}\xi^{2k}}. \tag{1.123}$$

All coefficients in Equation (1.123) can be parameterized in terms of A_0, $a_{2j} = a_{2j}(A_0)$, $b_{2j} = b_{2j}(A_0)$, and the PA becomes a one-parameter family of analytical approximations of the searching solution. Then, we compute the value of A_0 for which PA (1.123) decays to zero as ξ tends to infinity. It gives us conditions

$$a_{2j}(A_0) = 0, \quad b_{2j}(A_0) \neq 0. \tag{1.124}$$

One can compute the PA (1.123), then imposing the condition (1.124), the following convergent values of A_0 for varying orders $2N$ is obtained

N	1	2	3	4
A_0	$\pm\sqrt{3}$	± 2.20701	± 2.21121	± 2.21200

The numerical solution gives $A_0 \approx \pm 2.206208$, the difference between numerical and analytical solutions for $N = 4$ is 0.26%.

1.4.4 Hermite-Padé Approximations and Bifurcation Problem

PA can successfully work with functions having poles. However, it often becomes necessary to explore functions with branch points, and construct all their branches. In that case, one can use Hermite-Padé approximations [94], [220], [221]. Suppose it comes to a function with the expansion

$$f(\varepsilon) = \sum_{n=1}^{\infty} u_n \varepsilon^n, \tag{1.125}$$

and we managed to find the first few coefficients of this series

$$f_N(\varepsilon) = \sum_{n=1}^{N} u_n \varepsilon^n.$$

If it is known that this function has a branch point, we can try to transform the original series (1.125) in an implicit function

$$F(\varepsilon, f) = 0$$

and determine all required branches of it.

For this purpose we construct a polynomial $F_p(\varepsilon, f)$ of degree $p \geq 2$

$$F_p(\varepsilon, f) = \sum_{m=1}^{p} \sum_{k=0}^{m} C_{m-k,k} \varepsilon^{m-k} f^k.$$

It was assumed $C_{0.1} = 1$, and the remaining coefficients must be determined from the condition

$$F_p(\varepsilon, f_N(\varepsilon)) = O(\varepsilon^{N+1}) \text{ at } \varepsilon \to 0. \tag{1.126}$$

Polynomial F_p contains $0.5(p^2 + 3p - 2)$ unknowns, the condition (1.126) yields N linear algebraic equations, and hence, $N = 0.5(p^2 + 3p - 2)$. Once the polynomial F_p is found, one can easily find p branches of the solution from the equation

$$F_p = 0.$$

For the analysis of bifurcations of these solutions one can use Newton's polygon [237]. If a priori information about the searching function is known, it can be taken into account for constructing the polynomial F_p.

1.4.5 Estimates of Effective Characteristics of Composite Materials

We consider a macroscopically isotropic 2D composite material consisting of a matrix with inclusions. The aim is to determine the effective conductivity q from the known matrix (q_1) and inclusions (q_2) conductivities and the volume fraction (φ). As shown in [234], if we take $\varepsilon = (q_2/q_1) - 1$ as a small parameter value, the required effective conductivity can be written as follows

$$\frac{q}{q_1} = 1 + \varphi\varepsilon - 0.5\varphi(1 - \varphi)\varepsilon^2 + O(\varepsilon^3). \tag{1.127}$$

Using the first two terms of expansion (1.127), one obtains

$$\frac{1}{1 + \varphi\varepsilon} \leq \frac{q}{q_1} \leq 1 + \varphi\varepsilon,$$

hence the Wiener bounds follow

$$\left(\frac{1-\varphi}{q_1}+\frac{\varphi}{q_2}\right)^{-1}\leq q\leq(1-\varphi)q_1+\varphi q_2.$$

Three terms of the expansion (1.127) give a well-known Hashin-Shtrikman lower bound, originally based on variational principles

$$q_1+\frac{\varphi}{\frac{1}{q_2-q_1}+\frac{1-\varphi}{2q_1}}\leq q, \tag{1.128}$$

while $q_2 > q_1$; for $q_2 < q_1$ the inequalities in Equation (1.128) should be changed to the opposite.

Replacement $q_2 \leftrightarrow q_1$, $\varphi \leftrightarrow 1-\varphi$ gives an upper Hashin-Shtrikman bound

$$q\leq q_2+\frac{1-\varphi}{\frac{1}{q_1-q_2}+\frac{\varphi}{2q_2}}. \tag{1.129}$$

The first three terms in the expansion (1.127) do not depend on the specific geometry of the inclusions, so the estimates (1.128), (1.129) are the most common ones. Specifying the type of inclusions, we can construct the following terms in the expansion and using the PA or continued fractions, to get narrower and narrower bounds for the unknown effective conductivity.

1.4.6 Continualization

We study a chain of $n+2$ material points with the same masses m, located in equilibrium states in the points of the axis x with coordinates $jh (j = 0, 1, \ldots, n, n+1)$ and suspended by elastic couplings of stiffness c (Figure 1.7) [8].

According to Hooke's law the elastic force acting on the j-th mass is as follows

$$\sigma_j(t) = c[y_{j+1}(t) - y_j(t)] - c[y_j(t) - y_{j-1}(t)] = c[y_{j-1}(t) - 2y_j(t) + y_{j+1}(t)],$$

where $j = 1, 2, \ldots, n$ and $y_j(t)$ is the displacement of the j-th material point from its static equilibrium position.

Applying Newton's second law one obtains the following system of ODEs governing chain dynamics

$$m\sigma_{jtt}(t) = c(\sigma_{j+1} - 2\sigma_j + \sigma_{j-1}), j = 1, 2, \ldots, n. \tag{1.130}$$

Let us suppose the following BCs

$$\sigma_0(t) = \sigma_{n+1}(t) = 0. \tag{1.131}$$

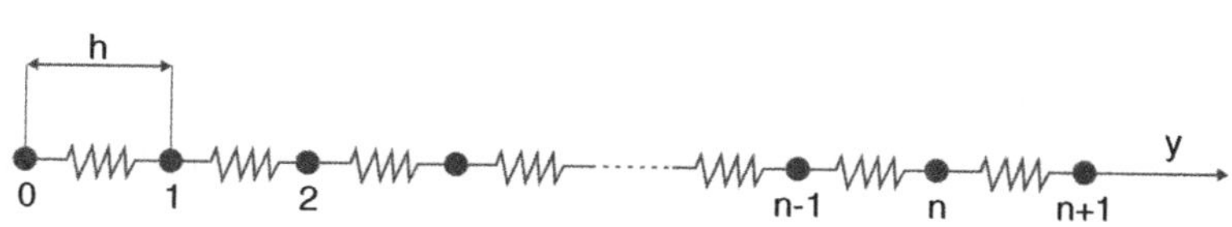

Figure 1.7 A chain of elastically coupled masses

For large values of n usually continuum approximation of discrete problem is applied. In our case it takes the form of

$$m\sigma_{tt}(x,t) = ch^2\sigma_{xx}(x,t), \tag{1.132}$$

$$\sigma(0,t) = \sigma(\ell,t) = 0. \tag{1.133}$$

Formally, one can rewrite Equation (1.130) as a pseudo-differential equation:

$$m\frac{\partial^2\sigma}{\partial t^2} + 4c\sin^2\left(-\frac{ih}{2}\frac{\partial}{\partial x}\right)\sigma = 0. \tag{1.134}$$

The pseudo-differential operator can be split into the Mclaurin series as follows

$$\sin^2\left(-\frac{ih}{2}\frac{\partial}{\partial x}\right) = -\frac{1}{2}\sum_{k=1}\frac{h^{2k}}{(2k)!}\frac{\partial^{2k}}{\partial x^{2k}}$$

$$= -\frac{h^2}{4}\frac{\partial^2}{\partial x^2}\left(1 + \frac{h^2}{12}\frac{\partial^2}{\partial x^2} + \frac{h^4}{360}\frac{\partial^4}{\partial x^4} + \frac{h^6}{10080}\frac{\partial^6}{\partial x^6}\right). \tag{1.135}$$

Retaining only the first term in the last line of Equation (1.135), one obtains a continuum approximation (1.132). Keeping the first three terms in Equation (1.135), the following model is obtained

$$m\frac{\partial^2\sigma}{\partial t^2} = ch^2\left(\frac{\partial^2}{\partial x^2} + \frac{h^2}{12}\frac{\partial^4}{\partial x^4} + \frac{h^4}{360}\frac{\partial^6}{\partial x^6}\right). \tag{1.136}$$

In the case of periodic BCs for a discrete chain one obtains the following BCs for Equation (1.136):

$$\sigma = \sigma_{xx} = \sigma_{xxxx} = 0 \ \text{ for } \ x = 0, \ell. \tag{1.137}$$

BVP (1.136), (1.137) is of the 6th order in spatial variables. Using PA we can obtain a modified continuum approximation of the 2nd order. If only two terms are left in the r.h.s. of Equation (1.135), then the PA can be cast into the following form

$$\frac{\partial^2}{\partial x^2} + \frac{h^2}{12}\frac{\partial^4}{\partial x^4} \approx \frac{\dfrac{\partial^2}{\partial x^2}}{1 - \dfrac{h^2}{12}\dfrac{\partial^2}{\partial x^2}}.$$

For justification of this procedure Fourier or Laplace transforms can be used. The corresponding so-called quasi-continuum model reads

$$m\left(1 - \frac{h^2}{12}\frac{\partial^2}{\partial x^2}\right)\sigma_{tt} - ch^2\sigma_{xx} = 0. \tag{1.138}$$

The BCs for Equation (1.138) have the form (1.133).

1.4.7 Rational Interpolation

In this subsection we follow [108]. A simple way to approximate a function is to choose a sequence of points

$$a = x_0 < x_1 < x_2 \cdots < x_n = b,$$

and to construct the interpolating polynomial $p_n(x)$

$$p_n(x_i) = f(x_i), \quad i = 0, 1, 2, \ldots, n.$$

However, as is well-known $p_n(x)$ may not be a good approximation to f, and for large $n \gg 1$ it can exhibit wild oscillations. If we are free to choose the distribution of the interpolation points x_i, one remedy is to cluster them near the end-points of the interval $[a, b]$, for example using various kinds of Chebyshev points.

A very popular alternative nowadays is to use splines (piecewise polynomials), which have become a standard tool for many kinds of interpolation and approximation algorithms, and for geometric modeling. However, it has been known for a long time that the use of rational functions can also lead to much better approximations than ordinary polynomials. In fact, both polynomial and rational interpolation, can exhibit exponential convergence when approximating analytic functions.

In "classical" rational interpolation, one chooses some M and N such that $M + N = n$ and fits a rational function of the form p_M/q_N to the values $f(x_i)$, where p_M and q_N are polynomials of degrees M and N, respectively. If n is even, it is typical to set $M + N = n/2$, and some authors have reported excellent results. The main drawback, though, is that there is no control over the occurrence of poles in the interval of interpolation.

In reference [51] it has been suggested that it might be possible to avoid poles by using rational functions of higher degree. Authors considered algorithms, which fit rational functions, whose numerator and denominator degrees can both be as high as n. This is a convenient class of rational interpolants because such an interpolant can be written in so-called barycentric form

$$r(x) = \frac{\sum\limits_{i=0}^{n} \frac{\lambda_i}{x-x_i} f(x_i)}{\sum\limits_{i=0}^{n} \frac{\lambda_i}{x-x_i} f(x_i)}$$

for some real values λ_i. Thus, it suffices to choose the weights λ_i in order to specify r, and the idea is to search for weights, which give interpolants r that have no poles and preferably good approximation properties. Various approaches are described in [108], in particular, one can choose $\lambda_i = (-1)^i, \ i = 0, 1, 2, \ldots, n$.

1.4.8 Some Other Applications

PA is widely used for the construction of solitons and other localized solutions of nonlinear problems, even in connection with the appeared term "padeon" (see [155], [156] for more details).

As a simple model, we consider the nonlinear BVP

$$y'' - y + 2y^3 = 0, \tag{1.139}$$

$$y(0) = 1, \quad y(\infty) = 0, \tag{1.140}$$

that has an exact localized solution

$$y = \cosh^{-1}(x). \tag{1.141}$$

Quasilinear asymptotics give a solution in the following form

$$y = Ce^{-x}(1 - 0.25C^2e^{-2x} + 0.0625C^4e^{-4x} + \ldots), \quad C = \text{const.} \tag{1.142}$$

It is easy to verify that with reconstructing the truncated series (1.142) in the PA, and with determining the constant C from the BCs (1.140), we arrive at the exact solution (1.141).

It is interesting also to use the PA to problems with the phenomenon of "blow-up", when the solution goes to infinity at a finite value of the argument. For example, the Cauchy problem

$$\frac{dx}{dt} = \alpha x + \varepsilon x^2, \quad x(0) = 1, \quad 0 < \varepsilon \ll \alpha \ll 1, \tag{1.143}$$

has the exact solution

$$x(t) = \frac{\alpha \exp(\alpha t)}{\alpha + \varepsilon - \varepsilon \exp(\alpha t)}, \tag{1.144}$$

which tends to infinity for $t \to \ln[(\alpha + \varepsilon)/\varepsilon]$.

Regular asymptotic expansion

$$x(t) \sim \exp(\alpha t) - \varepsilon\alpha^{-1} \exp(\alpha t)[1 - \exp(\alpha t)] + \ldots$$

cannot describe this phenomenon, but the use of the PA gives the exact solution (1.144).

PA allows us to expand the scope of the known approximate methods. For example, in the method of harmonic balance the representation of the solution of a rational function of the type

$$x(t) = \frac{\sum_{n=0}^{N} \{A_n \cos[(2n+1)\omega t] + B_n \sin[(2n+1)\omega t]\}}{1 + \sum_{n=0}^{N} \{C_m \cos(2m\omega t) + D_m \sin(2m\omega t)\}}$$

substantially increases the accuracy of approximation [128], [187].

PA can be used effectively to solve ill-posed problems. This could include reconstruction of functions in the presence of noise ([116]–[118]), various problems of dehomogenization (i.e., determining the components of a composite material on its homogenized characteristics) [81], etc. We must also mention 2D PA being illustrated and discussed in [86]. For other applications of PA see [22], [26], [27], [33], [40], [41], [45], [54], [82], [83], [85], [89], [90], [98], [102], [125], [126], [127], [135], [172], [177], [190], [249].

1.5 Matching of Limiting Asymptotic Expansions

1.5.1 Method of Asymptotically Equivalent Functions for Inversion of Laplace Transform

This method was originally proposed by Slepyan and Yakovlev for the treatment of integral transformations. Here is a description of this method, following [229]. Suppose that the Laplace transform of a function of a real variable $f(t)$ is

$$F(s) = \int_0^\infty f(t)e^{-st}\, ds.$$

In order to obtain an approximate expression for the inverse transform, it is necessary to clarify the behavior of the transform in vicinity of the points $s = 0$ and $s = \infty$, and to determine the nature and location of its singular points, as well as whether they lie on the exact boundary of the regularity or near it. Then the transform $F(s)$ is replaced by the function $F_0(s)$, allowing the exact inversion and satisfying the following conditions:

1. Functions $F_0(s)$ and $F(s)$ are asymptotically equivalent at $s \to \infty$ and $s \to 0$, i.e.

$$F_0(s) \sim F(s) \text{ at } s \to 0 \text{ and } s \to \infty.$$

2. Singular points of the functions $F_0(s)$ and $F(s)$, located on the exact boundary of the regularity, coincide.

Free parameters of the function $F_0(s)$ are chosen in such a way that they satisfy the conditions of the approximation of $F(s)$ in the sense of minimum relative error for all real values $s \geq 0$:

$$\min\left\{\max\left|\frac{F_0(s, \alpha_1, \alpha_2, \ldots, \alpha_k)}{F(s)} - 1\right|\right\}. \tag{1.145}$$

Condition (1.145) is achieved by variation of free parameters α_i. Often the implementation of equalities

$$\int_0^\infty F_0(s)\,\mathrm{d}s = \int_0^\infty F(s)\,\mathrm{d}s$$

or $F_0' \sim F_0'$ at $s \to 0$ leads to a rather precise fulfillment of requirements (1.145).

Constructed in such a way function is called asymptotically equivalent function (AEF).

Here is an example of constructing AEF. Find the inverse transform if the Laplace transform is given by the modified Bessel function [2], Chapter 9:

$$K_0(s) = -\ln(s/2)I_0(s) + \sum_{k=0}^{\infty} \frac{s^{2k}}{2^{2k}(k!)^2}\Psi(k+1) \tag{1.146}$$

where $\Psi(z)$ is the psi (digamma) function [2], Chapter 6.

For pure imaginary values of the argument $s(s = iy;\ 0 < |y| < \infty)$ function $K_0(s)$ has no singular points. Consequently, we can restrict the study of its behavior for $s \to 0$ and $s \to \infty$. The corresponding asymptotic expressions are [2], Chapter 9:

$$\begin{aligned} K_0(s) &= -\left[\ln\frac{s}{2} + \gamma\right] + O(s), & s \to 0, \\ K_0(s) &= \sqrt{\frac{\pi}{2s}}e^{-s}\left[1 + O\left(\frac{1}{s}\right)\right], & s \to \infty, \end{aligned} \tag{1.147}$$

where γ is the Euler's constant $(\gamma = 1.781\ldots)$[] (note the typo in the first formula (1.147) in [229]).

The analyzed Laplace transform has a branch point of the logarithmic type, branch point of an algebraic type, and an essential singularity. These singular points need to be stored in the structure of the zero approximation. The most simple way to obtain such a structure is to combine of two asymptotic representations (1.147), so that they mutually do not distort each other and contain free parameters, which could be disposed of in the future. As a result, we arrive at the zeroth order approximation

$$F_0(s) = e^{-s}\left[\ln\frac{s+\alpha}{s} + \sqrt{\frac{\pi}{2}}\frac{1}{\sqrt{s+\beta}}\right], \tag{1.148}$$

where α and β are the free parameters.

It is easy to see that expression (1.148) has the correct asymptotic behavior $s \to \infty$. The free parameters are determined from the condition of coincidence of the asymptotics of the functions $K_0(s)$ and $F_0(s)$ for $s \to 0$, and the equality of integrals

$$\int_0^{\infty} F_0(s)\, ds = \int_0^{\infty} K_0(s)\, ds.$$

As a result of calculations one obtains a system of transcendental equations

$$\ln\alpha + \sqrt{\frac{\pi}{2\beta}} = \ln 2 - \gamma,$$

$$\ln\alpha - e^{\alpha}\mathrm{Ei}(-\alpha) + \gamma + \frac{\pi}{\sqrt{2}}e^{\beta}[1 - \mathrm{erf}(\sqrt{2})] = \frac{\pi}{2},$$

where Ei(...) is the the sine integral [2], Chapter 7. erf(...) is the error function [2] (note typo in these formulas in [229]).

Solving the prescription system numerically, one finds $\alpha = 0.3192$, $\beta = 0.9927$.

Then the approximate inverse transform can be written as follows:

$$f_0(t) = \left\{\frac{1-\exp[-\alpha(t-1)]}{t-1} + \frac{\exp[-\beta(t-1)]}{\sqrt{2(t-1)}}\right\} H(t-1). \tag{1.149}$$

The exact expression for the function $f(t)$ is

$$f(t) = \frac{1}{\sqrt{t^2-1}} H(t-1). \tag{1.150}$$

Comparison of exact (1.150) (solid curve) and approximate (1.149) (dotted curve with circles) inversions is shown in Figure 1.8. As can be seen, a satisfactory result is obtained even in the zero approximation.

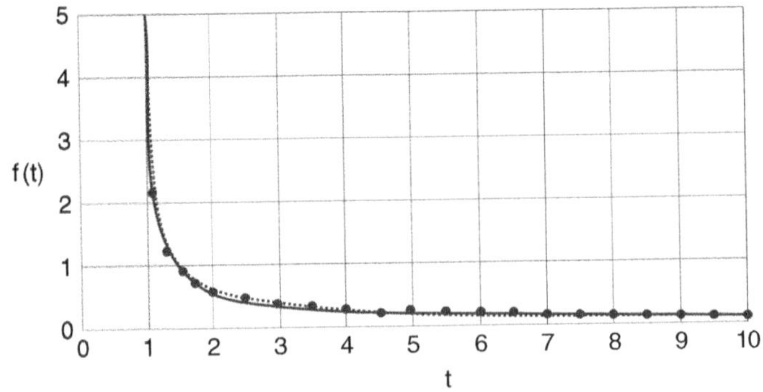

Figure 1.8 Comparison of the exact Laplace transform inversion with the treatment by the method of AEFs

Analogously, one can construct AEFs for inverse sine and cosine Fourier transforms, Hankel and other integral transforms.

1.5.2 Two-Point PA

The analysis of numerous examples confirms usually implemented a sort of "complementarity principle": if for $\varepsilon \to 0$ one can construct a physically meaningful asymptotics, there is a nontrivial asymptotics also for $\varepsilon \to \infty$. The most difficult in terms of the asymptotic approach is the intermediate case of $\varepsilon \sim 1$. In this domain numerical methods typically work well, however, if the task is to investigate the solution depending on the parameter ε, then it is inconvenient to use different solutions in different areas. Construction of an unified solution on the basis of limiting asymptotics is not a trivial task, which can be summarized as follows: we know the behavior of functions in zones I and III (Figure 1.9); we need to construct it in the zone II. For this purpose one can use a two-point PA (TPPA). We give the definition following [30].

Let

$$F(\varepsilon) = \sum_{i=0}^{\infty} c_i \varepsilon^i \quad \text{at } \varepsilon \to 0, \tag{1.151}$$

$$F(\varepsilon) = \sum_{i=0}^{\infty} c_i \varepsilon^{-i} \quad \text{at } \varepsilon \to \infty. \tag{1.152}$$

Its TPPA is a rational function of the form

$$f_{[n/m]}(\varepsilon) = \frac{a_0 + a_1\varepsilon + \cdots + a_n\varepsilon^n}{1 + b_1\varepsilon + \cdots + b_m\varepsilon^m}$$

with $k \leq m + n - 1$ coefficients, which are determined from the condition

$$(1 + b_1\varepsilon + \cdots + b_m\varepsilon^m)(c_0 + c_1\varepsilon + c_2\varepsilon^2 + \ldots) = a_0 + a_1\varepsilon + \cdots + a_n\varepsilon^n,$$

and the remaining $m + n - k$ coefficients of a similar condition for ε^{-1}.

As an example of TPPA using for matching of limiting asymptotics, consider the solution of the Van der Pol equation:

$$\ddot{x} + \varepsilon\dot{x}(x^2 - 1) + x = 0.$$

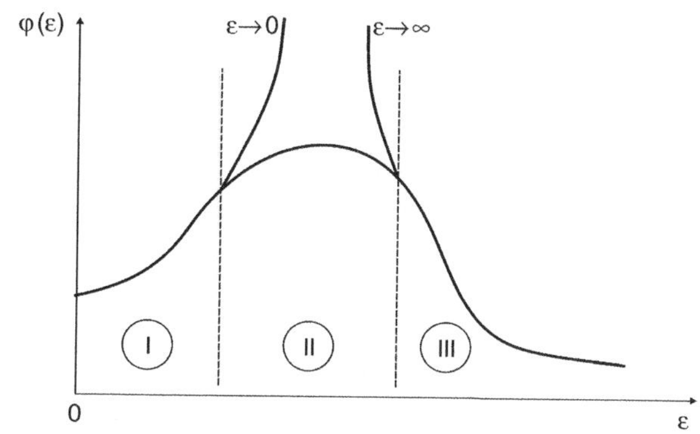

Figure 1.9 Matching of asymptotic solutions

Asymptotic expressions of the oscillation period for small and large values of ε are [3], [73], [133]:

$$T = 2\pi\left(1 + \frac{\varepsilon^2}{16} - \frac{5\varepsilon^4}{3072}\right) \quad \text{at } \varepsilon \to 0, \tag{1.153}$$

$$T = \varepsilon(3 - 2\ln 2) \quad \text{at } \varepsilon \to \infty. \tag{1.154}$$

For constructing TPPA we use the four conditions at $\varepsilon \to 0$ and the two conditions at $\varepsilon \to \infty$, then

$$T(\varepsilon) = \frac{a_0 + a_1\varepsilon + a_2\varepsilon^2 + a_3\varepsilon^3}{1 + b_1\varepsilon + b_2\varepsilon^2}, \tag{1.155}$$

where

$$a_0 = 2\pi, a_1 = \frac{\pi^2(3 - 2\ln 2)}{4(3 - 2\ln 2)^2 - \pi^2}, \quad a_2 = \frac{\pi(3 - 2\ln 2)^2}{2[4(3 - 2\ln 2)^2 - \pi^2]},$$

$$a_3 = \frac{\pi^2(3 - 2\ln 2)}{16[4(3 - 2\ln 2)^2 - \pi^2]}, \quad b_1 = \frac{\pi(3 - 2\ln 2)}{2[4(3 - 2\ln 2)^2 - \pi^2]},$$

$$b_2 = \frac{\pi^2}{16[4(3 - 2\ln 2)^2 - \pi^2]}.$$

Table 1.5 shows the results of the comparison of numerical values of the period, given in [3], [6] with the results calculated by formula (1.155).

Table 1.5 Comparison of numerical results and calculations using the TPPA

ε	T numerical	TPPA
1	6.66	6.61
2	7.63	7.37
3	8.86	8.40
4	10.20	9.55
5	11.61	10.81
6	13.06	12.15
7	14.54	13.54
8	16.04	14.96
9	17.55	16.42
10	19.08	17.89
20	34.68	33.30
30	50.54	49.13
40	66.50	65.10
50	82.51	81.14
60	98.54	97.20
70	114.60	113.29
80	130.67	129.40
90	146.75	145.49
100	162.84	161.61

Now we construct inverse Laplace transform with the TPPA. Let the original function be as follows:

$$f(t) = (1 + t^2)^{-0.5}. \tag{1.156}$$

Asymptotics of this function looks like

$$f(t) \cong \begin{cases} 1 - 0.5t^2 + \dots & \text{at} \quad t \to 0, \\ t^{-1} + \dots & \text{at} \quad t \to \infty. \end{cases}$$

TPPA in this case can be written as

$$f(t) = \frac{1 + 0.5t}{1 + 0.5t + 0.5t^2}. \tag{1.157}$$

Numerical results are shown in Figure 1.10. An approximate inversion (1.157) (upper curve) agrees well with the original (1.156) (lower curve) for all values of the argument.

Other examples on the effective use of the TPPA are reported in references [6], [75], [112], [258].

1.5.3 Other Methods of AEFs Construction

Unfortunately, the situations where both asymptotic limits have the form of power expansions are rarely encountered in practice, so we have to resort to other methods of AEFs constructing. Consider, for example, the following BVP

$$\varepsilon y_{xx} - xy = \varepsilon y, y(0) = 1, y(\infty) = 0, \varepsilon \ll 1. \tag{1.158}$$

Solution for small values of x can be written as follows

$$y = 1 - a\xi + \frac{1}{6}\xi^3 + O(\xi^4), \tag{1.159}$$

where $\xi = x\varepsilon^{-\frac{1}{3}}$, a is the arbitrary constant.

The solution for large values of x is constructed using the WKB method (see [62])

$$y = b\xi^{-\frac{1}{4}} \exp\left(-\frac{2}{3}\xi^{\frac{3}{2}}\right)\left[1 - \frac{5}{48}\xi^{-\frac{3}{2}} + O\left(\xi^{-3}\right)\right], \tag{1.160}$$

where b is an arbitrary constant.

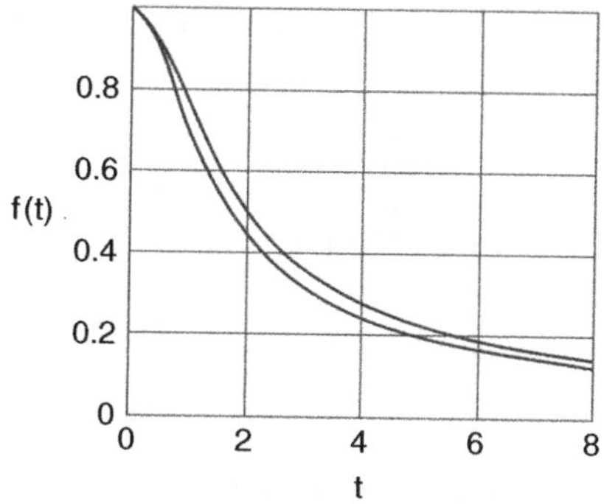

Figure 1.10 Exact and approximate Laplace transform inversions

Now we match these asymptotics. Because of the exponents occured in Equation (1.160), using TPPA in the original form is not possible. Therefore, we construct AEF based on the following considerations: for large values of the variable ξ the exponent from Equation (1.160) is taken into account in its original form, and for small values of the variable ξ it is expanded in a Maclaurin series. Constructed in this way AEF has the form

$$y_a = \frac{1 - a\xi + \frac{2}{3}\xi^{\frac{3}{2}} - \frac{2}{3}\xi^{\frac{5}{2}} + \frac{32}{5}a\xi^4}{1 + \frac{32}{5}\frac{a}{b}\xi^{\frac{17}{4}}} \exp\left(-\frac{2}{3}\xi^{\frac{3}{2}}\right). \tag{1.161}$$

The coefficients a and b in Equation (1.161) still remain undefined. For calculation of these constants one can use some integral relations, for example, obtained from Equation (1.161) by multiplying them with the weighting functions $1, x, x^2, \ldots$, and further integration over the interval $[0, \infty)$. In the end, such values of the constants are found

$$a = \frac{\sqrt[3]{3}\,\Gamma\left(\frac{2}{3}\right)}{\Gamma\left(\frac{2}{3}\right)}, \quad b = \frac{\sqrt[3]{9}\,\Gamma\left(\frac{2}{3}\right)}{2\sqrt{\pi}}. \tag{1.162}$$

Numerical calculations show that the formula (1.161) with constants (1.162) approximates the desired solution in the whole interval $[0, \infty)$ with an error not exceeding 1.5%.

When choosing the constants one can use other methods, in which case a lot depends on the skill of the researcher. Of course, it is necessary to ensure the correct qualitative behavior of AEFs, avoiding, for example zeros of the denominator, which do not correspond to the problem. To do this, one can vary the number of terms in the asymptotics and the numerator and denominator constructed uniformly suitable solutions. In general, the method of rational AEF can be described as follows [176]. Let us assume that function $f(z)$ has the following asymptotics:

$$f(z) = F(z) \text{ at } z \to \infty, \tag{1.163}$$

and

$$f(z) = \sum_{i=0}^{\infty} c_i z^i \text{ at } z \to 0. \tag{1.164}$$

Then the AEF can be produced from the Equations (1.165), (1.166) as follows

$$f(z) \approx \frac{\sum_{i=0}^{m} \alpha_i(z) z^i}{\sum_{i=0}^{n} \beta_i(z) z^i} \text{ at } z \to 0, \tag{1.165}$$

where α_i, β_i are considered not as constants but as some functions of z. Functions $\alpha_i(z)$ and $\beta_i(z)$ are chosen in such a way that:

1. The expansion of AEF (1.165) in powers of z for $z \to 0$ matches the PS (1.164);
2. The asymptotic behavior of AEF (1.165) for $z \to \infty$ coincides with the function $F(z)$ (1.163).

In the construction of AEFs a priori qualitative information is very important. For example, if from any considerations it is known that the unknown function is close to a power form, you can use the method of Sommerfeld [144]. Its essence is to replace a segment of the power series

$$f(x) = 1 + a_1 x + a_2 x^2 + \dots \tag{1.166}$$

by the function

$$f(x) \approx (1 + Ax)^{\mu}. \tag{1.167}$$

Expanding expression (1.167) in a MacLaurin series and comparing coefficients of this expansion with series (1.166), one obtains

$$A = \frac{a_1^2 - 2a_2}{a_1}; \mu = \frac{a_1^2}{a_1^2 - 2a_2}.$$

Numerical approaches also can be used for the construction of AEFs. For example, in reference [137] a computational technique for matching limiting asymptotics is described.

Sometimes it is possible to construct the so called composite equations, which can be treated as asymptotically equivalent equations. Let us emphasize, that the composite equations, due to [244], can be obtained in result of synthesis of the limiting cases. The principal idea of the method of the composite equations can be formulated in the following way [244]:

1. Identify the terms in the differential equations, whose neglect in the straightforward approximation is responsible for the nonuniformity.
2. Approximate those terms in so far as possible while retaining their essential character in the region of nonuniformity.

Let us dwell on the terminology. Here we use the term "asymptotically equivalent function". Other terms "reduced method of matched asymptotic expansions" [144], "quasifractional approximants" [75], "mimic function" [113] are also used.

1.5.4 Example: Schrödinger Equation

For the Schrödinger equation (1.71) with BCs (1.72) we obtained previously a solution for the exponent, little different from the two (1.79). In [56] the following asymptotic solutions for $N \to \infty$ are obtained:

$$E_0(N) = \frac{\pi^2}{4}(2N)^{-\frac{2}{N+1}}\Gamma\left(\frac{N}{N+1}\right)^2. \tag{1.168}$$

Using Ansatzes (1.79) and (1.168), we construct AEF

$$E_0(N) \sim \frac{\pi + \Gamma\left(\frac{N}{N+1}\right)^2}{4(2N + \alpha)^{\frac{2}{N+1}}}, \tag{1.169}$$

where $\alpha = \pi^2\Gamma(1.25) - 2 \approx 6.946$.

Numerical results are presented in Table 1.6. It is evident that formula (1.169) gives good results for all the values of N.

Table 1.6 Comparison of numerical and analytical results of the energy levels for the Schrödinger equation

N	E_0; numerical [56]	Equation (1.169)	Error,%
1	1.0000	1.0	0
2	1.0604	0.9974	5.9364
4	1.2258	1.17446	4.1882
10	1.5605	1.5398	1.33
50	1.1052	2.1035	0.079
200	2.3379	2.3376	0.006
500	2.4058	2.4058	≈ 0
1500	2.4431	2.4431	≈ 0
3500	2.4558	2.45558	≈ 0

1.5.5 Example: AEFs in the Theory of Composites

Now let us consider an application of the method of AEFs for the calculation of the effective heat conductivity of an infinite regular array of perfectly conducting spheres, embedded in a matrix with unit conductivity. The following expansion for the effective conductivity $\langle k\rangle$ has been reported in reference [215]

$$\langle k\rangle = 1 - \frac{3c}{-1 + c + a_1 c^{\frac{10}{3}} \frac{1+a_2 c^{\frac{11}{3}}}{1-a_3 c^{\frac{7}{3}}} + a_4 c^{\frac{14}{3}} + a_5 c^6 + a_6 c^{\frac{22}{3}} + O\left(c^{\frac{25}{3}}\right)}, \tag{1.170}$$

where c is the volume fraction of inclusions. Here we consider three types of space arrangement of spheres, namely, the simple cubic (SC), body centered cubic (BCC) and face centered cubic (FCC) arrays. The constants a_i for these arrays are given in Table 1.7.

In the case of perfectly conducting large spheres ($c \to c_{\max}$, where $c_{\max}$ is the maximum volume fraction for a sphere) the problem can be solved by means of a reasonable physical assumption that the heat flux occurs entirely in the region, where spheres are in a near contact. Thus, the effective conductivity is determined in the asymptotic form for the flux between two spheres, which is logarithmically singular in the width of a gap, justifying the assumption [185]:

$$\langle k\rangle = -M_1 \ln\ \chi - M_2 + O(\chi^{-1}), \tag{1.171}$$

where $\chi = 1 - (c/c_{\max})^{\frac{1}{3}}$ is the dimensionless width of a gap between the neighboring spheres, $\chi \to 0$ for $c \to c_m$, $M_1 = 0.5 c_{\max} p$, p is the number of contact points at the surface of a sphere;

Table 1.7 The constants $a_1, \ldots, a_6$ in Equation (1.170)

	a_1	a_2	a_3	a_4	a_5	a_6
SC array	1.305	0.231	0.405	0.0723	0.153	0.0105
BCC array	0.129	−0.413	0.764	0.257	0.0113	0.00562
FCC array	0.0753	0.697	−0.741	0.0420	0.0231	$9.14 \cdot 10^{-7}$

Table 1.8 The constants M_1, M_2 and c_{max}

	M_1	M_2	c_{max}
SC array	$\pi/2$	0.7	$\pi/6$
BCC array	$\sqrt{3}\pi/2$	2.4	$\sqrt{3}\pi/8$
FCC array	$0\sqrt{2}\pi$	7.1	$\sqrt{2}\pi/6$

M_2 is a constant, depending on the type of space arrangement of spheres. The values of M_1, M_2 and c_{max} for the three types of cubic arrays are given in Table 1.8.

On the basis of limiting solutions (1.170) and (1.171) we develop the AEF valid for all values of the volume fraction of inclusions $c \in [0, c_{max}]$:

$$\langle k \rangle = \frac{P_1(c) + P_2 c^{\frac{m+1}{3}} + P_3 \ln\ \chi}{Q(c)}. \tag{1.172}$$

Here the functions $P_1(c)$, $Q(c)$ and the constants P_2, P_3 are determined as follows:

$$Q(c) = 1 - c - a_1 c^{\frac{10}{3}}, \quad P_1(c) = \sum_{i=0}^{m} \alpha_1 c^{\frac{i}{3}}, \quad P_2 = 0 \quad \text{for } n = 1,$$

$$P_2 = \frac{-[P_1(c_{max}) + Q(c_{max})M_2]}{c_{max}^{\frac{m+1}{3}}} \quad \text{for } n = 2.$$

The AEF (1.172) takes into account leading terms of expansion (1.170) and leading terms of expansion (1.171), and the corresponding coefficients follow

$$\alpha_0 = 1, \quad \alpha_3 = 2 - \frac{Q(c_{max})M_1}{3c_{max}}, \quad \alpha_{10} = \alpha_1 - \frac{Q(c_{max})M_1}{10c_{max}^{\frac{10}{3}}},$$

$$\alpha_j = -\frac{Q(c_{max})M_1}{jc_{max}^{\frac{j}{3}}}, \quad j = 1, 2, \ldots, m - 1m, \quad j \neq 3, 10.$$

The increment of m and n leads to the growth of the accuracy of the obtained solution (1.172). Let us illustrate this dependence in the case of SC array. We calculated $\langle k \rangle$ for different values of m and n. In Figure 1.11 our analytical results are compared with experimental measurements from [186] (black dots); details of these data can be found in [184]. Finally, we restrict $m = 19$ and $n = 2$ for all types of arrays, as they provide a satisfactory agreement with numerical datas and a rather simple analytical form of the AEF (1.172).

Numerical results for the BCC and the FCC arrays are displayed in Figures 1.12 and 1.13, respectively. For BBC array the obtained AEF (1.172) is compared with the experimental results taken from [182] and [183]. For FCC array the experimental data are not available, therefore we are comparing with the numerical results obtained by [183] using the Rayleigh method. The agreement between the analytical solution (1.172) and the numerical results is quite satisfactory.

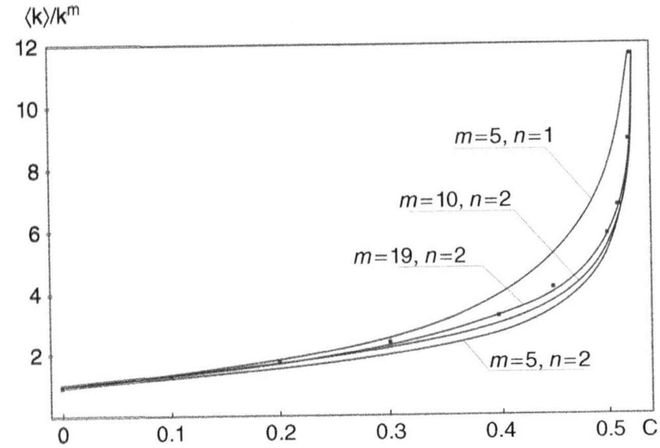

Figure 1.11 Effective conductivity $\langle k\rangle/k^m$ of the SC array vs. volume fraction of inclusions c

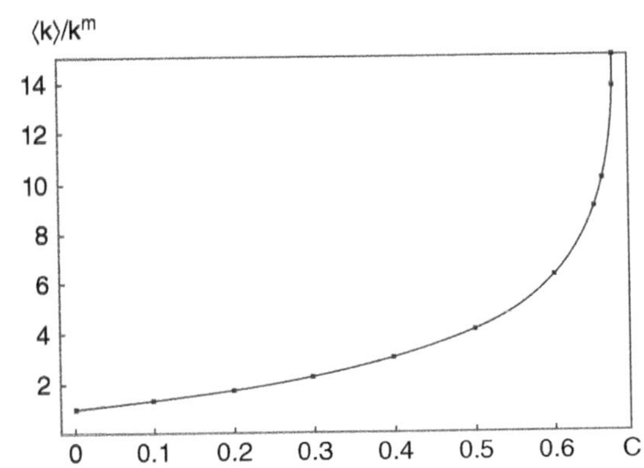

Figure 1.12 Effective conductivity $\langle k\rangle/k^m$ of the BCC array vs. volume fraction of inclusions c

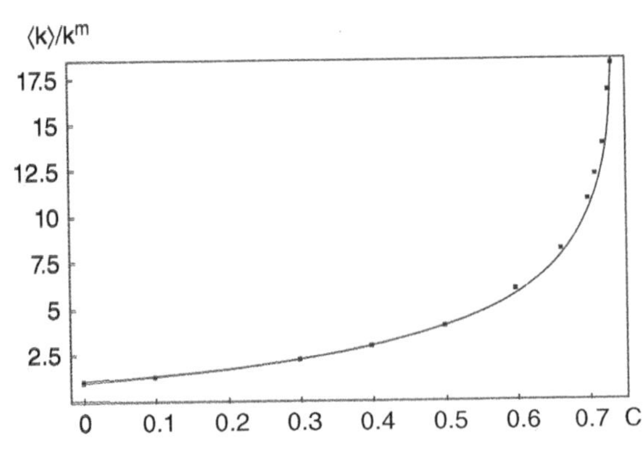

Figure 1.13 Effective conductivity $\langle k\rangle/k^m$ of the FCC array vs. volume fraction of inclusions c

1.6 Dynamical Edge Effect Method

1.6.1 Linear Vibrations of a Rod

We introduce the method of dynamical edge effect using a problem possessing an exact solution. We consider free vibrations of a rod of length L governed by the following PDE:

$$\frac{\partial^4 w}{\partial x^4} + a^2 \frac{\partial^2 w}{\partial t^2} = 0, \quad a^2 = \frac{\rho F}{EI}. \tag{1.173}$$

Consider two variants of the boundary conditions:

a) simple support

$$\text{for } x = 0, L \quad w = 0, \quad \frac{\partial^2 w}{\partial x^2} = 0; \tag{1.174}$$

b) rigid clamping

$$\text{for } x = 0, L \quad w = 0, \quad \frac{\partial w}{\partial x} = 0. \tag{1.175}$$

Since we consider natural vibrations, the process of looking for function $w(x, t)$ is assumed in the following form

$$w(x, t) = W(x)\exp(i\omega t).$$

The eigenfunction $W(x)$ is yielded by the following equation

$$\frac{d^4 W}{dx^4} - a^2\omega^2 W = 0. \tag{1.176}$$

Observe that for $\omega^2 \to \infty$ Equation (1.176) is reduced to trivial one, therefore an application of the boundary functions or of the matched asymptotic series does not belong to easy tasks.
A solution to Equation (1.176) with BCs (1.174) follows

$$W_m = \sin\left(\frac{m\pi}{L}x\right), \quad m = 1, 2, 3, \ldots; \tag{1.177}$$

$$\omega_m = \frac{1}{a}\left(\frac{m\pi}{L}\right)^2. \tag{1.178}$$

One may verify that the BVP (1.175), (1.176) does not allow us to obtain the solution (1.177). However, if an eigenfunction rapidly oscillates with respect to x, i.e. we consider a sufficiently high order vibration form, then we may expect that in this case a solution (1.177) holds for an internal region located far from the boundaries (Figure 1.14). Even though the boundary conditions are not satisfied, we may try to construct a solution compensating the occurred errors on the BC and being rapidly decaying while approaching the internal region, and hence the approximating formulas for eigenfunctions and frequencies can be derived.
We assume the following solution to Equation (1.176):

$$W_0 = \sin\frac{\pi(x - x_0)}{\lambda}, \tag{1.179}$$

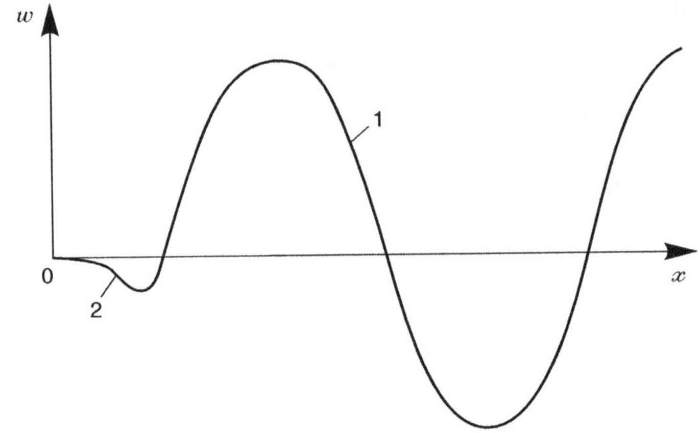

Figure 1.14 Curve 1 depicts the rapidly oscillating state, whereas curve 2 presents a sum of the fundamental state and dynamical edge effect

where x_0, λ correspond to shift and length of the wave, respectively. Values of x_0 and λ will be estimated in the process of construction of the dynamical edge effect.

Vibration frequency

$$\omega = \frac{1}{a}\left(\frac{\pi}{\lambda}\right)^2. \tag{1.180}$$

Let us present Equation (1.176) in the following way [240]:

$$\left(\frac{d^2}{dx^2} + a\omega\right)\left(\frac{d^2}{dx^2} - a\omega\right) W = 0. \tag{1.181}$$

Therefore, its general solution has the following form

$$W = W_1 + W_2,$$

where functions W_1 and W_2 are general solutions of the equations

$$\frac{d^2 W_1}{dx^2} + a\omega W_1 = 0, \tag{1.182}$$

$$\frac{d^2 W_2}{dx^2} - a\omega W_2 = 0. \tag{1.183}$$

When the eigenforms change rapidly ($a\omega \gg 1$), the following estimations hold

$$\frac{dW_1}{dx} \sim a\omega W_1, \qquad \frac{dW_2}{dx} \sim a\omega W_2.$$

Observe that W_1 represents rapidly oscillating function, whereas W_2 presents a sum of exponential functions with large exponents. Hence, here we deal rather with untypical situation, since we do not have small and large real values of the routes of the characteristic equation, which correspond to slow and fast changeable solution components. We deal here with splitting of two states, where one of them rapidly oscillates and the second is concentrated near the boundaries. In other words the characteristic equation has real and imaginary roots of the same moduli order.

In what follows we construct the edge effect described by Equation (1.183). Taking into account Equation (1.180), the following relations for the edge effects located in neighborhood of the edges $x = 0$ and $x = L$ are obtained:

$$W_{1cr} = C_1 \exp(-\pi\lambda^{-1}x),$$
$$W_{2cr} = C_2 \exp[-\pi\lambda^{-1}(x-L)]. \tag{1.184}$$

In order to define the eigenform and the associated frequency we need to find x_0, λ and arbitrary constants C_1, C_2 from the boundary conditions:

$$\text{for } x=0 \quad W_0 + W_{1cr} = 0, \qquad \frac{d}{dx}(W_0 + W_{1cr}) = 0; \tag{1.185}$$

$$\text{for } x=L \quad W_0 + W_{2cr} = 0, \qquad \frac{d}{dx}(W_0 + W_{2cr}) = 0. \tag{1.186}$$

Substituting Equations (1.179) and (1.184) into Equation (1.185) and (1.186) one gets

$$C_1 - \sin\frac{\pi x_0}{\lambda} = 0, \quad C_1 - \cos\frac{\pi x_0}{\lambda} = 0, \tag{1.187}$$

$$C_2 + \sin\left(\pi\frac{L-x_0}{\lambda}\right) = 0, \quad C_2 + \cos\left(\pi\frac{L-x_0}{\lambda}\right) = 0. \tag{1.188}$$

Furthermore, we obtain

$$\lambda = \frac{L}{m+0,5}, \quad m = 1,2,\ldots; \quad x_0 = \lambda(0,25+n), \quad n = 1,2,\ldots.$$

Finally, the formula for eigenfrequencies of a clamped rod is as follows:

$$\omega_m = \pi^2\frac{(m+0,5)^2}{aL^2}, \quad m = 1,2,\ldots. \tag{1.189}$$

Formula (1.189) for fundamental frequency gives the error of 1%.

The method described so far has been proposed by Bolotin [58], [59]. On the other hand, Keller and Rubinow [28], [146] have proposed the wave method for the Laplace equation. The latter one has been generalized into biharmonic equation [76], and next the equivalence of both Boltin's and Keller-Rubinow methods have been proved [77], [78].

1.6.2 Nonlinear Vibrations of a Rod

In order to present the main ideas of this method for a nonlinear case, we use the Kirchhoff equation:

$$EI\frac{\partial^4 w}{\partial x^4} - \frac{EF}{2L}\left[\int_0^L \left(\frac{\partial w}{\partial x}\right)^2 dx\right]\frac{\partial^2 w}{\partial x^2} + \rho F\frac{\partial^2 w}{\partial t^2} = 0. \tag{1.190}$$

Let the rod be elastically clamped, hence we have

$$\text{for } x = 0, L \quad w = 0, \quad \frac{\partial^2 w}{\partial x^2} - c^*\frac{\partial w}{\partial x} = 0, \tag{1.191}$$

where $c^* = c/(EI)$, and c is the coefficient of clamping.

Zero order solution is approximated by the following form

$$w_0 = A \sin \frac{\pi(x - x_0)}{\lambda} \xi(t). \tag{1.192}$$

Substituting Equation (1.192) into Equation (1.190) we get

$$\frac{d^2\xi}{dt^2} + \omega^2(1 + \gamma\xi^2)\xi = 0, \tag{1.193}$$

where

$$\omega^2 = EI\rho^{-1}\left(\frac{\pi}{\lambda}\right)^2, \quad \gamma = 0,25(1 + \lambda_1)\left(\frac{A}{r}\right)^2,$$

$$r = \sqrt{I/F}, \quad \lambda_1 = \frac{\lambda}{2\pi L}[\sin \frac{2\pi(L - x_0)}{\lambda} + \sin \frac{2\pi x_0}{\lambda}].$$

Equation (1.193) with the initial condition

$$\xi(0) = 1, \quad \frac{d\xi(0)}{dt} = 0 \tag{1.194}$$

has the following solution

$$\xi(t) = \text{cn}(\sigma t, k), \quad \sigma = \omega\sqrt{1 + \gamma}, \tag{1.195}$$

where cn(…,…) denotes Jacoby's cosine function of period T equal to $4K$; $K = \int_0^{\pi/2} (1 - k^2 \sin^2\varphi)^{-0,5} d\varphi$ is the full elliptic integral of the first kind with the modulus $k = \sqrt{0,5\gamma/(1 + \gamma)}$ ([2], chapter 16).

The solution of our problem far from edges is

$$w_0 = W_0(x)\text{cn}(\sigma t, k), \tag{1.196}$$

where $W_0(x) = A\sin(\pi(x - x_0))/\lambda$.

Solution (1.196) satisfies Equation (1.190) but it does not satisfy the BCs (1.191). In order to construct states localized in the vicinity of edges, we assume the following solution

$$w = w_0 + w_{cr}. \tag{1.197}$$

Substituting Equation (1.197) into Equation (1.190) yields

$$\frac{\partial^4}{\partial x^4}(w_0 + w_{cr}) - 0,5(r^2L)^{-1}\frac{\partial^2}{\partial x^2}(w_0 + w_{cr}) \int_0^L \left(\frac{\partial w_0}{\partial x} + \frac{\partial w_{cr}}{\partial x}\right)^2 dx$$
$$+ \frac{\rho}{EI}\frac{\partial^2}{\partial t^2}(w_0 + w_{cr}) = 0. \tag{1.198}$$

Observe that contrary to the earlier studied linear case, now functions w_0 and w_{cr} are coupled due to nonlinearity of the problem. On the other hand, the fundamental state as well as the edge effects differ strongly from the point of view of energy, since the latter one is localized in a boundary layer of the rod edges [15]. In what follows we are going to estimate orders of underintegral terms in Equation (1.198) with respect to $L/\lambda \gg 1$:

$$\int_0^L \left(\frac{\partial w_0}{\partial x}\right)^2 dx \sim \left(\frac{L}{\lambda}\right)^2; \quad \int_0^L \frac{\partial w_0}{\partial x}\frac{\partial w_{cr}}{\partial x} dx \sim \frac{L}{\lambda};$$

$$\int_0^L \left(\frac{\partial w_{cr}}{\partial x}\right)^2 dx \sim 1. \tag{1.199}$$

Taking into account only the term $(\pi/\lambda)^2$ in the first approximation in Equation (1.198), the latter equation is recast to the following form

$$\frac{\partial^4 w_0}{\partial x^4} - 0,5(r^2L)^{-1}\frac{\partial^2 w_0}{\partial x^2}\int_0^L \left(\frac{\partial w_0}{\partial x}\right)^2 dx + \frac{\rho}{EI}\frac{\partial^2 w_0}{\partial t^2}$$

$$+ \frac{\partial^4 w_{cr}}{\partial x^4} - 0,5(r^2L)^{-1}\frac{\partial^2 w_{cr}}{\partial x^2}\int_0^L \left(\frac{\partial w_0}{\partial x}\right)^2 dx + \frac{\rho}{EI}\frac{\partial^2 w_{cr}}{\partial t^2} = 0. \tag{1.200}$$

Substituting Equation (1.196) into Equation (1.200) yields the following equation

$$\frac{\partial^4 w_{cr}}{\partial x^4} - B\mathrm{cn}^2(\sigma t, k)\frac{\partial^2 w_{cr}}{\partial x^2} + \frac{\rho}{EI}\frac{\partial^2 w_{cr}}{\partial t^2} = 0, \tag{1.201}$$

where $B = \gamma(\frac{\pi}{\lambda})^2$.

Note that although we deal with the linear equation (1.201) but with time dependent coefficients. Since we cannot separate time and space dependent variables, we apply here Kantorovitch method [142]. Namely, we introduce the following approximation

$$w_{cr}(x,t) \cong W_{cr}(x)\mathrm{cn}(\sigma t, k). \tag{1.202}$$

Substituting Equation (1.202) into Equation (1.201) and reducing time the following ODE is obtained

$$\frac{d^4 W_{cr}}{dx^4} - B_1\frac{d^2 W_{cr}}{dx^2} - \left(\frac{\pi}{\lambda}\right)^2\left[\left(\frac{\pi}{\lambda}\right)^2 + B_1\right]W_{cr} = 0, \tag{1.203}$$

where

$$B_1 = A\left(\frac{2k^2-1}{2k^2} + \frac{\sqrt{1-k^2}}{2k\arcsin k}\right). \tag{1.204}$$

In relation (1.203) and further arcsin(…, …) is understood in the sense of its main value.

The characteristic equation of (1.203) has four roots, and two purely imaginary roots can be omitted here (they correspond to zero order solution). Real roots correspond to the following solution

$$W_{cr}(x) = C_1\exp\left[-\sqrt{\left(\frac{\pi}{\lambda}\right)^2 + B_1 x}\right] + C_2\exp\left[\sqrt{\left(\frac{\pi}{\lambda}\right)^2 + B_1 x}\right].$$

If the interaction of the rod edges can be neglected, then the condition of decaying of the boundary effect implies $C_2 = 0$ yielded by $x \to \infty$.

Satisfying the BC for $x = 0$ yields

$$W_0 + W_{cr} = 0, \quad \frac{d^2 W_0}{dx^2} + \frac{d^2 W_{cr}}{dx^2} = c^*\left(\frac{dW_0}{dx} + \frac{dW_{cr}}{dx}\right),$$

which allows to find the constant C_1:

$$C_1 = A \sin \frac{\pi x_0}{\lambda},$$

$$x_0 = \frac{\lambda}{\pi} \arctan \frac{\pi}{\lambda \left[\left(2(\pi/\lambda)^2 + B_1 \right) / c^* + \sqrt{(\pi/\lambda)^2 + B_1} \right]}. \quad (1.205)$$

Note that for $c^* \to 0$ and $c^* \to \infty$ formulas (1.204) yield also solution to the limiting cases of both simply supported and clamped rod edges.

Proceeding in the analogous way one may also construct the dynamics edge effect localized in the vicinity of $x = L$.

Forms of nonlinear rod vibrations can be separated into two groups with respect a symmetry type. In the case of symmetry regarding the point $x = L/2$, the condition

$$\frac{dW_0}{dx} = 0 \quad \text{for} \quad x = L/2$$

implies

$$L - 2x_0 = (2m+1)\pi, \quad m = 1, 2, \ldots. \quad (1.206)$$

In the case of antisymmetric forms, the condition

$$W_0 = 0 \quad \text{for} \quad x = L/2$$

yields

$$L - 2x_0 = 2n\pi, \quad n = 1, 2, \ldots. \quad (1.207)$$

Equations (1.206), (1.207) are reduced to the following one

$$L - 2x_0 = m\pi, \quad m = 1, 2, \ldots, \quad (1.208)$$

where odd (even) m values correspond to symmetric (antisymmetric) forms with respect to the point $x = L/2$.

Therefore, constants λ and x_0 are defined via Equations (1.204), (1.205) and (1.208).

1.6.3 *Nonlinear Vibrations of a Rectangular Plate*

We begin with Berger dynamic equation [76],[48]

$$D\nabla^4 w - T\nabla^2 w + \rho h \frac{\partial^2}{\partial t^2} w = 0, \quad (1.209)$$

$$Th^2 ab = 6D \int_0^b \int_0^a \left[\left(\frac{\partial w}{\partial x} \right)^2 + \left(\frac{\partial w}{\partial y} \right)^2 \right] dx\, dy. \quad (1.210)$$

Let the plate be ridigly clamped along its contour

$$\text{for} \quad x = 0, a \quad w = \frac{\partial w}{\partial x} = 0, \quad (1.211)$$

$$\text{for} \quad y = 0, b \quad w = \frac{\partial w}{\partial y} = 0. \quad (1.212)$$

The plate deflection follows

$$w_0(x, y, t) = A \sin k_1(x - x_1) \sin k_2(y - y_1)\xi(t), \tag{1.213}$$

where k_i, x_1, y_1 are constant quantities to be defined.

Substituting Ansatz (1.213) into Equations (1.209), (1.210) and reducing the space variables we obtain Equation (1.193) regarding time-dependent function $\xi(t)$ with the following coefficients

$$\omega^2 = D(\rho h)^{-1}(k_1^2 + k_2^2)^2,$$

$$\gamma = 1,5\left(\frac{A}{h}\right)^2 \frac{k_1^2(a_1 + \lambda_1)(a_2 - \lambda_2) + k_2^2(a_1 - \lambda_1)(a_2 + \lambda_2)}{a_1 a_2(k_1^2 + k_2^2)},$$

$$\lambda_1 = 0,5k_1^{-1}[\sin 2k_1(x - x_1)]|_0^a, \ \lambda_2 = 0,5k_2^{-1}[\sin 2k_2(y - y_1)]|_0^b.$$

A solution to this time-dependent equation takes the form (1.195) for initial conditions (1.194).

Therefore, the following zero order solution is found

$$w_0 = A \sin k_1(x - x_1) \sin k_2(y - y_1)\mathrm{cn}(\sigma t, k), \tag{1.214}$$

being valid in the internal plate part located sufficiently far from the plate boundaries.

Let us now proceed to construction of dynamic edge effects localized in vicinity of the plate contour. Substituting w in the form of (1.197) into Equations (1.209), (1.210) yields

$$D\nabla^4(w_0 + w_{cr}) - T\nabla^2(w_0 + w_{cr}) + \rho h\frac{\partial^2}{\partial t^2}(w_0 + w_{cr}) = 0, \tag{1.215}$$

$$Th^2ab = 6D\int_0^b\int_0^a\left[\left(\frac{\partial(w_0 + w_{cr})}{\partial x}\right)^2 + \left(\frac{\partial(w_0 + w_{cr})}{\partial y}\right)^2\right]dx\,dy. \tag{1.216}$$

Let us apply further the energy approach used in the case of a study of nonlinear vibrations of the rod. We estimate an order of the quantities standing on the r.h.s. of Equation (1.216) with respect to nondimensional parameters $ak_1 \sim bk_2 \gg 1$. The following relations hold:

$$\begin{aligned}
&\int_0^b\int_0^a\left(\frac{\partial w_0}{\partial x}\right)^2 dx\,dy \sim a^2k_1^2, &&\int_0^b\int_0^a\left(\frac{\partial w_0}{\partial y}\right)^2 dx\,dy \sim b^2k_2^2,\\
&\int_0^b\int_0^a\frac{\partial w_0}{\partial x}\frac{\partial w_{cr}}{\partial x}dx\,dy \sim ak_1, &&\int_0^b\int_0^a\frac{\partial w_0}{\partial y}\frac{\partial w_{cr}}{\partial y}dx\,dy \sim bk_2,\\
&\int_0^b\int_0^a\left(\frac{\partial w_{cr}}{\partial x}\right)^2 dx\,dy \sim 1, &&\int_0^b\int_0^a\left(\frac{\partial w_{cr}}{\partial y}\right)^2 dx\,dy \sim 1.
\end{aligned} \tag{1.217}$$

The given estimations imply that in relation (1.216) one may keep the solution components of order $a^2k_1^2 \sim b^2k_2^2 \gg 1$. In the first approximation, which depends only on the fundamental state, Equations (1.215), (1.216) are cast to the following form:

$$\nabla^4 w_{cr} - B\mathrm{cn}^2(\sigma t, k)\nabla^2 w_{cr} + \rho h D^{-1}\frac{\partial^2 w_{cr}}{\partial t^2} = 0, \tag{1.218}$$

where $B = \gamma(k_1^2 + k_2^2)$.

The linear PDE (1.218) with time-dependent coefficients serves for construction of four dynamic edge effects regarding the plate edges $x = 0, a$ and $y = 0, b$. We assume again that the plate sides possess sufficient length that their interaction can be neglected.

Let us consider an edge effect localized in vicinity of the edge $x = 0$ (the remaining cases can be solved in a similar way). We take

$$w_{cr}(x, y, t) = \Phi(x, t) \sin k_2(y - y_1),$$

and we obtain the following PDE

$$\frac{\partial^4 \Phi}{\partial x^4} - \left[B\mathrm{cn}^2(\sigma t, k) + 2k_2^2\right] \frac{\partial^2 \Phi}{\partial x^2} + k_2^2 \left[B\mathrm{cn}^2(\sigma t, k) + 2k_2^2\right] \Phi + \rho h D^{-1} \frac{\partial^2 \Phi}{\partial t^2} = 0. \qquad (1.219)$$

We apply Kantorovitch approach by assuming

$$\Phi(x, t) \cong \varphi(x)\mathrm{cn}(\sigma t, k).$$

Then, Equation (1.219) wields

$$\frac{d^4 \varphi}{dx^4} - (B_1 + 2k_2^2)\frac{d^2 \varphi}{dx^2} + \left[k_2^2(B_1 + 2k_2^2) - (k_1^2 + k_2^2)(B_1 + k_1^2 + k_2^2)\right] \varphi = 0. \qquad (1.220)$$

where

$$B_1 = B \left[\frac{2k^2 - 1}{2k^2} + \frac{\sqrt{1 - k^2}}{2k \arcsin k}\right].$$

Equation (1.220) is recast in the following equivalent form

$$\left(\frac{d^2}{dx^2} + k_1^2\right) \left(\frac{d^2}{dx^2} - k_1^2 - 2k_2^2 - B_1\right) \varphi = 0.$$

We neglect the first multiplier describing the fundamental state, and the second equation has a solution associated with the edge effect

$$\varphi(x) = C_1 \exp\left(-\sqrt{k_1^2 + 2k_2^2 + B_1 x}\right) + C_2 \exp\left(\sqrt{k_1^2 + 2k_2^2 + B_1 x}\right).$$

Finally, the edge effect in the neighborhood of the plate edge $x = 0$ can be written in the following form

$$w_{cr}^{(1)} = \left[C_1 \exp\left(-\sqrt{k_1^2 + 2k_2^2 + B_1 x}\right)\right.$$
$$\left. + C_2 \exp\left(\sqrt{k_1^2 + 2k_2^2 + B_1 x}\right)\right] \sin k_2(y - y_1)\mathrm{cn}(\sigma t, k). \qquad (1.221)$$

Proceeding in analogous way, the edge effect in vicinity of the plate edge $y = 0$ is governed by the following function

$$w_{cr}^{(2)} = \left[C_3 \exp\left(-\sqrt{k_1^2 + 2k_2^2 + B_1 y}\right)\right.$$
$$\left. + C_4 \exp\left(\sqrt{k_1^2 + 2k_2^2 + B_1 y}\right)\right] \sin k_1(x - x_1)\mathrm{cn}(\sigma t, k). \qquad (1.222)$$

We express BCs (1.211), (1.212), taking into account Equation (1.197), in the following form:

$$\text{for} \quad x=0 \qquad w_0 + w_{cr}^{(1)} = 0, \qquad \frac{\partial}{\partial x}\left(w_0 + w_{cr}^{(1)}\right) = 0; \tag{1.223}$$

$$\text{for} \quad y=0 \qquad w_0 + w_{cr}^{(2)} = 0, \qquad \frac{\partial}{\partial y}\left(w_0 + w_{cr}^{(2)}\right) = 0. \tag{1.224}$$

Conditions of decaging of the edge effects follow:

$$\text{for} \quad x \to \infty \quad w_{cr}^{(1)} \to 0, \tag{1.225}$$

$$\text{for} \quad y \to \infty \quad w_{cr}^{(2)} \to 0. \tag{1.226}$$

Then satisfaction of Equations (1.225), (1.226) implies $C_2 = C_4 = 0$, and next from Equations (1.223), (1.224) we get:

$$\begin{aligned} x_1 &= k_1^{-1} \arctan\left[k_1\left(k_1^2 + 2k_2^2 + B_1\right)^{-1/2}\right], \\ C_1 &= f k_1 \left[2\left(k_1^2 + k_2^2\right) + B_1\right]^{-1/2}; \end{aligned} \tag{1.227}$$

$$\begin{aligned} y_1 &= k_2^{-1} \arctan\left[k_2\left(k_1^2 + 2k_2^2 + B_1\right)^{-1/2}\right], \\ C_3 &= f k_2 \left[2\left(k_1^2 + k_2^2\right) + B_1\right]^{-1/2}. \end{aligned} \tag{1.228}$$

Let us split forms of plate vibrations into two groups with respect to a symmetry type regarding the lines $x = 0,5a$, $y = 0,5b$. In the case of symmetric form in both directions we have

$$\frac{\partial w_0}{\partial x} = 0 \quad \text{for} \quad x = a/2, \qquad \frac{\partial w_0}{\partial y} = 0 \quad \text{for} \quad y = b/2.$$

which implies

$$k_1(a - 2x_1) = (2m+1)\pi, \tag{1.229}$$

$$k_2(b - 2y_1) = (2n+1)\pi, \qquad m, n = 1, 2, \ldots. \tag{1.230}$$

Analogously, in the case of an antisymmetric form in both directions we have

$$w_0 = 0 \quad \text{for} \quad x = a/2, \qquad w_0 = 0 \quad \text{for} \quad y = b/2.$$

Relations (1.229), (1.230) yield

$$k_1(a - 2x_1) = 2p\pi, \tag{1.231}$$

$$k_2(b - 2y_1) = 2q\pi, \qquad p, q = 1, 2, \ldots. \tag{1.232}$$

Equations (1.229), (1.231) and (1.230), (1.234) can be cast into the following relations

$$k_1(a - 2x_1) = m\pi, \tag{1.233}$$

$$k_2(b - 2y_1) = n\pi, \qquad m, n = 1, 2, \ldots. \tag{1.234}$$

It means that one may get all possible vibration forms through proper choice of m and n.

1.6.4 Matching of Asymptotic and Variational Approaches

Asymptotic method of dynamical edge effect is generally devoted to determination of high frequencies and associated forms, but it yields also correct results for low frequency spectrum assuming that we deal with the kinematic boundary conditions. However, in the static problems the accuracy of low frequencies estimation in the object contour decreases. Application of the method described so far to estimate first fundamental frequency does not yield satisfactory results.

However, there exists a prospective direction to increase the efficiency of the dynamical edge effect method through matching it with one of the energetic approaches. The latter approaches allow us not only to improve the accuracy of the results obtained, but also to extend its area of application.

Let us consider the eigenvalue problem of a square ($0 \leq x, y \leq a$) of a plate simply supported along its contour. The governing equation is obtained from Equation (1.209) putting $N = 0$. The BCs have the following form:

$$w_{xx} + \nu w_{yy} = 0, \qquad w_{xxx} - 2(1-\nu)w_{xyy} = 0 \qquad \text{for} \quad x = 0, a; \tag{1.235}$$

$$w_{yy} + \nu w_{xx} = 0, \qquad w_{yyy} - 2(1-\nu)w_{yxx} = 0 \qquad \text{for} \quad y = 0, a. \tag{1.236}$$

In order to estimate the eigenfrequency the Rayleigh-Ritz method is applied [47], which uses the principle of virtual displacements. According to this principle, the work of internal and external forces acting on the plate on its virtual displacements is equal zero, which means that

$$U + V + R = 0. \tag{1.237}$$

Potential (U) and kinetic (V) energies of the plate are as follows

$$U = \frac{D}{2}\int_0^a\int_0^a (w_{xx}^2 + w_{yy}^2 + 2\nu w_{xx}^2 w_{yy}^2 + 2(1-\nu)w_{xy}^2)dx\, dy, \tag{1.238}$$

$$V = \frac{1}{2}\int_0^a\int_0^a \rho h w_t^2\, dx\, dy. \tag{1.239}$$

In this case, the work of external forces R is equal zero. Assuming the plate deflection in the following form

$$w(x, y, t) = W(x, y)\exp(i\omega t),$$

Equation (1.237) yields the plate frequency

$$\lambda^2 = \omega^2 a^4 \frac{\rho h}{D} = a^4 \left[\int_0^a\int_0^a \left(W_{xx}^2 + W_{yy}^2 + 2\nu W_{xx}^2 W_{yy}^2 + 2(1-\nu)W_{xy}^2\right) dx\, dy\right] \times \left[\int_0^a\int_0^a W^2\, dx\, dy\right]^{-1}. \tag{1.240}$$

Application of the dynamical edge effect method yields the following formula governing the plate deflection

$$W(x, y) = W_0(x, y) + W_1(x)\sin(\beta_2 y + l_2) + W_2(y)\sin(\beta_1 x + l_1),$$

where

$$W_0(x) = \sin(\beta_1 x + l_1)\sin(\beta_2 + l_2), \tag{1.241}$$

$$W_1(x) = C_{11}\exp[\alpha_1(x-a)] + C_{12}\exp(-\alpha_1 x),$$

$$W_2(x) = C_{21}\exp[\alpha_2(y-a)] + C_{22}\exp(-\alpha_2 y). \tag{1.242}$$

Taking into account the boundary conditions, wave numbers are defined through the following system of transcendental equations:

$$\beta_i a = 2l_i + m\pi, \quad i = 1,2, \quad m = 0,1,2,\dots, \tag{1.243}$$

where

$$l_i = \arctan\left[\frac{\beta_i}{\alpha_i}\left(\frac{\beta_i^2 + (2-\nu)\beta_k^2}{\beta_i^2 + \nu\beta_k^2}\right)^2\right],$$

$$\alpha_i = (\beta_i^2 + 2\beta_k^2)^{1/2}, \quad i = 1,2, \quad k = 1,2, i \neq k.$$

constants C_{ij} in Equation (1.242) are defined as follows

$$C_{i1} = \frac{\alpha_i^2 \sin l_i}{\alpha_i^2 - \nu\beta_k^2}, \quad C_{i2} = \frac{\alpha_i^2 \sin(\beta_i a + l_i)}{\alpha_i^2 - \nu\beta_k^2}, \quad i = 1,2, \quad k = 1,2, \quad i \neq k. \tag{1.244}$$

Formula (1.240), taking into account Equations (1.241)–(1.244), allows to define the plate frequency.

In Table 1.9 a comparison of the obtained nondimensional frequency λ of the square plate being free along its contour for $\nu = 0.225$ is reported (Rayleigh-Bolotin method (MRB) with values obtained by Rayleigh-Ritz [121] as well as the traditional Bolotin asymptotic method (AMB) are applied). We do not consider vibrations with respect to the cylindrical surface, since in the latter case one may get the exact solution. This is why numbers corresponding to the associated vibration forms are omitted in Table 1.9.

Results obtained via MRR of higher order approximations have the high order of accuracy in the interval of lower eigenfrequencies. However, increasing the number of a vibration form implies decrease of the obtained accuracy. Comparison of data obtained via different methods shows that error of first frequency estimation through MRB (2.7%) is essentially less than that using the traditional method of dynamical effect (13.6%). Increasing vibration form number

Table 1.9

Number of vibrations	λ, MRR[121]	λ, MRB	Error, %	λ, AMB	Error, %
1	14, 10	14, 48	2, 7	12, 41	13, 6
3	35, 96	36, 68	2, 0	34, 60	3, 9
5	65, 24	66, 33	1, 7	63, 44	2, 8
6	74, 45	75, 28	1, 1	73, 59	2, 5
7	109, 30	109, 10	0, 2	106, 30	2, 8

both asymptotic solutions approach the exact value (MRB yield upper, whereas edge effect lower bands estimation).

Consider now the problem of stability of a square clamped isotropic plate subjected to action of uniformly distributed compressing load N_0 applied to the middle plate surface. BCs are governed by (1.211), (1.212) ($b = a$), and in Equation (1.209) one needs to substitute T by $-T_0$. We apply MRB in order to find the critical compressing load. In this case the kinetic energy of the plate equals zero, whereas the work of active forces follows

$$R = -\frac{T_0}{2}\int_0^a \int_0^a (w_x^2 + w_y^2)\, dx\, dy.$$

Equation (1.237) yields the following nondimensional critical value of the force

$$P = \left(\frac{a}{\pi}\right)^2 \left[\int_0^a \int_0^a (w_{xx}^2 + 2w_{xx}^2 w_{yy}^2 + w_{yy}^2)\, dx\, dy\right]\left[\int_0^a \int_0^a (w_x^2 + w_y^2)\, dx\, dy\right]^{-1}, \quad (1.245)$$

where $P = T_0 a^2/(D\pi^2)$.

A solution to the problem is taken in the form of (1.241). Due to the symmetry of the problem, the wave numbers are detected from the following equation

$$\beta a = 2\arctan\,[\mathrm{th}(0,5\beta a)] + \pi.$$

The constants l_i, C_i are defined as follows

$$l_i = 0,5(\pi - \beta a), \quad C_{i1} = C_{i2} = -\cos(0.5\beta a)/\mathrm{ch}(0.5\beta a), \quad i = 1, 2.$$

This found solution is substituted into Equation (1.245), and then P is defined. The critical compressing force obtained via MRR [253] is $P = 5,31$. The computational error associated with the AMB (MRB) application reaches the value of 18% (8%).

The approach described so far also allows us to apply the method of dynamical edge effect for plates of either complicated forms (rectangular, circle [17], [18]) or design (for instance ribbed plates [16]).

1.6.5 On the Normal Forms of Nonlinear Vibrations of Continuous Systems

It is known that while investigating linear vibrations of discrete systems with finite number of degrees of freedom a key role is played by normal vibrations. Kauderer [141] has shown that also in a nonlinear system there exist solutions playing a similar role to that exhibited by normal vibrations of linear systems. He called them main vibrations and proposed a way of constructing their trajectories in a configuration state. Rosenberg [213], in order to define normal forms of vibrations of nonlinear systems with a finite degrees of freedom, formulated the problem in the configuration space in an approximate way and succeded in finding a few classes of nonlinear systems having solutions exhibited by lines teajectories (see [168], [171], [170], [239]). Attempts to generalize the introduced concepts into continuous systems are associated with separation of finite and space variables ([57], [171], [225]), i.e. they rely on the possibility of the following representation

$$U(\mathbf{x}, t) = X(\mathbf{x})T(t).$$

It is clear that the latter approach can be validated only for some boundary conditions. Applying a concept of the dynamical edge effect one may introduce the following definition. We say that a function $U(x, t)$ is called a normal form of nonlinear vibration of a continuous system if

$$U(\mathbf{x}, t) = X(\mathbf{x})T(t) + Y(\mathbf{x}, t),$$

where $T(t)$ stands for a periodic function, $Y(\mathbf{x}, t)$ is a quasi-periodic function, and $Y(\mathbf{x}, t)$ is small in comparison to $X(\mathbf{x})T(t)$ in a certain energetic norm. The latter condition can be either verified by a priori or posteriori approach.

1.7 Continualization

1.7.1 Discrete and Continuum Models in Mechanics

In mechanics of continuum media we deal with continuous objects described by continuous functions. Although this is reliable, it is a common article of faith that somehow the average of the microsystems is exactly described by the equations of fluid mechanics, but no one has proved this for a *realistic* model of the fine structure: the best proofs are for idealized models of a rarefied monatomic gas ([208] p. 217). A transition from a real discrete nature to its continuum model requires the introduction of a certain averaging. On the other hand, the key factors of the theory of elasticity, stress, strain and the rest are formally *defined* as limits, by considering arbitrarily small parts of the body, but are only *meaningful* as representing average behavior over regions that are large in comparison to atoms ([208], p. 286). The usually applied methods of averaging and homogenization are described in chapter 6. Since continualization of discrete relations has its own peculiarities, we have decided to illustrate them in this chapter. It is appropriate to introduce the reader to this matter using relatively simple examples, since the construction of mechanics of a continuum medium using only "first principles" and applied so far only to molecular theory belongs to more difficult branches of physics [260], [150], [152].

The average approach works reasonably well for the determination of global characteristics. As has been pointed out by Ulam ([236], p. 89, 90), the simplest problems involving an actual infinity of particles in distribution of matter already appear in classical mechanics. A discussion on these will permit us to introduce more general schemes which may possibly be useful in future physical theories.

Strictly speaking, one has to consider a true infinity in the distribution of matter in all problems of the physics of continua. In the classical treatment, as usually given in textbooks of hydrodynamics and field theory, this is, however, not really essential, and in most theories serves merely as a convenient limiting model of *finite* systems enabling one to use the algorithms of the calculus. The usual introduction of the continuum leaves much to be discussed and examined critically. The derivation of the equations of motion for fluids, for example, runs somewhat as follows. One images a very large number N of particles, say with equal masses constituting a net approximating the continuum, which are to be studied. The forces between these particles are assumed to be given, and one writes Lagrange equations for the motion of N particles. The finite system of ODEs becomes in the limit $N = \infty$ one or several *partial* DEs. The Newtonian laws of conservation of energy and momentum are seemingly correctly formulated for the limiting case of the continuum. There appears at once, however, at least a possible objection to the unrestricted validity of this formulation. For the very fact that the limiting equations imply tacitly the continuity and differentiability of the functions describing the motion of the continuum seems to impose various *constraints* on the possible motions of the approximating finite systems. Indeed, at any stage of the limiting process, it is quite conceivable for two neighboring particles to be moving in opposite directions with a relative velocity which does not need to tend to zero as N becomes infinite, whereas the continuity imposed on the solution of the limiting continuum excludes such a situation. There are, therefore, constraints on the class of possible motions which are not explicitly recognized. This means that a viscosity or other type of constraints must be introduced initially, singling out "smooth" motions from the totality of all possible ones. In some cases, therefore, the usual DEs of hydrodynamics may constitute a misleading description of the physical process.

On the other hand, nowadays development in technology and industry requires inclusion of micro-structural effects, which may play a crucial role when the characteristic magnitude of an excitation is of order of the characteristic size of the analyzed micro-structure object. In particular, we mention here modeling of crystal, polymer and composite materials, nano-materials, dynamics of cracks, description of hysteretic effects, mechanics of failures, fractals theory of phase transition and theory of plasticity [24], [55], [66], [91], [93], [111], [150], [199], [201], [203].

Micro-structural effects can be investigated within the frame of discrete models [152], however, even modern computers do not allow us to get reasonably validated results matched with the reasonably low computational time. Therefore, continuum description of micro- and nano-effects belongs to a challenging research topic. In addition, in many cases, one may apply modeling of mixed discrete-continuum systems, where its one part is continuous and the other is discrete.

1.7.2 Chain of Elastically Coupled Masses

Let us consider a simple example of a chain consisting of $n+2$ particles of the same masses m lying in the rest in the points of axis x with the coordinates jh $(j = 0, 1, \ldots, n, n+1)$ and linked by elastic couplings of stiffness c (Figure 1.7).

According to Hook's law the elastic force acting on the j-th mass is as follows:

$$\begin{aligned}\sigma_j(t) &= c[y_{j+1}(t) - y_j(t)] - c[y_j(t) - y_{j-1}(t)]\\ &= c[y_{j-1}(t) - 2y_j(t) + y_{j+1}(t)], \quad j = 1, 2, \ldots, n,\end{aligned}$$

where $y_j(t)$ is the displacement of j-th point with regard to the equilibrium position.

Applying Newton second law the following ODEs are derived

$$my_{jtt}(t) = c[y_{j-1}(t) - 2y_j(t) + y_{j+1}(t)], \quad j = 1, 2, \ldots, n. \tag{1.246}$$

System (1.246) can be recast to the following form:

$$m\sigma_{jtt}(t) = c(\sigma_{j+1} - 2\sigma_j + \sigma_{j-1}), \quad j = 1, \ldots, n. \tag{1.247}$$

Let the chain ends be fixed, then

$$y_0(t) = y_{n+1}(t) = 0. \tag{1.248}$$

In general, the initial conditions have the following form

$$y_j(t) = \varphi_j^{(0)}, \quad y_{jt}(t) = \varphi_j^{(1)} \quad \text{for} \quad t = 0. \tag{1.249}$$

This model has been proposed by Newton in estimating sound velocity [70]. He assumed that sound in air moves in the same way as an elastic wave moves along the masses chain. Equation (1.247) has been studied by J. Bernoulli ([49], [205]) who considered the problem of massless finite elastic string composed of particles of equal masses uniformily located along the string.

As has been shown in [193], for an arbitrary solution to the problem (1.246), (1.248), (1.249) the full chain energy is constant. Besides, the solutions to the problems so far stated are asymptotically stable in the Lyapunov sense.

A solution to the problem (1.246), (1.248), (1.249) can be expressed via elementary functions with the help of a discrete variant of the method of variables separation. Therefore, normal oscillation forms are constructed

$$y_j(t) = C_j T(t), \quad j = 1, \ldots, n,$$

where constants C_j present the solution to the eigenvalue problem:

$$-\lambda C_j = C_{j+1} - 2C_j + C_{j-1}, \quad j = 1, \ldots, n, \quad C_0 = C_{n+1} = 0, \tag{1.250}$$

and the function $T(t)$ satisfies the following equation

$$mT_{tt} + c\lambda T = 0. \tag{1.251}$$

Solution to the eigenvalue problem (1.250) takes the following form [193]:

$$C_k = A \sin \frac{k\pi}{n+1}, \quad \lambda_k = 4\sin^2 \frac{k\pi}{2(n+1)}, \quad k = 1, 2, \ldots, n. \tag{1.252}$$

We assume a solution to Equation (1.251) in the form $T = A\exp(i\omega t)$. Relations (1.251), (1.252) yield Lagrange formula for determination of frequencies ω_k of the discrete system:

$$\omega_k = 2\sqrt{\frac{c}{m}} \sin \frac{k\pi}{2(n+1)}, \quad k = 1, 2, \ldots, n. \tag{1.253}$$

Since all values of λ_k are different, they are simple, and each of them is associated with one eigenvalue for $\mathbf{C}_k(C_1^{(k)}, C_2^{(k)}, \ldots, C_n^{(k)})$:

$$\mathbf{C}_k = \operatorname{cosec} \frac{k\pi}{n+1} \left(\sin \frac{k\pi}{n+1}, \sin \frac{2k\pi}{n+1}, \ldots, \sin \frac{nk\pi}{n+1} \right),$$
$$k = 1, 2, \ldots, n. \tag{1.254}$$

Eigenvectors are mutually orthogonal, and a square of the moduli of eigenvector follows:

$$|\mathbf{C}_k|^2 = \frac{n+1}{2} \operatorname{cosec}^2 \frac{k\pi}{n+1}, \quad k = 1, 2, \ldots, n. \tag{1.255}$$

Each eigenfrequency (1.253) is associated with the normal oscillations form:

$$y_j^{(k)}(t) = C_j^{(k)} [A_k \cos(\omega_k t) + B_k \sin(\omega_k t)], \quad k = 1, 2, \ldots, n. \tag{1.256}$$

A general solution to problem (1.247)–(1.249) is described by a sum of normal oscillations:

$$y_j(t) = \sum_{k=1}^{n} C_j^{(k)} [A_k \cos(\omega_k t) + B_k \sin(\omega_k t)], \quad j = 1, \ldots, n. \tag{1.257}$$

Let us consider now the problem of masses chain movement, when a constant unit force acts on particles with zero number. A solution to this system is described by Equation (1.247) with the following boundary and initial conditions;

$$\sigma_0(t) = 1, \quad \sigma_{n+1}(t) = 0, \tag{1.258}$$

$$\sigma_j(t) = \sigma_{jt}(t) = 0 \quad \text{for} \quad t = 0. \tag{1.259}$$

The nonhomogenous BVP (1.247), (1.258), (1.259) is transformed to the BVP with respect to Equation (1.247) with the homogenous boundary and nonhomogenous initial condition via the following relationship

$$\sigma_j(t) = 1 - \frac{jl}{n+1} + \sigma_{0j}(t).$$

In order to find function $\sigma_j^{(0)}(t)$, we apply the normal form method:

$$\sigma_j^{(k)}(t) = C_j^{(k)}[A_k \cos(\omega_k t) + B_k \sin(\omega_k t)], \quad k = 1, 2, \ldots, n.$$

As a result, the following exact solution of the problem (1.246), (1.258), (1.259) is obtained:

$$\sigma_j(t) = 1 - \frac{jl}{n+1} - \frac{1}{n+1}\sum_{k=1}^{n} \sin\frac{\pi kj}{n+1}\text{ctan}\frac{\pi k}{2(n+1)}\cos(\omega_k t),$$

$$j = 1, 2, \ldots, n. \tag{1.260}$$

1.7.3 *Classical Continuum Approximation*

For large values n we usually apply the continuum approximation of the discrete problem, which in our case (Equations (1.247), (1.258), (1.259)) takes the following form:

$$m\sigma_{tt}(x,t) = \text{ch}^2\sigma_{xx}(x,t), \tag{1.261}$$

$$\sigma(0,t) = 1, \quad \sigma(l,t) = 0, \tag{1.262}$$

$$\sigma(x,0) = \sigma_t(x,0) = 0, \tag{1.263}$$

where $l = (n+1)h$.

Having at hand a solution to the BVPs (1.261)–(1.263), one may transit to a solution of the discrete medium according to the formulas

$$\sigma_j(t) = \sigma(jh,t), \quad j = 0, 1, \ldots, n, n+1. \tag{1.264}$$

Formally, this approximation can be obtained in the following way. Let us denote by D the difference operator occurred in Equation (1.247):

$$m\sigma_{jtt}(t) = cD\sigma(t). \tag{1.265}$$

Using the operator $\exp(h\frac{\partial}{\partial x})$ one gets [152]:

$$D = \exp\left(h\frac{\partial}{\partial x}\right) + \exp\left(-h\frac{\partial}{\partial x}\right) - 2 = -4\sin^2\left(-\frac{ih}{2}\frac{\partial}{\partial x}\right). \tag{1.266}$$

In what follows we explain the obtained relation. The Maclaurin formula for the infinitely differentiated function $F(x)$ takes the form

$$F(x+1) = \left[1 + \frac{\partial}{\partial x} + \frac{1}{2!}\frac{\partial^2}{\partial x^2} + \ldots\right]F(x) = \exp\left(\frac{\partial}{\partial x}\right)F(x). \tag{1.267}$$

Expressions of the form $\exp(\partial/\partial x)$ are called pseudo-differential operators. Using conditions (1.265)–(1.267), we recast the system (1.247) in the form of pseudo-differential equation [178]:

$$m\frac{\partial^2\sigma}{\partial t^2} + 4c\sin^2\left(-\frac{ih}{2}\frac{\partial}{\partial x}\right)\sigma = 0. \tag{1.268}$$

Development of the pseudo-differential operator into the Maclaurin series yields:

$$\sin^2\left(-\frac{ih}{2}\frac{\partial}{\partial x}\right) = -\left(\frac{h^2}{4}\frac{\partial^2}{\partial x^2} + \frac{h^4}{48}\frac{\partial^4}{\partial x^4} + \frac{h^6}{1\,440}\frac{\partial^6}{\partial x^6} + \dots\right). \tag{1.269}$$

Keeping only the first term in the series (1.269), we obtain a continuum approximation (1.261). Application of the Maclaurin series requires a small difference in displacements of the neighboring particles. Physically it means that we are investigating vibrations of a few particles located on the space period (see Figure 1.15), i.e. we proceed within the so-called long wave approximation. Note that the vertical axis corresponds to displacements in direction x, and we deal with the one-dimensional problem.

The continuum system (1.261) has the following infinite spectrum:

$$\alpha_k = \pi\sqrt{\frac{c}{m}}\frac{k}{n+1}, \quad k = 1, 2, \dots. \tag{1.270}$$

Although formulas (1.270) approximate reasonably good low frequencies of vibrations of the discrete system (1.253), but the n-th frequency α_k of continuum system differs from the n-th frequency of discrete system ω_k more than 50%. Accuracy of approximation (1.270) can be increased, but the following general conclusion follows. Frequencies of the continuum system $\omega_{n+1}, \omega_{n+2}, \dots$ do not have any relations to those of the discrete system (see [200, chapter 20]).

Observe that L.I. Mandelsshtam criticized the described method [167]; however today it is rigorously approved mathematically with the help of the Fourier transform [152].

1.7.4 "Splashes"

It is not difficult to derive the exact solution to problem (1.261)–(1.263) using the D'Alembert method and operational calculus [154]:

$$\sigma(x,t) = H\left(nh\arcsin\left|\sin\left(\frac{\pi}{2n}\sqrt{\frac{c}{m}}t\right)\right| - x\right), \tag{1.271}$$

where $H(\dots)$ stands for Heriside's function.

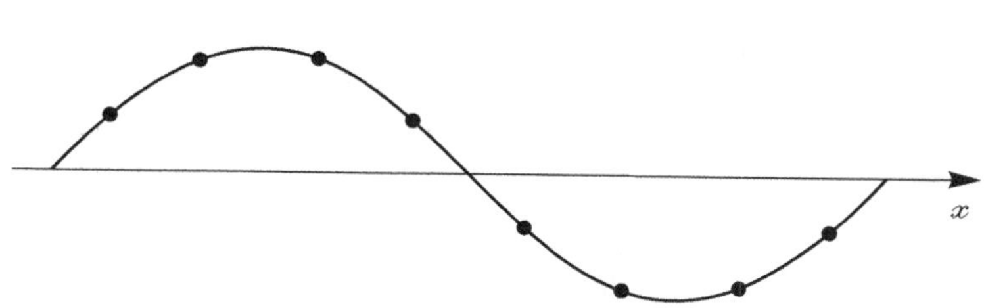

Figure 1.15 Solution form $\sigma = \sigma(x,t)$ in the fixed time instant $t = const$ (points correspond to a discrete system whereas solid curve represents a continuum system)

Table 1.10 Splashes

n	8	16	32	64	128	256	$n \to \infty$
P_n	1,7561	2,0645	2,3468	2,6271	2,9078	3,1887	$P_n \to \infty$

Formula (1.271) implies that for all time instants the following estimation holds

$$|\sigma(x, t)| \leq 1. \tag{1.272}$$

It is tempting to extend the estimation (1.272) into a discrete system [261] using relation (1.264). However, numerical and analytical investigations [154], [193], [105], [106], [107] have shown that one needs to include a difference between the global and local characteristics of a discrete system. Investigation the low spectrum of a discrete systems allows for a smooth transition into an averaged description. However, in the case of external excitations solutions of discrete systems do not smoothly transit into the wave equation solutions for $h \to 0$ [178]. It has been numerically shown [154], [193], [105], [106], [107] that for certain particles in the discrete chain the quantity $P_j = |\sigma_j(t)|$ can essentially overcome the bounded value 1 [105] (see Table 1.10).

It is interesting to note that the magnitude of splashes does not depend on the parameter m/c.

Amplitude of chain oscillations becomes arbitrary large with increase of N, but the system energy is constant and does not depend on N. However we do not deal here with a paradox, since oscillations amplitude is of order of the sum of $\sigma_j(t)$, whereas the potential energy is of the order of square of those quantities [193].

Amplitude of vibrations of a particle with the fixed number is bounded for $N \to \infty$, but the amplitude of vibrations of a particle with a certain number increases with the increase of N and tends to infinity for $N \to \infty$ in a way to that of $\ln N$ [193].

Note that a rigorous prove of the above observations is achieved assuming that $N + 1$ is either a simple number or a power of two. However, this result is of negligible meaning [193].

In the language of mechanics what we just said means that when analyzing the so-called "local properties" of a one-dimensional continuous medium, one cannot treat the medium as the limiting case of a linear chain of point masses, obtained when the number of points increases without limit [154]. Physically, this phenomenon can be interpreted in a rather simple way. Excited vibrations include both low and high harmonics, and the latter ones are defined via the continuum approximation with relatively large errors.

It is tempting to construct an improved theory of continuum media including splash effects. The mentioned theory should reasonably good describe harmonics of a solution with respect to an arbitrary period associated with the problem. In mathematical sense, the problem is reduced to that of approximation of nonlocal (difference) operator by the local (differential) one.

1.7.5 Envelope Continualization

It has been observed that asymptotics appear in pairs. Classical continuum approximation yields reliable results with respect to low part of the spectrum of the finite chain of particles.

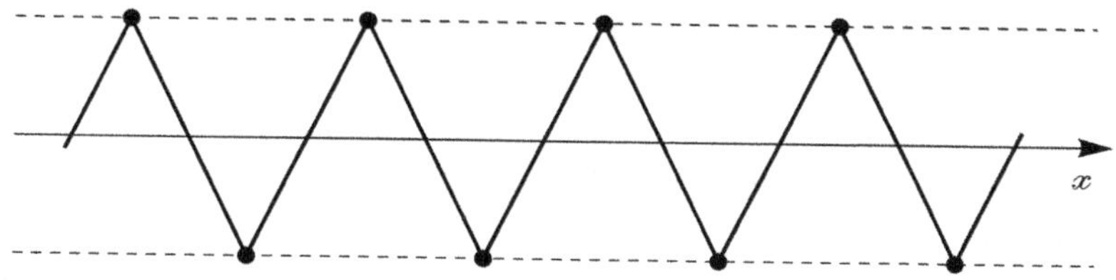

Figure 1.16 Saw-tooth chain vibrations

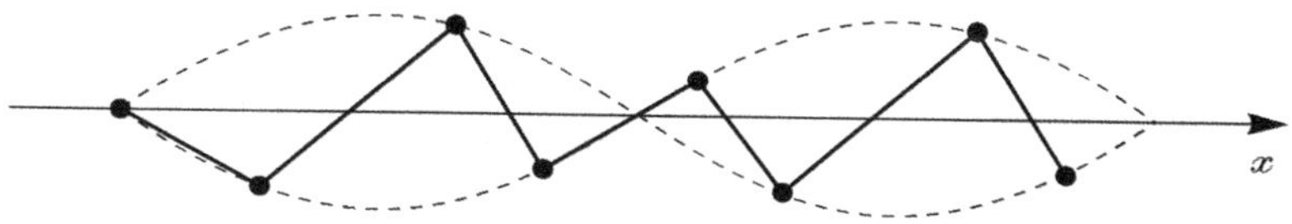

Figure 1.17 Envelope continualization

On the other hand, the maximum frequency of vibrations occurrs in the case of the saw-tooth vibrations (Figure 1.16)

In this case $\sigma_k = (-1)^k\Omega$, and the equation yielding Ω has the following form

$$m\Omega_{tt} + 4c\Omega = 0.$$

In the case of oscillations close to a saw-tooth shape, one may use the so called short-wave approximation (envelope continualization) [149] (Figure 1.17). A change of variables

$$\sigma_k = (-1)^k\Omega_k \tag{1.273}$$

allows us to transit from Equations (1.247), (1.258), (1.259) to the following BVP:

$$m\Omega_{ktt} + c(4\Omega_k + \Omega_{k-1} - 2\Omega_k + \Omega_{k+1}) = 0, \tag{1.274}$$

$$\Omega_0 = 1, \quad \Omega_{n+1} = 0, \tag{1.275}$$

$$\Omega_k = \Omega_{kt} = 0 \quad \text{for} \quad t = 0, \quad k = 0, 1, \ldots, n+1. \tag{1.276}$$

Further, the following relation is used:

$$\Omega_{k-1} - 2\Omega_k + \Omega_{k+1} = -4\sin^2\left(-\frac{ih}{2}\frac{\partial}{\partial x}\right)\Omega =$$

$$\left(h^2\frac{\partial^2}{\partial x^2} + \frac{h^4}{12}\frac{\partial^4}{\partial x^4} + \frac{h^6}{320}\frac{\partial^6}{\partial x^6} + \ldots\right)\Omega. \tag{1.277}$$

Substituting Equation (1.277) into Equation (1.274) considering h^2 as the small parameter, and taking into account only terms of zero and first orders with respect to h^2, we get

$$m\Omega_{tt} + 4c\Omega + ch^2\Omega_{xx} = 0. \tag{1.278}$$

It is not difficult to derive the boundary and initial conditions to Equation (1.278):

$$\Omega = 1 \quad \text{for} \quad x = 0, \qquad \Omega = 0 \quad \text{for} \quad x = l,$$

$$\Omega = \Omega_t = 0 \quad \text{for} \quad t = 0.$$

The approaches described so far (classical continualization and envelope continualization) can be treated as a discrete model with reasonably good accuracy, and it exhibits two continual approximations, i.e. for a chain and for an envelope.

1.7.6 Improvement Continuum Approximations

In what follows we discuss the problem of the improved continuum approximations. If in the series (1.269) we keep three first terms, then the following equation is obtained:

$$m\frac{\partial^2 \sigma}{\partial t^2} = ch^2\left(\frac{\partial^2}{\partial x^2} + \frac{h^2}{12}\frac{\partial^4}{\partial x^4} + \frac{h^4}{360}\frac{\partial^6}{\partial x^6}\right)\sigma. \tag{1.279}$$

The problem regarding boundary conditions of Equation (1.279) does not belong to trivial ones (see the interesting discussion by Raman and Bohr regarding periodic conditions for the chain [61]). One may define them only if we determine the chain movement for $k = -1, -2, -3, k = N+2, N+3, N+4$. In other words, the boundary is substituted by the boundary domain [152]. In particular, in the case of periodic extension (simple support), we get

$$\sigma = \sigma_{xx} = \sigma_{xxxx} = 0 \quad \text{for} \quad x = 0, l. \tag{1.280}$$

If we take $\sigma_k(t) = 0$ for $k = -1, -2, -3, k = N+2, N+3, N+4$, then BCs (1.280) refer to clamping

$$\sigma = \sigma_x = \sigma_{xxx} = 0 \quad \text{for} \quad x = 0, l.$$

Comparison of the n-th frequency of the continuum system (1.279), (1.280) with the corresponding frequency of a discrete system exhibits the essential increase of accuracy (we have 2.1 instead of 2 in the exact solution, which yields the error of 5%). Note that the estimation of the continuum approximation error with respect to maximal frequency of the discrete chains is somehow conventional, but most simple.

In the general case, keeping in (1.269) N terms, one gets the so-called intermediate continuum models [105]. It is assumed that for $N = \infty$ the exact input equation is obtained:

$$m\frac{\partial^2 \sigma}{\partial t^2} = 2c\sum_{k=1}^{N}\frac{h^{2k}}{(2k)!}\frac{\partial^{2k}\sigma}{\partial x^{2k}}. \tag{1.281}$$

BCs for Equation (1.281) have the following form:

$$\frac{\partial^{2k}\sigma}{\partial x^{2k}} = 0 \quad \text{for} \quad x = 0, l, \quad k = 0, 1, \ldots, N-1 \tag{1.282}$$

or

$$\sigma = 0, \quad \frac{\partial^{2k-1}\sigma}{\partial x^{2k-1}} = 0 \quad \text{for} \quad x = 0, l, \quad k = 1, \ldots, N-1. \tag{1.283}$$

The corresponding BVPs are correct (they are also stable during the numerical realization) for odd N. In this case Equation (1.281) is of the hyperbolic type [106]. Application of the intermediate continuum models allow us to determine the splash effects [105].

Observe that analogous ideas presented so far are used in the method of differential approximations to estimate errors of difference systems [227].

Construction of intermediate continuum models is based on the development of the difference operator into the Taylor series. It seems that the more effective ones are continuum models relying on the Padé approximations, which are called quasi-continuum approximations [84], [211], [212]. In order to approximate the operator (1.269) within the Padé algorithm, one may also apply either Fourier or Laplace transforms. If one keeps only three first terms in the series (1.269), then the Padé approximation follows

$$\frac{\partial^2}{\partial x^2} + \frac{h^2}{12}\frac{\partial^4}{\partial x^4} \approx \frac{\frac{\partial^2}{\partial x^2}}{1 - \frac{h^2}{12}\frac{\partial^2}{\partial x^2}}.$$

The corresponding quasi-continuum model takes the form

$$m\left(1 - \frac{h^2}{12}\frac{\partial^2}{\partial x^2}\right)\sigma_{tt} - ch^2\sigma_{xx} = 0. \tag{1.284}$$

BCs for Equation (1.284) have the following form

$$\sigma = 0 \quad \text{for} \quad x = 0, l. \tag{1.285}$$

Error estimation of the n-th frequency in comparison to the discrete chain is of the amount of 16.5%. Equation (1.284) has lower dimension in comparison to approximation (1.279).

Now, having in hand both long- and short-wave asymptotes, one may apply two points Padé approximation (see chapter 9.2). Let us construct two-point Padé approximation of difference operator, using the first term of series (1.269). Besides, we require that n-th frequency of vibrations of the continuum system should coincide with the corresponding frequency of the discrete system $\omega_n = 2\sqrt{c/m}\sin[n\pi/2(n+1)]$. For large values of n one may apply the following approximation

$$\alpha_n \approx 2\sqrt{c/m}. \tag{1.286}$$

Continuum approximation is governed by the following equation

$$m\left(1 - a^2h^2\frac{\partial^2}{\partial x^2}\right)\sigma_{tt} - ch^2\sigma_{xx} = 0, \tag{1.287}$$

with the BCs (1.285).

Frequencies of vibrations yielded by the BVP (1.287), (1.285) follow

$$\alpha_k = \pi\sqrt{\frac{c}{m}}\frac{k}{\sqrt{(n+1)^2 + \alpha^2k^2}}, \quad k = 1, 2, \ldots. \tag{1.288}$$

Application of formula (1.286) yields $a^2 = 0,25 - \pi^{-2}$. The largest error in estimation of the eigenfrequencies is achieved for $k = [0,5(n+1)]$ and does not overcome 3%. Approximation (1.287) allows us to include the splash phenomenon.

1.7.7 Forced Oscillations

Let us begin with the classical continuum approximations. A solutions to the Equation (1.261) is as follows

$$\sigma = 1 - \frac{x}{l} + u(x, t),$$

and functions $u(x, t)$ are defined via the following formulas

$$m\frac{\partial^2 u}{\partial t^2} = \mathrm{ch}^2\frac{\partial^2 u}{\partial x^2}, \tag{1.289}$$

$$u(0, t) = u(l, t) = 0, \tag{1.290}$$

$$u(x, 0) = -1 + \frac{x}{l}, \quad u_t(x, 0) = 0. \tag{1.291}$$

A solution to the BVP (1.289)–(1.291) is found using the Fourier method, and has the following form

$$\sigma = 1 - \frac{x}{l} - \frac{2}{\pi}\sum_{k=1}^{\infty}\frac{1}{k}\sin\left(\frac{k\pi x}{l}\right)\cos(\alpha_k t), \tag{1.292}$$

where α_k, $k = 1, 2, 3, \ldots$.

Formula (1.292) describes oscillations either of a string or a rod. If we are aimed on approximation of the particle chain oscillations, then we keep only n first harmonics in the infinite sum, since the remaining ones have no relations to the chain movements:

$$\sigma = 1 - \frac{x}{l} - \frac{2}{\pi}\sum_{k=1}^{n}\frac{1}{k}\sin\left(\frac{k\pi x}{l}\right)\cos(\alpha_k t). \tag{1.293}$$

Observe that solution (1.260) and (1.293) differ from each other not only regarding frequencies α_k and ω_k (formulas (1.253) and (1.270), respectively). In order to overcome this drawback one may use either equations (1.279), (1.284) or (1.286). However, also the coefficients of the series (1.292) and (1.293) understood as projections onto the normal oscillation forms at a discrete and continuum system $\frac{1}{n+1}\mathrm{ctan}\frac{\pi k}{2(n+1)}$ and $\frac{2}{\pi k}$ differ strongly for $k \gg 1$. This phenomenon appears due to the development into series regarding normal forms for a discrete system using the summation of k from 1 to n, whereas integration with respect to x is carried out from 0 to l in the case of a continuous system. One may improve the results using the Euler-Maclaurin formulas [100], [101]:

$$\sum_{k=0}^{n+1} f(k) = \int_0^{n+1} f(x)dx + \frac{1}{2}[f(0) + f(n+1)] + \tag{1.294}$$

$$\sum_{j=1}^{\infty}\frac{(-1)^{j+1}}{j+1}B_j\left[\frac{d^j f(n+1)}{dx^j} - \frac{d^j f(0)}{dx^j}\right].$$

Here B_i are Bernoulli numbers, where $B_0 = 1$, $B_1 = -1/2$, $B_2 = 1/6$, $B_3 = 0$.

One may also apply the following recurrence formula:

$$B_n = -\frac{1}{n+1}\sum_{k=1}^{n} C_{n+1}^{k+1} B_{n-k}.$$

In order to introduce the development regarding normal forms of a discrete system, one may apply formulas 4.4.2.6, 4.4.1.5 and 4.4.1.7 taken from handbook [209]

$$\sum_{k=0}^{n+1}\sin^2\frac{k\pi j}{n+1} = \frac{n+1}{2}, \tag{1.295}$$

$$\sum_{k=0}^{n+1}\left(1-\frac{j}{n+1}\right)\sin\frac{k\pi j}{n+1}=\frac{1}{n+1}\text{cotan}\frac{j\pi}{2(n+1)}. \tag{1.296}$$

The corresponding integrals follow

$$\int_0^{n+1}\sin^2\frac{\pi jx}{n+1}dx=\frac{n+1}{2}, \tag{1.297}$$

$$\int_0^{n+1}\left(1-\frac{x}{l}\right)\sin\frac{\pi jx}{n+1}dx=\frac{2}{\pi j}. \tag{1.298}$$

The values of sum (1.295) and integral (1.297) coincide. Applying the Euler-Mclaurin formula one may get the value of the continuum projection more closer to the sum value (1.296):

$$\sum_{k=0}^{n+1}\left(1-\frac{j}{n+1}\right)\sin\frac{k\pi j}{n+1}=$$

$$\int_0^{n+1}\left(1-\frac{x}{l}\right)\sin\frac{\pi jx}{n+1}dx+\frac{1}{2}[\sin 0+\sin(j\pi)]-\frac{j\pi}{6(n+1)}\cos 0+\ldots=$$

$$\frac{2}{\pi j}\left[1-\frac{\pi^2 j^2}{12(n+1)^2}\right]. \tag{1.299}$$

According to formula (1.299), one may construct a simple relation relatively well, approximating the sum (1.296) for arbitrary values of j from $j=1$ to $j=n$.

For this purpose, the second term in the r.h.s. of Equation (1.299) should be substituted by the following approximation

$$\sum_{k=0}^{n+1}\left(1-\frac{j}{n+1}\right)\sin\frac{k\pi j}{n+1}\approx\frac{2}{\pi j}\left[1-\frac{j^2}{(n+1)^2}\right].$$

1.8 Averaging and Homogenization

We begin with a terminology background. We understand by averaging the process applied to nonlinear problems in mechanics, whereas homogenization deals with the averaging process regarding DEs with quickly changing coefficients. In both approaches the same idea of splitting of fast and slow solution components is applied (see Figure 1.18).

1.8.1 Averaging via Multiscale Method

One may apply different forms of averaging procedure beginning with the Van der Pol method of slowly changeable amplitudes [63] up to the Hilbert transformation [238]. However, these methods cannot be interpreted in a simple way. Namely, the following question appears: why should the averaging with respect to fast time be carried out and why should the slow time t be frozen as well as the function of t while applying averaging with respect to fast time? The multiscale method allows us to clarify this averaging approach ([257], p. 130).

In what follows we apply the multiscales method. It should be noted that all of the asymptotic methods yield the same equation, which has been pointed out by N.N. Moiseev [189].

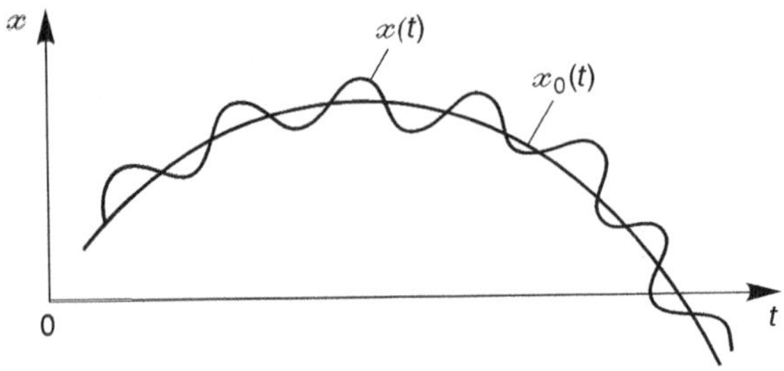

Figure 1.18 Splitting of the input solution into a sum of slow (averaged) and fast changeable components

The multiscale method presents the most regular approach to find higher order approximations. On the other hand, this method in many cases yields a solution in the form of convergent series ([257], p. 144).

We begin with the Duffing equation with small nonlinearity:

$$\ddot{x} + x + \varepsilon x^3 = 0, \quad \varepsilon \ll 1. \tag{1.300}$$

Linear equation ($\varepsilon = 0$) has the following general solution:

$$x_0 = A\cos t + B\sin t, \tag{1.301}$$

where $A, B = const.$

It is tempting to assume that for $0 < \varepsilon \ll 1$ a solution to Equation (1.300) can be presented in the form (1.301), where A and B are functions slowly changed in time t.

According to the multiscale method two scales we introduced the slow time $\tau = \varepsilon t$ keeping notation t for the "fast time," and hence

$$\frac{d}{dt} = \frac{\partial}{\partial t} + \varepsilon\frac{\partial}{\partial \tau}.$$

A solution x is presented in the following series form

$$x = x_0(t,\tau) + \varepsilon x_1(t,\tau) + \ldots,$$

and we have

$$\frac{dx}{dt} = \frac{\partial x_0}{\partial t} + \varepsilon\left(\frac{\partial x_0}{\partial \tau} + \frac{\partial x_1}{\partial t}\right) + \ldots,$$

$$\frac{d^2x}{dt^2} = \frac{\partial^2 x_0}{\partial t^2} + \varepsilon\left(2\frac{\partial^2 x_0}{\partial \tau\,\partial t} + \frac{\partial^2 x_1}{\partial t^2}\right) + \ldots.$$

After splitting with respect to ε, the following recurrent series is obtained

$$\frac{\partial^2 x_0}{\partial t^2} + x_0 = 0, \tag{1.302}$$

$$\frac{\partial^2 x_1}{\partial \tau^2} + x_1 = 2\frac{\partial^2 x_0}{\partial \tau\,\partial t} - x_0^3, \tag{1.303}$$

....

A solution to Equation (1.302) can be written in the following form:

$$x_0 = A(\tau)\cos t + B(\tau)\sin t. \tag{1.304}$$

Equation (1.303) is cast to the form:

$$\frac{\partial^2 x_1}{\partial \tau^2} + x_1 = P(t,\tau) = 2\frac{dA}{d\tau}\sin t - 2\frac{dB}{d\tau}\cos t - (A\cos t + B\sin t)^3. \tag{1.305}$$

Lack of secular terms in solution to Equation (1.303) requires

$$\int_0^{2\pi} P(t,\tau)\cos t\,dt = 0, \quad \int_0^{2\pi} P(t,\tau)\sin t\,dt = 0. \tag{1.306}$$

Condition (1.306) implies the following system of two first order ODEs with respect to functions $A(\tau)$, $B(\tau)$:

$$\frac{dA}{d\tau} = \frac{3}{8}(A^2 + B^2)B, \quad \frac{dB}{d\tau} = -\frac{3}{8}(A^2 + B^2)A.$$

Analogously, one may derive a system of averaged equations if the zero order solution of Equation (1.300) has the following form:

$$x = a(t)\cos(t + \theta(\tau)) \tag{1.307}$$

under the condition

$$\frac{dx}{dt} = -a(t)\sin(t + \theta(\tau)). \tag{1.308}$$

Now functions $a(t)$ and $\theta(\tau)$ have the meaning of an amplitude and phase of vibrations, respectively.

Differentiating formula (1.307) regarding time t yields:

$$\frac{dx}{dt} = -a\sin(t+\theta) + \frac{da}{dt}\cos(t+\theta) - \frac{d\theta}{dt}a\sin(t+\theta). \tag{1.309}$$

Substituting Equation (1.309) into Equation (1.300) and using Equation (1.308), one gets:

$$\frac{da}{dt}\sin(t+\theta) + a\frac{d\theta}{dt}\cos(t+\theta) = \varepsilon a^3\cos^3(t+\theta). \tag{1.310}$$

Solving Equations (1.309) and (1.310) with respect to $\frac{da}{dt}$ and $\frac{d\theta}{dt}$, one obtains:

$$\frac{da}{dt} = \varepsilon a^3\cos^3(t+\theta)\sin(t+\theta), \tag{1.311}$$

$$\frac{d\theta}{dt} = \varepsilon a^2\cos^4(t+\theta). \tag{1.312}$$

Since a and θ are slowly changed functions in time (ε is small) then their changes within the time $T = 2\pi$, being a period of the right-hand sides, is small. Averaging of the r.h.s. of Equations (1.311) and (1.312) on the interval $[t, t+T]$, where the quantities a and θ appearing in the right-hand sides of equations are assumed to be constant, yields

$$\frac{1}{T}\int_0^T \cos^3(t+\theta)\sin(t+\theta)dt = 0, \quad \frac{1}{2\pi}\int_0^{2\pi}\cos^4(t+\theta)dt = \frac{3}{8}.$$

It further follows that

$$\frac{da}{dt} = 0, \tag{1.313}$$

$$\frac{d\theta}{dt} = \frac{3}{8}\varepsilon a^2. \tag{1.314}$$

Equation (1.313) implies that a is constant, whereas relation (1.314) yields $\theta = \frac{3}{8}\varepsilon a_0^2 t + \theta_0$. Hence, in the first order approximation we get

$$u = a_0 \cos\left(1 + \frac{3}{8}\varepsilon A_0^2\right) t + O(\varepsilon).$$

It is clear that zero order solutions presented by expressions (1.304) and (1.307) are equivalent.

Improvement term to the vibration frequency coincides with that obtained via other methods (for instance via Lindstedt-Poincaré method).

Observe that not only linear equations may serve as a zero order approximation. For example, in reference [79], nonlinear DEs are taken from the beginning (zero order approximation) allowing to achieve final solutions in the form of elliptic functions. Although the procedure is more complicated, but the nonlinear effects are taken already in the first equation of the successive series of equations.

1.8.2 Frozing in Viscoelastic Problems

The viscoelastic problems are associated with the integro-differential equations. As a typical example one may consider equation governing vibrations of the viscoelastic rectangular plate taking into account geometric nonlinearity:

$$\Gamma(\nabla^4 w) - \Gamma(N\nabla^2 w) + \rho_1 w_{tt} = 0,$$

where $\Gamma(\varphi) = \varphi + \int_0^t R(t-\tau)\varphi(\tau)d\tau$, $N = \frac{6}{abh^2}\int_0^a \int_0^b (w_x^2 + w_y^2)dx\, dy$, $\rho_1 = \frac{\rho h}{D}$, and R stands for the relaxation kernel.

If the plate is simply supported, then one may separate the variables, and applying

$$w = A(t)\sin\frac{m\pi x}{a}\sin\frac{n\pi y}{b}$$

the following nonlinear integro-differential equation regarding $A(t)$ is obtained

$$B(m,n)\Gamma(A + 1,5h^{-2}A^3) + \rho_1 A_{tt} = 0, \tag{1.315}$$

where

$$B(m,n) = \pi^4\left[\left(\frac{m}{a}\right)^2 + \left(\frac{n}{b}\right)^2\right]^2.$$

Since an exact solution of Equation (1.315) cannot be achieved, therefore we are going to simplify equation with respect to the low frequencies. Let us begin with a study of the following integral

$$I(t) = \int_0^t A(\tau)R(t-\tau)d\tau. \tag{1.316}$$

If the variable $A(\tau)$ slowly changes with respect to the change of the kernel relaxation $R(t-\tau)$, then one may introduce freezing for $t=\tau$ assuming that

$$I(t) \approx A(t)I_1(t),$$

where $I_1(t) = \int_0^t R(t-\tau)d\tau$.

The so far described approach is known as the method of freezing [103], [104]. Applying this method to Equation (1.315) yields

$$B[1 + I_1(t)](A + 1,5h^{-2}A^3) + \rho_1 A_{tt} = 0. \tag{1.317}$$

Note that Equation (1.317) is an ODE one with variable coefficients. Now we apply the averaging to solve it. Assuming

$$I_2(t) \approx \lim_{T\to\infty} \frac{1}{T}\int_0^T I_1(t)dt,$$

the following ODE with constant coefficients is obtained

$$B[1 + I_2](A + 1,5h^{-2}A^3) + \rho_1 A_{tt} = 0.$$

1.8.3 The WKB Method

The origin of this method requires further investigations [133]. However, we follow the traditional approach referring to Wentzel, Kramers and Brillouin. We consider the longitudinal vibrations of a rod ($0 \le x \le l$) with nonconstant stiffness $EF\varphi_1(x)$ and density $\rho F\varphi_2(x)$, where E, F, ρ are constants [34], [235].

Input equation and BCs are as follows

$$EF\frac{d}{dx}\left[\varphi_1(x)\frac{du}{dx}\right] + \rho\varphi_2(x)F\omega^2 u = 0, \tag{1.318}$$

$$u(0) = u(L) = 0. \tag{1.319}$$

In the nondimensional variables, we get

$$\frac{d^2u}{d\xi^2} + q(\xi)\frac{du}{d\xi} + q_1(\xi)\lambda^2 u = 0, \tag{1.320}$$

where $q(\xi) = \frac{\varphi_1'(\xi)}{\varphi_1(\xi)}$, $q_1(\xi) = \frac{\varphi_2(\xi)}{\varphi_1(\xi)}$, $\lambda^2 = \frac{\rho\omega^2}{EL^2}$, $\xi = \frac{x}{L}$.

Let us construct a solution with respect to high frequencies $\lambda^2 \gg 1$. For this purpose we apply the following change of variables

$$u = \exp(\lambda\psi(\xi)). \tag{1.321}$$

Substituting Ansatz (1.321) into Equations (1.320) one gets

$$\lambda^2\psi'^2 + \lambda\psi'\psi'' + \lambda q\psi' + q_1\lambda^2 = 0. \tag{1.322}$$

We are looking for a solution to Equation (1.322) in the series form with respect to λ^{-1}:

$$\psi = \psi_0 + \lambda^{-1}\psi_1 + \lambda^{-2}\psi_2 + \dots,$$

Terms of this series satisfy the following recurrent system of equations

$$\psi_0'^2 + q_1 = 0, \tag{1.323}$$

$$2\psi_1' + \psi_0'' + q = 0, \tag{1.324}$$

$$\dots\dots\dots$$

Equation (1.323) yields

$$\psi_0 = \pm i\sqrt{q_1(\tau)}.$$

Approximating solution of Equation (1.318) follows

$$u \approx C_1 \sin\left(\int_0^\xi \sqrt{q_1(\tau)}d\tau\right) + C_2 \cos\left(\int_0^\xi \sqrt{q_1(\tau)}d\tau\right).$$

BCs (1.319) imply

$$C_2 = 0, \sin\left(\lambda \int_0^1 \sqrt{q_1(\tau)}d\tau\right) = 0,$$

and hence

$$\lambda \int_0^1 \sqrt{q_1(\tau)}d\tau = \pi n, \quad n = 1, 2, 3, \dots.$$

Finally, the following formula allows us to find a frequency

$$\omega = \frac{\pi n L\sqrt{E}}{\sqrt{\rho}\int_0^\xi \sqrt{q_1(\tau)}d\tau}.$$

In what follows we show other modification of the WKB method using the example of vibrations of a rod with variable transversal crossection, governed by the following equation

$$\frac{d^2}{dx^2}\left[EI\varphi_1(x)\frac{d^2 w}{dx^2}\right] - \rho\varphi_2(x)F\omega^2 w = 0. \tag{1.325}$$

We study the clamped rod faces, and hence

$$w = \frac{dw}{dx} = 0 \quad \text{for} \quad x = 0, L. \tag{1.326}$$

The nondimensional form follows

$$\varepsilon^4 \frac{d^2}{d\xi^2}\left[\varphi_1 \frac{d^2 w}{d\xi^2}\right] - \varphi_2 w = 0, \tag{1.327}$$

$$w = \frac{dw}{d\xi} = 0 \quad \text{for} \quad \xi = 0, 1, \tag{1.328}$$

where $\varepsilon^4 = \frac{EI}{\omega^2 \rho F L^4}$, $\xi = \frac{x}{L}$.

A solution to the Equation (1.327) is sought in the form

$$w = \exp\left(\varepsilon^{-1}\int_0^\xi \psi(\tau)d\tau\right)[u_0(\xi) + \varepsilon u_1(\xi) + \varepsilon^2 u_2(\xi) + \dots]. \tag{1.329}$$

Substituting the Ansatz (1.329) into Equation (1.327), after splitting with regard to ε, the following recurrent system of equations is obtained:

$$(\varphi_1\psi^4 - \varphi_2)u_0 = 0, \tag{1.330}$$

$$4\varphi_1\psi^3 u_0' + 6\varphi_1\psi^2\psi' u_0 + 2\psi^3\varphi_1' u_0 = 0, \tag{1.331}$$

$$\dots\dots\dots$$

Equation (1.330) gives the following solution:

$$\psi_{1,2} = \pm\left(\frac{\varphi_2}{\varphi_1}\right)^{1/4}, \quad \psi_{1,2} = \pm i\left(\frac{\varphi_2}{\varphi_1}\right)^{1/4}.$$

Equation (1.331) yields

$$u_0(\xi) = \frac{1}{\psi^{3/2}\varphi_1^{1/2}}.$$

General solution to Equation (1.327) has the following form (in the first order approximation):

$$w = C_1 \sin\left(\varepsilon^{-1}\int_0^\xi \psi(\tau)d\tau\right)u_0(\xi) + C_2\cos\left(\varepsilon^{-1}\int_0^\xi \psi(\tau)d\tau\right)u_0(\xi) +$$
$$C_3\exp(-\varepsilon^{-1}\psi(0)\xi)u_0(0) + C_4\exp(-\varepsilon^{-1}\psi(1)(1-\xi))u_0(1). \tag{1.332}$$

In expression (1.332), for quickly decaging components, the function $u_0(\xi)$ is frozen on one of the interval ends.

BCs (1.327) allow us to define the vibration frequency

$$\omega = \pi^2(n+0,5)^2\sqrt{\frac{EI}{\rho F L^4}}\left[\int_0^l\left[\frac{\varphi_2(x)}{\varphi_1(x)}\right]^{1/4}dx\right]^{-2}, \quad n = 1,2,3,\dots. \tag{1.333}$$

Formula (1.333) for $\varphi_1 = \varphi_2 = 1$ coincides with Bolotin formula (4.1.17). In other words, the WKB method generalizes the method of Bolotin into the problems with variable coefficients.

1.8.4 Method of Kuzmak-Whitham (Nonlinear WKB Method)

Efficient generalization of the averaging method for ODEs has been proposed by Kuzmak [153] and for PDEs by Whitham [256]. According to the latter method, a quickly oscillating solution to a PDE has the following form:

$$w = w_0(\tau,\mathbf{x},t) + \varepsilon w_1(\tau,\mathbf{x},t) + \dots, \quad \tau = \varepsilon^{-1}S(\mathbf{x},t). \tag{1.334}$$

Functions $w_i(\tau,\mathbf{x},t)$ are periodic with respect to τ. The Kuzmak-Whitham method can be treated as a variant of the multiscale method, where τ plays a role of a fast variable. Substituting Ansatz (1.334) into the input PDE, and after splitting with respect to ε, the problem is reduced to ODEs, where the first equation is nonlinear one, and the remaining equations are linear and nonhomogenous. Solvability conditions of these equations are reduced to a system of nonlinear PDEs yielding a phase $S(\mathbf{x}, t)$ and dependence of the functions being sought on slow variables $\mathbf{x}, t$. Therefore, the Kuzmak-Whitham method generalizes the WKB method into a nonlinear case. We apply the method illustrated so far to construct simplified nonlinear evolution equations of shallow shells, which yields Berger equation as the particular case (see also [124]).

In what follows we consider the shallow shell with the curvatures R_1 and R_2 and dimensions a and b. The governing nondimensional equations follow:

$$\frac{1}{12(1-\nu^2)}\varepsilon^2\nabla^2\tilde{w} - \nabla_k\tilde{F} - \varepsilon(\tilde{F}_{\xi\xi}\tilde{w}_{\eta\eta} - 2\tilde{F}_{\xi\eta}\tilde{w}_{\xi\eta} + \tilde{F}_{\eta\eta}\tilde{w}_{\xi\xi}) + \tilde{w}_{\tau\tau} = 0,$$

$$\nabla^4\tilde{F} + \nabla_k\tilde{w} + \varepsilon(\tilde{w}_{\xi\xi}\tilde{w}_{\eta\eta} - \tilde{w}_{\xi\eta}^2) = 0, \quad (1.335)$$

$$\tilde{F}_{\eta\eta} = \frac{1}{1-\nu^2}[\tilde{u}_\xi - \varepsilon_{10}\tilde{w} + 0,5\varepsilon\tilde{w}_\xi^2 + \nu(\tilde{v}_\eta - \varepsilon_{20}\tilde{w} + 0.5\varepsilon\tilde{w}_\eta^2)],$$

$$\tilde{F}_{\xi\xi} = \frac{1}{1-\nu^2}[\tilde{v}_\eta - \varepsilon_{20}\tilde{w} + 0,5\varepsilon\tilde{w}_\eta^2 + \nu(\tilde{u}_\xi - \varepsilon_{10}\tilde{w} + 0.5\varepsilon\tilde{w}_\xi^2)], \quad (1.336)$$

$$\tilde{F}_{\xi\eta} = -\frac{1}{2(1+\nu)}(\tilde{u}_\eta + \tilde{v}_\xi + \varepsilon\tilde{w}_\xi\tilde{w}_\eta).$$

Here $\tau = \sqrt{\frac{\rho(1-\nu^2)}{E}}at$, $\nabla_k \equiv \varepsilon_{20}\frac{\partial^2}{\partial\xi^2} + \varepsilon_{10}\frac{\partial^2}{\partial\eta^2}$, $\varepsilon = \frac{h}{a}$, $\varepsilon_{i0} = \frac{a}{R_i}$, $\{\xi, \eta\} = \frac{1}{a}\{x, y\}$, $\tilde{F} = \frac{F}{Eha}$, $\{\tilde{u}, \tilde{v}, \tilde{w}\} = \frac{1}{h}\{u, v, w\}$, F is the Airy function.

Let us introduce the fast variable $\varepsilon^\alpha\theta(\xi, \eta)$ (parameter $\alpha < 0$ will be further estimated), and therefore

$$\frac{\partial}{\partial\xi} = \frac{\partial}{\partial\xi} + \varepsilon^\alpha\theta_\xi\frac{\partial}{\partial\theta}, \quad \frac{\partial}{\partial\eta} = \frac{\partial}{\partial\eta} + \varepsilon^\alpha\theta_\eta\frac{\partial}{\partial\theta}.$$

Displacement and Airy functions are being sought in the series forms of a slow and fast components, where the latter ones treated as periodic regarding θ of a period T, follow:

$$\tilde{F} = F^0(\xi, \eta, \tau) + \varepsilon^{\beta_1}F^1(\xi, \eta, \varepsilon^\alpha\theta, \tau), \qquad \tilde{w} = w^0(\xi, \eta, \tau) + \varepsilon^{\beta_2}w^1(\xi, \eta, \varepsilon^\alpha\theta, \tau),$$

$$\tilde{u} = u^0(\xi, \eta, \tau) + \varepsilon^{\beta_3}u^1(\xi, \eta, \varepsilon^\alpha\theta, \tau), \qquad \tilde{v} = v^0(\xi, \eta, \tau) + \varepsilon^{\beta_4}v^1(\xi, \eta, \varepsilon^\alpha\theta, \tau).$$

Let us also introduce parameters of asymptotic integrations γ_i and δ with the following scaling:

$$F^0 \sim \varepsilon^{\gamma_1}w^0, \quad w^0 \sim \varepsilon^{\gamma_2}, \quad u^0 \sim \varepsilon^{\gamma_3}, \quad v^0 \sim \varepsilon^{\gamma_4}, \quad \frac{\partial}{\partial\tau}(\ldots) \sim \varepsilon^\delta(\ldots).$$

We carry out the asymptotic analysis of systems (1.335), (1.336) assuming $\varepsilon_{10} \sim \varepsilon_{20} \sim 1$. As a result, we obtain the following estimations for the parameters α, β_i, γ_i, δ: $\alpha = -0,5$, $\beta_1 = 0$, $\beta_2 < 0$, $\beta_3 \geq 0$, $\beta_4 \geq -0,5$, $\gamma_1 = 1$, $\gamma_2 = 0$, $\gamma_3 > 0$, $\gamma_4 > 0$, $\delta = 0$.

The corresponding limiting system is

$$\frac{1}{12(1-\nu^2)}w^1_{\theta\theta\theta\theta}(\theta_\xi^2 + \theta_\eta^2) - (\varepsilon_{20}\theta_\xi^2 + \varepsilon_{10}\theta_\eta^2)F^1_{\theta\theta} -$$

$$(F^0_{\xi\xi}\theta_\eta^2 + 2F^0_{\xi\eta}\theta_\xi\theta_\eta + F^0_{\eta\eta}\theta_\xi^2)w^1_{\theta\theta} + w^1_{\tau\tau} = 0, \quad (1.337)$$

$$F^1_{\theta\theta\theta\theta}(\theta_\xi^2 + \theta_\eta^2)^2 - (\varepsilon_{20}\theta_\xi^2 + \varepsilon_{10}\theta_\eta^2)w^1_{\theta\theta} = 0, \quad (1.338)$$

$$\varepsilon^{-1}F^1_{\theta\theta}\theta_\xi\theta_\eta + F^0_{\eta\eta} = \frac{1}{2(1-\nu^2)}(w^1_\theta)^2\theta_\xi\theta_\eta, \tag{1.339}$$

$$\varepsilon^{-1}F^1_{\theta\theta}\theta_\xi\theta_\eta + F^0_{\xi\xi} = \frac{1}{2(1-\nu^2)}(w^1_\theta)^2\theta_\xi\theta_\eta, \tag{1.340}$$

$$\varepsilon^{-1}F^1_{\theta\theta}\theta_\xi\theta_\eta + F^0_{\xi\eta} = -\frac{1}{2(1+\nu)}(w^1_\theta)^2\theta_\xi\theta_\eta. \tag{1.341}$$

Derivatives of F^0 appeared on Equation (1.337) are defined via Equations (1.339)–(1.341) after averaging with respect to θ. According to periodicity of function F^1_θ, the following relation holds:

$$\int_0^T F^1_{\theta\theta}d\theta = 0.$$

Finally, we get

$$-(F^0_{\xi\xi}\theta^2_\eta + 2F^0_{\xi\eta}\theta_\xi\theta_\eta + F^0_{\eta\eta}\theta^2_\xi)w^1_{\theta\theta} = \frac{1}{2(1-\nu^2)}\int_0^T [(w^1_\theta)^2(\theta^2_\xi + \theta^2_\eta)^2 -$$

$$2w^1[\varepsilon_{10}(\theta^2_\xi + \nu\theta^2_\eta) + \varepsilon_{20}(\theta_\eta + \theta_\xi)]]d\theta. \tag{1.342}$$

Taking into account relation (1.342), the following simplified equations are obtained:

$$\frac{D}{h}\nabla^4 w - \left(\frac{1}{R_2}\frac{\partial^2}{\partial x^2} + \frac{1}{R_1}\frac{\partial^2}{\partial y^2}\right)F - \frac{E}{ab(1-\nu^2)}\times$$

$$\left\{0.5\nabla^2 w\int_a^b\int_a^b\left[w_x^2 + w_y^2\right]dx\,dy - w_{xx}\int_0^a\int_0^b\left(\frac{\nu}{R_1} + \frac{1}{R_2}\right)w\,dy\,dx -\right.$$

$$\left. w_{yy}\int_0^a\int_0^b\left(\frac{1}{R_1} + \frac{\nu}{R_2}\right)w\,dy\,dx\right\} + \rho w_{tt} = 0, \tag{1.343}$$

$$\frac{D}{h}\nabla^4 F - \left(\frac{1}{R_2}\frac{\partial^2}{\partial x^2} + \frac{1}{R_1}\frac{\partial^2}{\partial y^2}\right)w = 0. \tag{1.344}$$

Observe that for $R_1 \to \infty$, $R_2 \to \infty$ Equation (1.343) transits into Berger equation, and in the one-dimensional case we obtain the Kirchhoff equation.

1.8.5 Differential Equations with Quickly Changing Coefficients

In what follows we introduce the homogenization method using the following simple 1D problem [46], [138], [139], [140], [148], [169]:

$$\frac{d}{dx}\left[a\left(\frac{x}{\varepsilon}\right)\frac{du}{dx}\right] = q(x), \tag{1.345}$$

$$u = 0 \quad \text{for} \quad x = 0, L. \tag{1.346}$$

Here $a(x/\varepsilon)$ is periodic with respect to x, and has the period ε.

Variation of the r.h.s. of Equation (1.345) is small, but the coefficient $a(x/\varepsilon)$ changes quickly. Therefore, one may apply the method of two scales, and introduce fast $\eta = x/\varepsilon$ and slow $y = x$ variables. Then, the derivative follows:

$$\frac{d}{dx} = \frac{\partial}{\partial y} + \varepsilon^{-1}\frac{\partial}{\partial \eta}, \tag{1.347}$$

and instead of the input ODE we get a PDE.

Its solution is assumed to have the following form:

$$u = u_0(\eta, y) + \varepsilon u_1(\eta, y) + \dots, \tag{1.348}$$

where $u_0, u_1, \dots$ are periodic functions with respect to η of period 1.

Substituting relations (1.347), (1.348) into input Equation (1.345) and BCs (1.346), and comparing the terms standing by the same powers of ε, the following recurrent system of equations is obtained

$$\frac{\partial}{\partial \eta}\left[a(\eta)\frac{\partial u_0}{\partial \eta}\right] = 0, \tag{1.349}$$

$$\frac{\partial}{\partial \eta}\left[a(\eta)\frac{\partial u_0}{\partial y}\right] + a(\eta)\frac{\partial^2 u_0}{\partial y\,\partial \eta} + \frac{\partial}{\partial \eta}\left[a(\eta)\frac{\partial u_1}{\partial \eta}\right] = 0, \tag{1.350}$$

$$\frac{\partial}{\partial \eta}\left[a(\eta)\frac{\partial u_2}{\partial \eta}\right] + a(\eta)\frac{\partial^2 u_0}{\partial y^2} + \frac{\partial}{\partial \eta}\left[a(\eta)\frac{\partial u_1}{\partial y}\right] + a(\eta)\frac{\partial^2 u_1}{\partial y\,\partial \eta} = q(y), \tag{1.351}$$

....

$$u_j = 0 \quad \text{for} \quad y = 0, L, \quad \eta = 0, L/\varepsilon, \quad j = 1, 2, 3, \dots. \tag{1.352}$$

Equation (1.349), according to periodicity of functions u_0 with respect to η, yields $u_0 = u_0(y)$. It means that u_0 stands for a certain averaged part of function u and does not depend on a fast variable. In a series of physical problems, the existence of averaged part is already implied from the problem statement, and hence the first term of series (1.348) can be treated as not dependent on the fast variable. Equation (1.350) takes the following form

$$\frac{\partial}{\partial \eta}\left[a(\eta)\frac{\partial u_1}{\partial \eta}\right] = -\frac{\partial a(\eta)}{\partial \eta}\frac{du_0}{dy}. \tag{1.353}$$

This equation is considered on the period ($0 \le \eta \le 1$) and hence it is referred to as a cell or local governing equation. Solution to one cell problem is essentially simpler in comparison to the whole space solution. In this case we obtain

$$\frac{\partial u_1}{\partial \eta} = -\frac{\partial u_0}{\partial y} + \frac{C(y)}{a}. \tag{1.354}$$

Periodicity conditions of the first improvement term to the homogenized solution $u_1|_0^1 = 0$ allows us to determine the constant $C(y)$:

$$C = \hat{a}\frac{du_0}{dy}, \hat{a} = \left[\int_0^1 a^{-1} d\eta\right]^{-1}.$$

Excluding $\partial u_1/\partial \eta$ from (1.351) yields

$$\frac{\partial}{\partial \eta}\left(a\frac{\partial u_2}{\partial \eta}\right)+\frac{\partial}{\partial \eta}\left(a\frac{\partial u_1}{\partial y}\right)+\hat{a}\frac{d^2u_0}{dy^2}=q(y). \tag{1.355}$$

Now, in order to withdraw slow components from Equation (1.355), we apply the homogenization procedure acting on each equation term by the averaging operator $\int_0^1(\ldots)d\eta$. First two terms, in result of averaging, are equal to zero due to the periodicity condition, and hence Equation (1.355) takes the form

$$\hat{a}\frac{d^2u_0}{dy^2}=q(y). \tag{1.356}$$

We apply the following BCs for Equation (1.356):

$$u_0=0 \quad \text{for} \quad y=0,L. \tag{1.357}$$

In what follows we define the functions u_1 from conditions (1.354) :

$$u_1=\frac{du_0}{dy}\left(\hat{a}\int_0^1 a^{-1}d\eta-\eta\right),0\le\eta\le 1.$$

Further, function u_1 is periodically extended with respect to coordinates with the period of 1. The found value u_1 does not satisfy, in general, BCs (1.346), and the associated errors are of the order ε. In order to remove them, the following problem is solved:

$$\frac{d}{dx}\left[a\left(\frac{x}{\varepsilon}\right)\frac{du}{dx}\right]=0,$$

$$u|_{x=0}=A=u_1|_{y=\eta=0},u|_{x=l}=B=u_1|_{y=L,\ \eta=L/\varepsilon}.$$

Applying to this problem the homogenization approach again, the following first approximation is obtained:

$$\hat{a}\frac{d^2u_{01}}{dy^2}=0, u_{01}|_{y=0}=A, u_{01}|_{y=L}=B.$$

It is then tempting to apply the following solutions form:

$$u=u_0(y)+\varepsilon[u_{01}(y)+\varepsilon u_{02}(y)+\varepsilon^2u_{03}(y)+\ldots]+$$
$$\varepsilon[u_1(\eta,y)+\varepsilon u_2(\eta,y)+\varepsilon^2u_3(\eta,y)+\ldots]. \tag{1.358}$$

where $u_i(\eta,y)$ are a function having averaged values with respect to a period equal to zero.

Let us consider one more example of the following nonlinear equation:

$$\frac{d}{dx}\left[a\left(\frac{x}{\varepsilon}\right)\frac{du}{dx}\right]+b\left(\frac{x}{\varepsilon}\right)u^3=q(x), \tag{1.359}$$

$$u=0 \quad \text{for} \quad x=0,L. \tag{1.360}$$

Introducing the fast and slow variables η and y, and approximating the function u in the form (1.348), the following recurrent relations are obtained:

$$\frac{\partial}{\partial \eta}\left[a(\eta)\frac{\partial u_1}{\partial \eta}\right]+\frac{da(\eta)}{d\eta}\frac{du_0}{dy}=0, \tag{1.361}$$

$$\frac{\partial}{\partial \eta}\left[a(\eta)\frac{\partial u_2}{\partial \eta}\right]+\frac{\partial}{\partial \eta}\left[a(\eta)\frac{\partial u_1}{\partial y}\right]+$$

$$a(\eta)\frac{\partial^2 u_1}{\partial y\,\partial \eta}+a(\eta)\frac{d^2 u_0}{dy^2}+b(\eta)u_0^3=q(y), \tag{1.362}$$

$$\dots\dots\dots\ \dots$$

$$u_0=0 \quad \text{for} \quad y=0,l,$$

$$u_1=0 \quad \text{for} \quad y=0,l, \quad \eta=0,L/\varepsilon, \tag{1.363}$$

$$\dots\dots\dots\dots$$

Equation (1.361) coincides with Equation (1.350), and the local problem is not changed, when new terms are added without a change of higher order derivatives. Using the solution (1.354), the following homogenized equation is derived:

$$\hat{a}\frac{d^2 u_0}{dy^2}+\hat{b}u^3=q(y), \quad \hat{b}=\int_0^1 b(\eta)d\eta. \tag{1.364}$$

BCs for Equation (1.364) have the form (1.363). It should be emphasized that

$$u=u_0+O(\varepsilon), \quad \text{but} \quad \frac{du}{dx}=\frac{du_0}{dy}+\frac{\partial u_1}{\partial \eta}+O(\varepsilon).$$

In other words, although the solution u_0 to the homogenized equation approximates the function u with accuracy up to the terms of order ε, in the relations for the derivative one has to keep terms with u_1. Their occurrence generates problems in the process of numerical computations of the solution due to errors introduced by differentiations.

Let us now discuss the physical aspects of the coefficients of the homogenized Equation (1.364). It is clear that both coefficients b and $1/a$ are averaged. Sometimes averaging of the stiffness is referred to as Voigt averaging [60], [252], whereas the averaging of compliance is called Reuss averaging [60], [210]. These estimations present averaged arithmetic and averaged harmonic characteristics of the matrices and inclusions for composites. For a wide range of problems true values of the averaged coefficients of the homogenized Equation (1.364) $\tilde{a}_{ij}$ are located in between the averaged coefficients of Voigt ($\overline{a}_{ij}$) and Reuss ($\hat{a}_{ij}$):

$$\hat{a}_{ij}\le\tilde{a}_{ij}\le\overline{a}_{ij}. \tag{1.365}$$

Estimation (1.365) is known as the Voigt-Reuss pitchfork or Hill pitchfork, although it has been obtained first by Wiener [259]. However, the interval estimated in the mentioned pitchfork is relatively large.

In Figure 1.19, as an example, results obtained via computation of the homogenized conductivity d of the composite material composed of the matrices and square inclusions are reported. Input problem is governed by Laplace equations associated with a periodically nonhomogeneous medium.

A cell of periodicity presents a square of the side 1, whereas inclusions of the size $1/3$ are located symmetrically with respect to the square center, whereas the ratio of conductivities of the matrix and inclusion is denoted by d_0. Dotted (dashed) curve corresponds to Voigt's (Reuss) estimation. Solid curve corresponds to results of homogenization using numerical solution to

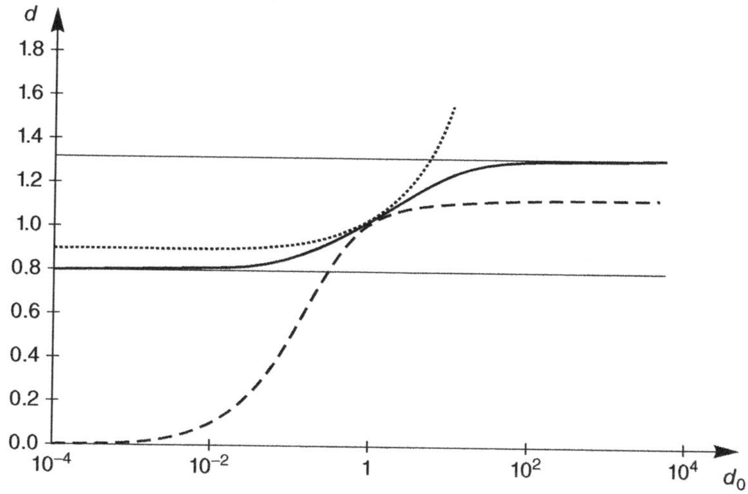

Figure 1.19 Results of homogenization of Laplace equation associated with a periodically nonhomogenous medium, with the Voigt (dot curve) and Reuss (dashed) curve estimation

the problem on a cell [32], [63]. Figure 5.1 gives insight into application of estimation (1.365) with respect to practical problems.

Let us consider now the eigenvalue problem

$$\frac{d}{dx}\left[a\left(\frac{x}{\varepsilon}\right)\frac{du}{dx}\right] + \lambda u = 0, \qquad u = 0 \quad \text{for} \quad x = 0, L. \tag{1.366}$$

We present the eigenform in the form of (1.358), and the eigenvalue λ is presented by the series

$$\lambda = \lambda_0 + \varepsilon\lambda_1 + \varepsilon^2\lambda_2 + \dots. \tag{1.367}$$

Substituting series (1.358), (1.367) into the input BVP (1.366) and taking into account formulas for the derivative (1.347), the following recurrent set of equations is obtained

$$\frac{\partial a}{\partial \eta}\frac{du_0}{dy} + \frac{\partial}{\partial \eta}\left[a\frac{\partial u_1}{\partial \eta}\right] = 0, \tag{1.368}$$

$$\frac{\partial}{\partial \eta}\left(a\frac{\partial u_2}{\partial \eta}\right) + \frac{\partial}{\partial \eta}\left(a\frac{\partial u_1}{\partial y}\right) + a\frac{\partial^2 u_1}{\partial y\,\partial \eta} + \frac{\partial a}{\partial \eta}\frac{du_{01}}{dy} + a\frac{d^2 u_0}{dy^2} + \lambda_0 u_0 = 0, \tag{1.369}$$

$$\frac{\partial}{\partial \eta}\left(a\frac{\partial u_3}{\partial \eta}\right) + \frac{\partial}{\partial \eta}\left(a\frac{\partial u_2}{\partial \eta}\right) + \frac{\partial a}{\partial \eta}\frac{du_{02}}{dy} + a\frac{\partial^2 u_2}{\partial y\,\partial \eta} + a\frac{d^2 u_{01}}{dy^2} + \lambda_1 u_0 + \lambda_0(u_{01} + u_1) = 0, \tag{1.370}$$

$$\dots\ \dots\ \dots\ \dots$$

$$u_0 = 0 \quad \text{for} \quad y = 0, L, \tag{1.371}$$

$$u_1 + u_{01} = 0 \quad \text{for} \quad \eta = 0, L/\varepsilon, \tag{1.372}$$

$$\dots\ \dots\ \dots\ \dots$$

The yielded value $\partial u_1/\partial\eta$ by Equation (1.368) is substituted into Equation (1.369) and into BCs (1.371), and after averaging the following BVP with respect to u_0, λ_0 is defined:

$$\hat{a}\frac{d^2u_0}{dy^2} + \lambda_0 u_0 = 0, \qquad u_0 = 0 \quad \text{for} \quad y = 0, L.$$

Now, Equation (1.369) yields

$$\frac{\partial u_2}{\partial\eta} = -\frac{\partial u_1}{\partial y} - \frac{du_{01}}{dy} + \frac{C_1(y)}{a}.$$

Due to periodicity condition of the function u_2 with respect to η, we get

$$C_1 = \hat{a}\frac{du_{01}}{dy} + \hat{a}\frac{\partial\hat{u}_1}{\partial y}, \quad \text{for} \quad \hat{u}_1 = \int_0^1 u_1 d\eta.$$

Substituting the already found values u_1, u_2 into Equation (1.370), and applying the averaging procedure, we obtain

$$\hat{a}\frac{d^2u_{01}}{dy^2} + \lambda_0 u_{01} + \hat{a}\frac{\partial^2\hat{u}_1}{\partial y^2} + \lambda_0\hat{u}_1 + \lambda_1 u_0 = 0. \tag{1.373}$$

BC for Equations (1.373) are obtained from BCs (1.372), and they have the following form:

$$u_{01} = -\hat{u}_1 \quad \text{for} \quad y = 0, L. \tag{1.374}$$

Improvement term to frequency λ_1 is defined via perturbation method, and then the slow improving term to the homogenized solution u_{01} is yielded by a solution to the BVPs (1.373), (1.374).

The approach described so far allows us to determine a solution in an arbitrary approximation regarding ε. Its most attractive advantage is generality. Indeed, having found a solution to a local problem, one may also define a solution to the input problem, as well as solve the eigenvalue problem. If one adds into the equation nonlinear terms in a way not disturbing higher order derivatives, then a construction of homogenized relations can be carried out in the similar way. The local problem remains the same as in the linear case, as well as the higher order approximations being linear. The whole nonlinearity is located in homogenized BVPs with smooth coefficients, which can be easily solved either via numerical or variational methods.

1.8.6 Differential Equation with Periodically Discontinuous Coefficients

We consider an application of the homogenization method to solve problems of DEs with periodically discontinuous coefficients. They are also known as problems with periodic barriers [251]. As an example we consider deformation of a membrane reinforced by threads.

Equations of equilibrium in intervals $kl < y_1 < (k+1)l$ are formulated as follows

$$\frac{\partial^2 u}{\partial x_1^2} + \frac{\partial^2 u}{\partial y_1^2} = Q(x_1, y_1). \tag{1.375}$$

Condition of linking neighborhood parts, being also known as the matching of jump conditions [251], have the following form

$$\lim_{y_1 \to kl+0} u \equiv u^+ = \lim_{y_1 \to kl-0} u \equiv u^-, \quad k = 0, \pm 1, \pm 2, \ldots,$$

$$\left(\frac{\partial u}{\partial y_1}\right)^+ - \left(\frac{\partial u}{\partial y_1}\right)^- = p\frac{\partial^2 u}{\partial x_1^2}, \tag{1.376}$$

where p is the parameter characterizing relative thread stiffness.

BCs for $x_1 = 0, H$ follow

$$u = 0. \tag{1.377}$$

Let us assume, that an external load is periodic with respect to y_1, and its period L is essentially larger than the distance between threads. Then, one may apply homogenization approach by taking $\varepsilon = l/L$ as the perturbation/small parameter. Introducing, instead of y_1, fast ($\eta = y_1/l$) and slow ($y = y_1/L$) variables yields

$$\frac{\partial}{\partial y_1} = \frac{1}{L}\left(\frac{\partial}{\partial y} + \varepsilon^{-1}\frac{\partial}{\partial \eta}\right). \tag{1.378}$$

Function u is approximated by the series

$$u = u_0(x, y) + \varepsilon^{\alpha}[u_{01}(x, y) + u_1(x, y, \eta)] +$$
$$\varepsilon^{\alpha_1}[u_{02}(x, y) + u_2(x, y, \eta)] + \ldots, \tag{1.379}$$

where $0 < \alpha < \alpha_1 < \ldots, x = x_1/L$.

Substituting Ansatz (1.379) into Equation (1.375) and into condition (1.376), and taking into account the formula (1.378), we get

$$\nabla^2 u_0 + \varepsilon^{\alpha-2}\frac{\partial^2 u_1}{\partial \eta^2} + 2\varepsilon^{\alpha-1}\frac{\partial^2 u_1}{\partial y\, \partial \eta} + \varepsilon^{\alpha_1-2}\frac{\partial^2 u_2}{\partial \eta^2} +$$
$$2\varepsilon^{\alpha_1-1}\frac{\partial^2 u_2}{\partial y\, \partial \eta} + O(\varepsilon^{\alpha}) = q(x, y), \tag{1.380}$$

$$[u_0 + \varepsilon^{\alpha}(u_{01} + u_1) + \ldots]^+ = [u_0 + \varepsilon^{\alpha}(u_{01} + u_1) + \ldots]^-, \tag{1.381}$$
$$\varepsilon^{\alpha-1}\left[\left(\frac{\partial u_1}{\partial \eta}\right)^+ - \left(\frac{\partial u_1}{\partial \eta}\right)^-\right] + O(\varepsilon^{\alpha}) = p_1\left[\frac{\partial^2 u_0}{\partial x^2} + O(\varepsilon^{\alpha})\right],$$

where $q = L^2 Q$, $p_1 = p/L$, $\nabla^2 u_0 = \frac{\partial^2 u_0}{\partial x^2} + \frac{\partial^2 u_0}{\partial y^2}$, $(\ldots)^{\pm} = \lim_{\eta \to k \pm 0} u$.

It should be emphasized that a majority of the works devoted to homogenization of periodic systems, in particular those purely mathematical, are carried out using the following implicit statements: the occurred system parameters are of the same order. However, in our case a way

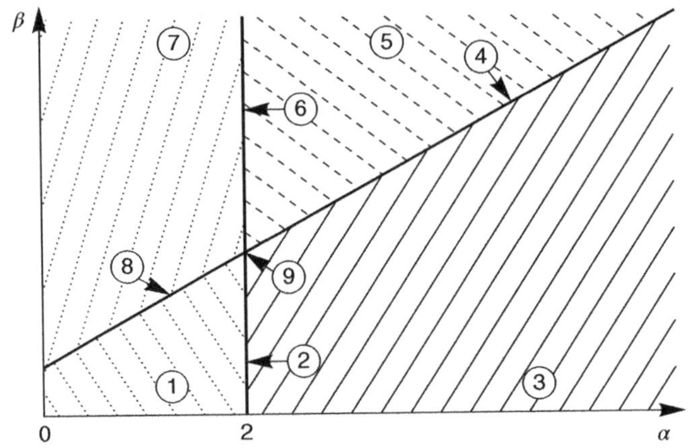

Figure 1.20 Partition of the parameter quadrant plane into nine zones with different asymptotics

of construction of asymptotics essentially depends on the order of the relative thread stiffness p_1 in comparison to parameter ε. In what follows we introduce the parameter β characterizing this order ($p_1 \sim \varepsilon^{\beta}$) and we study a possible character of limiting systems depending on α, β.

Owing to Equation (1.380), the following different limiting system occurs for $0 < \alpha < 2$, $\alpha = 2$ and $\alpha > 2$:

$$\text{for} \quad 0 < \alpha < 2 \quad \frac{\partial^2 u_1}{\partial \eta^2} = 0, \tag{1.382}$$

$$\text{for} \quad \alpha = 2 \quad \nabla^2 u_0 + \frac{\partial^2 u_1}{\partial \eta^2} = q, \tag{1.383}$$

$$\text{for} \quad \alpha > 2 \quad \nabla^2 u_0 = q. \tag{1.384}$$

Limiting relations are obtained from relation (1.381) for $\varepsilon \to 0$, and they have the following form

$$\text{for} \quad \beta < \alpha - 1 \quad \frac{\partial^2 u_0}{\partial x^2} = 0, \tag{1.385}$$

$$\text{for} \quad \beta = \alpha - 1 \quad \left(\frac{\partial u_1}{\partial \eta}\right)^+ - \left(\frac{\partial u_1}{\partial \eta}\right)^- = p_1 \varepsilon^{1-\alpha} \frac{\partial^2 u_0}{\partial x^2}, \tag{1.386}$$

$$\text{for} \quad \beta > \alpha - 1 \quad \left(\frac{\partial u_1}{\partial \eta}\right)^+ = \left(\frac{\partial u_1}{\partial \eta}\right)^-. \tag{1.387}$$

The quadrant plane of the parameters $\beta > 0$, $\alpha > 0$ have nine different areas (Figure 1.20).

In what follows we are going to study them in some detail. Let $\beta < \alpha - 1$, which means physically that the threads are stiff. Equation (1.385) yields $u_0 = 0$ and hence we cannot apply the homogenization approach here. For zones 1–3, we have the following governing limiting equation

$$\frac{\partial^2 u_1}{\partial \eta^2} = q. \tag{1.388}$$

The case $\beta > \alpha - 1$, corresponds to zones 4–6. Physically it means that we deal with weak threads, and the limiting case is governed by (1.384).

Zones 7 and 8 are described by equations out of the physical meaning. A key role plays zone 9 ($\alpha = 2$, $\beta = 1$) associated with averaged thread stiffness. The limiting system is composed of Equations (1.383), (1.386), and the transition conditions take the form

$$u_1^+ = u_1^-, \tag{1.389}$$

$$\left(\frac{\partial u_1}{\partial \eta}\right)^+ - \left(\frac{\partial u_1}{\partial \eta}\right)^- = p_2 \frac{\partial^2 u_0}{\partial x^2}, \tag{1.390}$$

where $p_2 = p/l$.

Equation (1.382) implies

$$u_1 = 0,5(q - \nabla^2 u_0)\eta^2 + C(x, y)\eta + C_1(x, y).$$

Constant $C_1(x, y)$ is associated with the term u_{01}, which is defined through homogenized equation of the successive approximations. Conditions (1.389) yield

$$C(x, y) = -0,5(q - \nabla^2 u_0)\, L. \tag{1.391}$$

We have to satisfy one more condition (1.390), but there is a lack of constants. However, condition (1.390) implies the looked for homogenization equation. Indeed, substituting the found value u_1 into Equation (1.390), we obtain

$$\nabla^2 u_0 + p_2 \frac{\partial^2 u_0}{\partial x^2} = q. \tag{1.392}$$

Equation (1.392) should be integrated taking into account the following BCs:

$$u_0 = 0 \quad \text{for} \quad x = 0, H/L.$$

Physically, a transition into Equation (1.392) is associated with a "smeared" of the threads stiffness (transition into structurally-orthotropic theory). Finally, function u_1 can be approximated in the following way:

$$u_1 = 0.5 p_2 \frac{\partial^2 u_0}{\partial x^2} \eta(\eta - 1).$$

In general, BCs are not satisfied. Boundary error quickly changes with respect to η, and yields occurrence of a boundary layer u_b. We construct the latter through introduction of the variable $\xi = x_1/l$ via the following series:

$$u_b = \varepsilon^{\gamma_1} u_{b1}(x, y, \xi, \eta) + \varepsilon^{\gamma_2} u_{b2}(x, y, \xi, \eta) + \dots,$$

where $0 < \gamma_1 < \gamma_2 < \dots$.

Equations yielding the function u_{b1} have the following form:

$$\frac{\partial^2 u_{b1}}{\partial \xi^2} + \frac{\partial^2 u_{b1}}{\partial \eta^2} = 0,$$

$$u_{b1}|_{\eta=k} = 0, \quad k = 0, \pm 1, \dots.$$

BCs (we consider only one edge) for $x = \xi = 0$ have the following form:

$$u_{b1} = -u_1.$$

In order to construct boundary layer, one may apply the Kantorovitch method, taking u_{b1} in the following form:

$$u_{b1} = \Phi(\xi)\eta(\eta - 1).$$

and now the boundary conditions $\eta = 0, 1$ are satified. Furthermore, the standard Kantorovitch technique can be applied [142].

Let us now describe more rigorously the notion of fast and slow changes of the load. Function $f(\varepsilon, \theta)$ is called oscillating with velocity ε^{-1} on the period 2π, if [251]

$$0 < C_1 \le \int_0^{2\pi} |f(\varepsilon, \theta)|^2 d\theta \le C_2 < \infty, \quad \left| \int_0^{\alpha} f(\varepsilon, \theta) d\theta \right| \le C\varepsilon, \quad 0 \le \alpha \le 2\pi,$$

where C, C_1, C_2 are certain constants.

1.8.7 Periodically Perforated Domain

We consider the Poisson equation, which describes membrane deformation

$$\nabla^2 u = f(x, y) \tag{1.393}$$

in the multi-connected domain Ω (Figure 1.21) [122], [123]. Small parameter ε characterizes a ratio of the characteristics size of the repeated part (cell) and the characteristic domain dimension.

On the boundary of holes Neuman BCs are given

$$\frac{\partial u}{\partial \mathbf{n}_i} = 0 \quad \text{on} \quad \partial\Omega_i, \tag{1.394}$$

where $\mathbf{n}_i$ denotes an external normal to the contour of i-th hole.

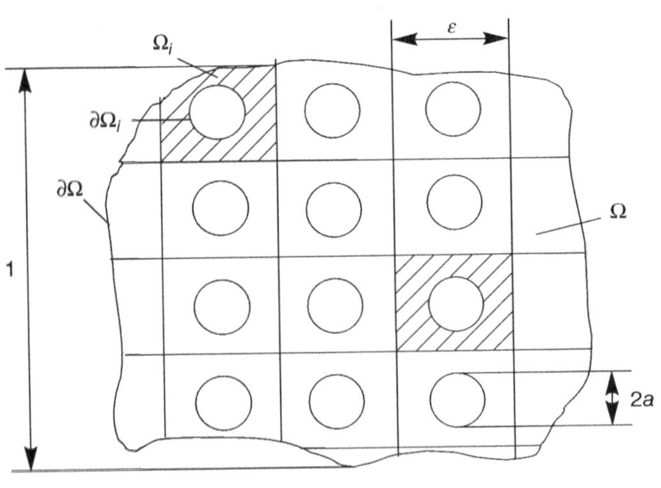

Figure 1.21 Perforated medium

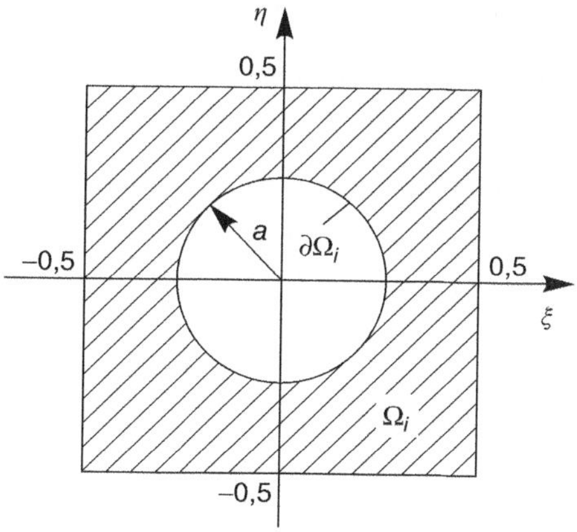

Figure 1.22 Periodically repeated cell

Membrane edges are clamped

$$u = 0 \quad \text{on} \quad \partial\Omega. \tag{1.395}$$

We introduce fast variables $\xi = x/\varepsilon$, $\eta = y/\varepsilon$. The solution is assumed to be of the form

$$u = u_0(x,y) + \varepsilon u_1(x,y,\xi,\eta) + \varepsilon^2 u_2(x,y,\xi,\eta) + \ldots, \tag{1.396}$$

where u_j ($j = 1, 2, \ldots$) are periodic functions with period 1 with respect to ξ, η.
Partial derivatives follow:

$$\frac{\partial}{\partial x} = \frac{\partial}{\partial x} + \varepsilon^{-1}\frac{\partial}{\partial \xi}, \quad \frac{\partial}{\partial y} = \frac{\partial}{\partial y} + \varepsilon^{-1}\frac{\partial}{\partial \eta}. \tag{1.397}$$

A periodically repeated cell in fast variables is shown in Figure 1.22.
Substituting Ansatz (1.396) into BVP (1.393)–(1.395), and taking into account Equations (1.397), the splitting procedure with respect to ε yields the following recurrent sequence of the BVPs:

$$\frac{\partial^2 u_1}{\partial \xi^2} + \frac{\partial^2 u_1}{\partial \eta^2} = 0 \quad \text{in } \Omega_i, \tag{1.398}$$

$$\frac{\partial u_1}{\partial \mathbf{k}} + \frac{\partial u_0}{\partial \mathbf{n}} = 0 \quad \text{on} \quad \partial\Omega_i, \tag{1.399}$$

$$\frac{\partial^2 u_0}{\partial x^2} + \frac{\partial^2 u_0}{\partial y^2} + 2\left(\frac{\partial^2 u_1}{\partial x\,\partial \xi} + \frac{\partial^2 u_1}{\partial y\,\partial \eta}\right) + \frac{\partial^2 u_2}{\partial \xi^2} + \frac{\partial^2 u_2}{\partial \eta^2} = f \quad \text{in} \quad \Omega_i, \tag{1.400}$$

$$\frac{\partial u_2}{\partial \mathbf{k}} + \frac{\partial u_1}{\partial \mathbf{n}} = 0 \text{ on } \partial\Omega_i, \tag{1.401}$$

....

$$u_i = 0, \quad i = 0, 1, 2, \ldots \quad \text{on} \quad \partial\Omega. \tag{1.402}$$

Here **k** denotes a normal to the hole contour in fast variables.

We define the averaging operator in the following way:

$$\tilde{\Phi}(x,y) = \iint\limits_{\Omega_i} \Phi(x,y,\xi,\eta)\, d\xi\, d\eta. \tag{1.403}$$

After application of the averaging operator (1.256), Equation (1.400) yields

$$\left(\frac{\partial^2 u_0}{\partial x^2} + \frac{\partial^2 u_0}{\partial y^2}\right)(1-\pi a^2) + \iint\limits_{\Omega_i} \left(\frac{\partial^2 u_1}{\partial x\, \partial \xi} + \frac{\partial^2 u_1}{\partial y\, \partial \eta}\right) d\xi\, d\eta =$$

$$(1-\pi a^2) f. \tag{1.404}$$

Homogenized BC takes the following form

$$u_0 = 0 \quad \text{on } \partial\Omega. \tag{1.405}$$

Now we proceed to the problem on cell (1.398), (1.399) taking into account the condition of a periodic continuation, i.e. conditions of equality of the function u_1 and its first order derivatives regarding the respective coordinates lying on contrary located cell sides are satisfied.

Reduction of the periodic problems to those of BVPs has been described, for instance in monograph ([32], chapter 6). In both cases displacements on two contrary located external cell boundaries as well as normal derivatives on two remaining sides are equal zero.

Assume that the opening diameter $2a$ is small in comparison to the cell dimensions. In the first approximation u_1 ($u_1 \approx u_1^{(1)}$), one may transit into the infinite plane problem of the opening

$$\frac{\partial^2 u_1^{(1)}}{\partial \xi^2} + \frac{\partial^2 u_1^{(1)}}{\partial \eta^2} = 0, \tag{1.406}$$

$$\frac{\partial u_1^{(1)}}{\partial \mathbf{k}} + \frac{\partial u_0}{\partial \mathbf{n}} = 0 \quad \text{on } \partial\Omega_i, \tag{1.407}$$

$$u_1^{(1)} \to 0 \quad \text{for } \xi^2 + \eta^2 \to \infty. \tag{1.408}$$

In polar coordinates the BVP (1.406)–(1.408) can be cast into the following form

$$\frac{\partial^2 u_1^{(1)}}{\partial r^2} + \frac{1}{r}\frac{\partial u_1^{(1)}}{\partial r} + \frac{1}{r^2}\frac{\partial^2 u_1^{(1)}}{\partial \theta^2} = 0, \tag{1.409}$$

$$\left.\frac{\partial u_1^{(1)}}{\partial r}\right|_{r=a} = -\frac{\partial u_0}{\partial x}\cos\theta - \frac{\partial u_0}{\partial y}\sin\theta, \tag{1.410}$$

$$u_1^{(1)} \to 0 \quad \text{for } r \to \infty. \tag{1.411}$$

A solution to the BVP (1.409)–(1.411) is

$$u_1^{(1)} = \frac{a^2}{r}\left(\frac{\partial u_0}{\partial x}\cos\theta + \frac{\partial u_0}{\partial y}\sin\theta\right). \tag{1.412}$$

Observe that functions $u_1^{(1)}$ do not satisfy the periodicity conditions. In order to avoid compensating the discipancy, we obtain in the second approximation ($u_1 \approx u_1^{(1)} + u_1^{(2)}$) the following BVP:

$$\Delta u_1^{(2)} = 0 \text{ in } \Omega_i^*,$$

$$u_1^{(2)}(0,5,\eta) - u_1^{(2)}(-0,5,\eta) = u_1^{(1)}(-0,5,\eta) - u_1^{(1)}(0,5,\eta),$$

$$u_1^{(2)}(\xi,0,5)-u_1^{(2)}(\xi,-0,5)=u_1^{(1)}(\xi,-0,5)-u_1^{(1)}(\xi,0,5),$$
$$u_{1\xi}^{(2)}(0,5,\eta)-u_{1\xi}^{(2)}(-0,5,\eta)=u_{1\xi}^{(1)}(-0,5,\eta)-u_{1\xi}^{(1)}(0,5,\eta),$$
$$u_{1\eta}^{(2)}(\xi,0,5)-u_{1\eta}^{(2)}(\xi,-0,5)=u_{1\eta}^{(1)}(\xi,-0,5)-u_{1\eta}^{(1)}(\xi,0,5).$$

Let us present $u_1^{(2)}$ in the form

$$u_1^{(2)}=u_1^{(12)}+u_1^{(22)}, \tag{1.413}$$

where functions $u_1^{(12)}$ satisfy homogenous BCs with respect to ξ and nonhomogenous ones with respect to η. The function $u_1^{(22)}$ can be obtained from $u_1^{(12)}$ via change of the variables $(\xi\leftrightarrow\eta, x\leftrightarrow y)$.

The following BVP for $u_1^{(12)}$ estimation is obtained:

$$\Delta u_1^{(12)}=0 \text{ in } \Omega_i^*, \tag{1.414}$$

$$u_1^{(12)}(0,5,\eta)=u_1^{(12)}(-0,5,\eta), u_{1\xi}^{(12)}(0,5,\eta)=u_{1\xi}^{(12)}(-0,5,\eta), \tag{1.415}$$

$$\begin{gathered}u_1^{(12)}(\xi,0,5)-u_1^{(12)}(\xi,-0,5)=u_1^{(1)}(\xi,-0,5)-u_1^{(1)}(\xi,0,5),\\ u_{1\eta}^{(12)}(\xi,0,5)-u_{1\eta}^{(12)}(\xi,-0,5)=u_{1\eta}^{(1)}(\xi,-0,5)-u_{1\eta}^{(1)}(\xi,0,5).\end{gathered} \tag{1.416}$$

A general solution to Equation (1.414) takes the following form:

$$\begin{aligned}u_1^{(12)}=A_0+B_0\eta+\sum_{n=1}^{\infty}[(A_n\mathrm{ch}(2\pi n\eta)+B_n\mathrm{sh}(2\pi n\eta))\cos(2\pi n\xi)+\\ (C_n\mathrm{ch}(2\pi n\eta)+D_n\mathrm{sh}(2\pi n\eta))\sin(2\pi n\xi)],\end{aligned} \tag{1.417}$$

where A_n, B_n, C_n, D_n are arbitrary constants.

Let us present now BCs (1.416) in the following form:

$$u_1^{(12)}(\xi,0,5)-u_1^{(12)}(\xi,-0,5)=-\frac{\partial u_0}{\partial y}a^2(\xi^2+0,25)^{-1}, \tag{1.418}$$

$$u_{1\eta}^{(12)}(\xi,0,5)-u_{1\eta}^{(12)}(\xi,-0,5)=2\frac{\partial u_0}{\partial x}a^2\xi(\xi^2+0,25)^{-2}. \tag{1.419}$$

Developing r.h.s. of relations (1.418), (1.419) into Fourier series and substituting solution (1.417) into Equations (1.418), (1.419), we obtain

$$A_n=D_n=0, n=0,1,\ldots, B_0=-\frac{\partial u_0}{\partial y}\pi a^2=\frac{\partial u_0}{\partial y}B_0^*,$$

$$B_n=-\frac{\partial u_0}{\partial y}\frac{2a^2}{\mathrm{sh}\pi n}[e^{-\pi n}\mathrm{Im}E_1(\pi n(i-1))-e^{\pi n}\mathrm{Im}E_1(\pi n(i+1))]=\frac{\partial u_0}{\partial y}B_n^*,$$

$$C_n=B_n \text{ including the change } \frac{\partial u_0}{\partial y}\Rightarrow\frac{\partial u_0}{\partial x}, n=0,1,2,\ldots.$$

Here $E_1(\ldots)$ denotes integral exponential function ([2], chapter 5), $i = \sqrt{-1}$.

Finally, we have

$$\overline{u}_1^{(2)} = \frac{\partial u_0}{\partial y} B_0^* \eta +$$

$$\sum_{n=1}^{\infty} B_n^* \left(\frac{\partial u_0}{\partial y} \mathrm{sh}(2\pi n\eta) \cos(2\pi n\xi) + \frac{\partial u_0}{\partial x} \mathrm{ch}(2\pi n\eta) \sin(2\pi n\xi) \right).$$

Function $u_1^{(22)}$ is constructed in the analogous way.

Substitution of formula $u_1 = u_1^{(1)} + u_1^{(2)}$ into Equation (1.404) yields the following homogenized equation:

$$q \left(\frac{\partial^2 u_0}{\partial x^2} + \frac{\partial^2 u_0}{\partial y^2} \right) = f, \tag{1.420}$$

where

$$q = 1 - \pi a^2 + \frac{8\pi^2 a^4}{1 - \pi a^2} \times$$

$$\sum_{n=1}^{\infty} \frac{n}{\mathrm{sh}\pi n} (e^{-\pi n} \mathrm{Im} E_1(\pi n(i-1)) - e^{\pi n} \mathrm{Im} E_1(\pi n(i+1))). \tag{1.421}$$

Series appeared in (1.421) is absolutely convergent with fastly decreasing terms $|a_{n+1}/a_n| \to \exp(-\pi)$.

Homogenized BC for Equation (1.420) takes the form of (1.405).

In what follows we briefly discuss a paradox reported by Bakhvalov and Eglit [31]. They considered two following cases. In the first case homogenization has been carried out for a medium with holes. In the second case a certain medium with inclusions has been homogenized, and then in the homogenized relations the characteristics of inclusions have been set to zero. The corresponding limiting systems have not coincided. However, r.h.s. of the studied Poisson equations regarding inclusions have been homogenized with respect to the whole cell area, whereas in the case of the medium with holes – only on the cell area without opening. The discussed paradox can be simply omitted. The r.h.s. for openings should be homogenized regarding the cell area without holes, whereas coefficients of the l.h.s. can be obtained via the limiting transition applied to the problem on inclusions.

1.8.8 Waves in Periodically Nonhomogenous Media

The homogenization method for the problems on waves distribution in a periodically nonhomogeneous media is often defined via representation of its solution through a scalar product of a periodic function and a certain modulated function. Mathematicians call this approach the Floquet method [70], whereas among physicists it is known as the Bloch method [86], [143].

In what follows we recall how to find a solution to differential equation with periodic coefficients. Consider the following Hill equation:

$$\frac{d^2 y(x)}{dx^2} - \varphi(x) y(x) = 0,$$

where $\varphi(x)$ is the periodic function of period a.

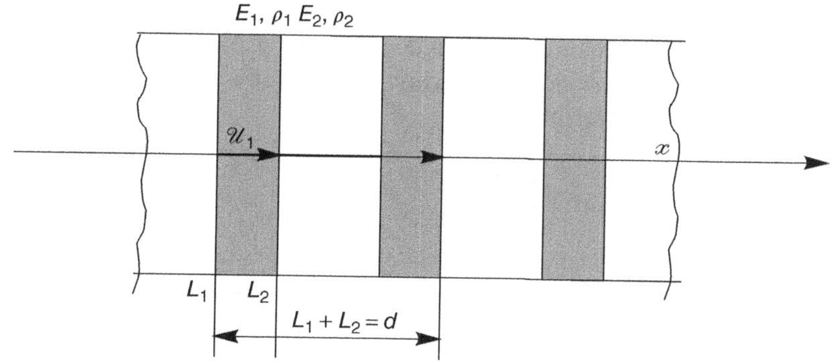

Figure 1.23 Scheme of 1D composite material

A particular solution to Hill equation, owing to the Floquet theorem [70], has the following form

$$y_1(x) = \Phi(x)\exp(i\mu x),$$

where $\Phi(x)$ is a periodic function of period a, and μ is a complex characteristic exponent.

We demonstrate a use of the homogenization method via an example of the 1D lagered composite (Figure 1.23) [80]. An equation governing motion of the neighborhood composite parts follows:

$$E_k u_{kxx} - \rho_k u_{ktt} = 0, \quad k = 1, 2. \tag{1.422}$$

On the contact boundaries the following conditions hold:

$$u_1 = u_2, \quad E_1 u_{1x} = E_2 u_{2x}. \tag{1.423}$$

In addition, the following quasi-periodicity conditions should be satisfied

$$u_k(x + d, t) = u_k(x, t)\exp(ird), \tag{1.424}$$

where $k = 1, 2$, r - wave number, $r = 2\pi/L$, L wave length.

The solution of the composite parts has the following form

$$u_k(x, t) = A_k \exp[i(p_k x + \omega t)] + B_k \exp[i(-p_k x + \omega t)], \tag{1.425}$$

where $p_k = \omega/C_k$, $C_k = \sqrt{E_k/\rho_k}$, $k = 1, 2$.

Substituting Ansatz (1.425) into BCs (1.423), (1.424), one gets a system of four linear homogenous algebraic equations regarding unknown coefficients A_k, B_k. Comparing to zero the system determinant yields the following Equation [37], [206]

$$\cos(rd) = \cos\Omega\cos(\Omega a) - \frac{b^2+1}{2b}\sin\Omega\sin(\Omega a), \tag{1.426}$$

where $\Omega = \frac{\omega L_1}{C_1}$, $a = \frac{L_2 C_1}{L_1 C_2}$, $b = \frac{\sqrt{E_1\rho_1}}{\sqrt{E_2\rho_2}}$.

In Equation (1.426) parameter b presents a rotation of impedances of composite components, whereas the parameter a stands for a time ratio associated with a wave transition.

Homogenized solution can be obtained from Equation (1.426) for small Ω and small wave number rd (long-wave approximation), and it is assumed that parameters a and b are of order 1. Developing the right- and left-hand sides of Equation (1.426) into a Maclaurin series, and retaining only the first terms, we get

$$\Omega = rd\left[(1+a)^2 + \frac{(b-1)^2 a}{b}\right]^{-1/2}. \tag{1.427}$$

In order to solve the transcendental Equations (1.426) one may apply perturbation method different from the homogenization method. Let, for instance $b = 1 + \varepsilon$, $\varepsilon \ll 1$, then Equation (1.426) can be cast into the form

$$\cos(rd) = \cos[\Omega(1+a)] - \varepsilon_1 \sin\Omega \sin(\Omega a), \tag{1.428}$$

where $\varepsilon_1 = 0,5\varepsilon^2/(1+\varepsilon)$.

In zero order approximation we have

$$\cos(rd) = \cos[\Omega(1+a)],$$

and hence

$$\Omega_0 = (rd + 2\pi k)/(1+a).$$

Representing further the solution of Equation (1.428) in the form

$$\Omega = \Omega_0 + \varepsilon_1\Omega_1 + \ldots,$$

we get

$$\Omega_1 = -\frac{\sin\Omega_0 \sin(\Omega_0 a)}{(1+a)\sin(rd)}.$$

In the case when neighborhood composite parts differ strongly with respect to stiffness, i.e. $b \ll 1$, one may introduce a small parameter $\varepsilon_2 = 1/b$, and hence Equation (1.426) can be presented in the following form:

$$\varepsilon_2 \cos rd = \varepsilon_2 \cos\Omega \cos(\Omega a) - \frac{1}{2}(1+\varepsilon_2^2)\sin\Omega \sin(\Omega a). \tag{1.429}$$

Possible simplification of Equation (1.429) may depend on an order of quantity a. If $a \sim 1$, then $L_1/L_2 \sim \varepsilon_2$ (length of one composite part is essentially less than the length of the second one), then a solution to Equation (1.429) can be predicted in the following form:

$$\Omega = \sqrt{\varepsilon_2}\Omega_0 + \varepsilon_2\Omega_1 + \ldots. \tag{1.430}$$

Substituting series (1.430) into Equation (1.429) one gets (first approximation)

$$\cos rd = 1 - \frac{1}{2\varepsilon_2}\sin(\sqrt{\varepsilon_2}\Omega_0)\sin(\sqrt{\varepsilon_2}\Omega_0\tau). \tag{1.431}$$

Developing the r.h.s. of formula (1.431) into series regarding Ω_0 and keeping terms of second and fourth orders, one may approximate with relatively high accuracy a chain of two periodically repeated masses coupled via the same springs [192].

References

[1] Abassy T.A., El-Tawil M.A., Saleh H.K.: The solution of Burgers' and good Boussinesq equations using ADM-Padé technique. *Chaos, Solitons and Fractals*, 32, 1008–26 (2007).

[2] Abramowitz M., Stegun I.A.: *Handbook of Mathematical Functions, with Formulas, Graphs, and Mathematical Tables*. Dover Publications, New York, 1965.

[3] Andersen C.M., Dadfar M.B., Geer J.F.: Perturbation analysis of the limit cycle of the Van der Pol equation. *SIAM J. Appl. Math.*, 44(5), 881–95 (1984).

[4] Andersen C.M., Geer J.F.: Power series expansion for the frequency and period of the limit cycle of the Van der Pol equation. *SIAM J. Appl. Math.*, 42(3), 678–93 (1982).

[5] Andrianov I.V.: Application of Padé approximants in perturbation methods. *Adv. Mech.*, 14(2), 3–25 (1991).

[6] Andrianov I.V., Awrejcewicz J.: New trends in asymptotic approaches: summation and interpolation methods. *Appl. Mech. Rev.*, 54(1), 69–92 (2001).

[7] Andrianov I.V., Awrejcewicz J., Barantsev R.G.: Asymptotic approaches in mechanics: new parameters and procedures. *Appl. Mech. Rev.*, 56(1), 87–110 (2003).

[8] Andrianov I.V., Awrejcewicz J., Weichert D.: Improved continuous models for discrete media. *Mathematical Problems in Engineering*, ID 986242 (2010).

[9] Andrianov I.V., Bulanova N.S.: Using Padé approximants to estimate the domain of applicability of the small-parameter method. *J. Sov. Math.*, 63(5), 532–34 (1993).

[10] Andrianov I.V., Danishevskyy V.: Asymptotic investigation of the nonlinear dynamic boundary value problem for rod. *Technische Mechanik*, 15(1), 5–35 (1995).

[11] Andrianov I.V., Danishevskyy V.: Asymptotic approach for nonlinear periodical vibrations of continuous structures. *J. Sound Vibr.*, 249(3), 465–81 (2002).

[12] Andrianov I.V., Gristchak V.Z., Ivankov A.O.: New asymptotic method for the natural, free and forced oscillations of rectangular plates with mixed boundary conditions. *Technische Mechanik*, 14(3/4), 185–93 (1994).

[13] Andrianov I.V., Ivankov A.O.: Application of the Padé approximant in the method of introducing a parameter when investigating biharmonic equations with complex boundary conditions. *USSR Comp. Math. Math. Phys.*, 27(1), 193–7 (1987).

[14] Andrianov I.V., Ivankov A.O.: New asymptotic method for solving of mixed boundary value problem. *International Series of Numerical Mathematics*, 106, 39–45 (1992).

[15] Andrianov I.V., Ivankov A.O.: On the solution of plate bending mixed problems using modified technique of boundary conditions perturbation. *ZAMM*, 73(2), 120–2 (1993).

[16] Andrianov I.V., Krizhevskiy G.A.: Modifications of the asymptotic method in problems of dynamics of the reinforced constructions. *Stroitielnaya Mekhanika*, 2, 66–8 (1987) (in Russian).

[17] Andrianov I.V., Krizhevskiy G.A.: Computation of vibration modes of parallelogram plate via an approximating method. *Izvestia VUZ, Stroitelstvo i Architektura*, 2, 48–9 (1988) (in Russian).

[18] Andrianov I.V., Krizhevskiy G.A.: Analytical investigation of geometrically non-linear vibrations of sectorial plates reinforced by radial ribbs. *DAN USSR A*, 11, 30–3 (1989) (in Russian).

[19] Andrianov I.V., Lesnichaya V.A., Manevitch L.I.: *Homogenization Methods in Statics and Dynamics of Ribbed Shells*. Nauka, Moscow, 1985 (in Russian).

[20] Andrianov I.V., Manevitch L.I.: *Asymptotology: Ideas, Methods and Applications*. Kluwer Academic Publishers, Dordrecht, Boston, London, 2002.

[21] Andrianov I.V., Mikolenko V.A., Kholod E.G.: Nonlinear dynamics of a plane fibrous composite taking into account the width of the fiber. *Mech. Solids*, 34(2), 71–5 (1999).

[22] Apresyan L.A.: Padé approximants. *Radiophysics and Quantum Electronics*, 22(6), 449–66 (1979).

[23] Argatov I.I.: *Introduction to the Asymptotic Modeling in Mechanics*. Politechnika, Sankt-Petersburg, 2004 (in Russian).

[24] Askes H., Metrikine A.V.: One-dimensional dynamically consistent gradient elasticity models derived from a discrete microstructure. Part 2: Static and dynamic response. *European J. Mechanics A/Solids*, 21, 573–88 (2002).

[25] Awrejcewicz J., Andrianov I.V., Manevitch L.I.: *Asymptotic Approaches in the Nonlinear Dynamics: New Trends and Applications*. Springer-Verlag, Berlin, Heidelberg, 1998.

[26] Azrar L., Benamar R., Potier-Ferry M.: An asymptotic-numerical method for large-amplitude free vibrations of thin elastic plates. *J. Sound Vibr.*, 220(4), 695–727 (1999).

[27] Azrar L., Cochelin B., Damil N., Potier-Ferry M.: An asymptotic numerical method to compute the postbuckling behaviour of elastic plates and shells. *Int. J. Num. Meth. Eng.*, 36, 1187–1213 (1994).

[28] Babich V.M., Buldyryev V.S.: *Asymptotic Methods in Short Wave Diffraction Problems*. Nauka, Moscow, 1972 (in Russian).

[29] Baker G.A.: *Essential of Padé Approximation*. Academic Press, New York, 1975.

[30] Baker G.A., Graves-Morris P.: *Padé Approximants*. Cambridge University Press, Cambridge, New York, 1996.

[31] Bakhvalov N.S., Eglit M.E.: On limiting behavior of periodic media with soft-moduli inclusions. *Comp. Math. Math. Physics*, 35(6), 719–29 (1995).

[32] Bakhvalov N.S., Panasenko G.P.: *Averaging Processes in Periodic Media: Mathematical Problems in Mechanics of Composite Materials*. Kluwer, Dordrecht, 1989.

[33] Basdevant J.L.: The Padé approximation and its physical applications. *Fort. Physic*, 20(5), 284–328 (1977).

[34] Bauer S.M., Filippov S.B., Smirnov A.L., Tovstik P.E.: Asymptotic methods in mechanics with applications to thin shells and plates. *Asymptotic Methods in Mechanics. CRM Proc. Lect. Notes*, 3, 3–140 (1994).

[35] Beckermann B., Kaliaguine V.: The diagonal of the Padé table and the approximation of the Weyl function of the second-order difference operator. *Constr. Appr.*, 13, 481–510 (1997).

[36] Beckermann B., Matos A.S., Wielonsky F.: Reduction of the Gibbs phenomenon for smooth functions with jumps by the ε-algorithm. *J. Comp. Appl. Math.*, 219, 329–49 (2008).

[37] Bedford A., Drumheller D.S.: *Introduction to Elastic Wave Propagation*. John Wiley & Sons, Ltd., Chichester, 1996.

[38] Bellman R.: Perturbation methods applied to nonlinear dynamics. *J. Appl. Math.*, 22, 500–2 (1955).

[39] Bellman R.: *Perturbation Techniques in Mathematics, Physics and Engineering*. Holt, Rinehart and Winston, New York, 1964.

[40] Bender C.M., Boetcher S.: Continued fraction as a discrete nonlinear transform. *J. Math. Phys.*, 35(1), 364–7 (1994).

[41] Bender C.M., Boetcher S.: Determination of $f(\infty)$ from the asymptotic series for $f(x)$ about $x = 0$. *J. Math. Phys.*, 35(4), 1914–21 (1994).

[42] Bender C.M., Boetcher S., Milton K.A.: A new perturbative approach to nonlinear partial differential equations. *J. Math. Phys.*, 32(11), 3031–8 (1991).

[43] Bender C.M., Milton K.A., Pinsky S.S., Simmons L.M. Jr,.: A new perturbation approach to nonlinear problems. *J. Math. Phys.*, 30(7), 1447–55 (1989).

[44] Bender C.M., Orszag S.A.: *Advanced Mathematical Methods for Scientists and Engineers*. McGraw-Hill, New York, 1978.

[45] Bender C.M., Vinson J.P.: Summations of power series by continued exponentials. *J. Math. Phys.*, 37(8), 4103–19 (1996).

[46] Bensoussan A., Lions J.-L., Papanicolaou G.: *Asymptotic Analysis for Periodic Structures*. North-Holland, Amsterdam, 1978.

[47] Berdichevskiiy V.L.: *Variational Principles of Mechanics of Continuous Media*. Nauka, Moscow, 1983 (in Russian).

[48] Berger H.M.: A new approach to analysis of large deflections of plates. *J. Appl. Mech.*, 22(4), 465–72 (1955).

[49] Bernoulli J.: Dechordis vibrantibls. *Commentarii Academiae Scientiarum Imperialis Petropolitanae*, 3, 13–28 (1728).

[50] Berry M.V.: Asymptotics, superasymptotics, hyperasymptotics. *NATO ASI Series*, 284, 1–14 (1992).

[51] Berrut J.-P., Mittelmann H.D.: Lebesque constant minimizing linear rational interpolation of continuous functions over the interval. *Computers and Mathematics with Applications*, 33, 77–86 (1997).

[52] Berry M.V., Howls C.J.: Hyperasymptotics. *Proc. Roy. Soc. A*, 430, 653–68 (1990).

[53] Berry M.V., Howls C.J.: Hyperasymptotics for integrals with saddles. *Proc. Roy. Soc. A*, 434, 657–75 (1991).

[54] Bessis D.: Padé approximants in noise filtering. *J. Comp. Appl. Math.*, 66, 85–8 (1996).

[55] Blanc X., Bris C.Le, Lions P.-L.: From molecular models to continuum mechanics. *Archive Rational Mech. Anal.*, 164, 341–81 (2002).

[56] Boettcher S., Bender C.M.: Nonperturbative square-well approximation to a quantum theory. *J. Math. Phys.*, 31(11), 2579–85 (1990).

[57] Bogoliubov N.N. Mitropolsky Yu.A.: *Asymptotic Methods in the Theory of Nonlinear Oscillations*. Gordon and Breach, London, 1985.

[58] Bolotin V.V.: The edge effect in the oscillations of elastic shells. *J. Appl. Math. Mech.*, 24(5), 1257–72 (1960).

[59] Bolotin V.V.: *Random Vibrations of Elastic Systems*. Martinus Nijhoff Publ., the Hague, Boston, 1984.

[60] Bolotin V.V., Novichkov Y.N.: *Mechanics of Multi-Layer Structures*. Mashinostroyeniye, Moscow, 1980 (in Russian).

[61] Born M., Huang K.: *Dynamical Theory of Crystal Lattices*. Oxford University Press, Oxford, 1954.

[62] Bosley D.L.: A technique for the numerical verification of asymptotic expansions. *SIAM Rev.*, 38(1), 128–35 (1996).

[63] Bourgat J.F.: Numerical experiments of the homogenization method for operators with periodic coefficients. *Lecture Notes Math.*, 704, 330–56 (1979).

[64] Boyd J.P.: *Weakly Nonlocal Solitary Waves and Beyond-All-Orders Asymptotics. Generalized Solutions and Hyperasymptotic Perturbation Theory*. Kluwer, Dordrecht, 1998.

[65] Boyd J.P.: The Devil's invention: asymptotic, superasymptotic and hyperasymptotic series. *Acta Applicandae Mathematicae*, 56, 1–98 (1999).

[66] Braides A., Gelli M.S.: The passage from discrete to continuous variational problems: a nonlinear homogenization process. *Nonlinear Homogenization and its Appl. to Composites, Polycristals and Smart Materials*. Eds. Ponte Castaneda P., Telega J.J., Gambin B. Dordrecht: Kluwer, 2004, 45–64.

[67] Brezinski C.: Composite sequence transformations. *Numerical Mathematics*, 46, 311–21 (1985).

[68] Brezinski C.: Convergence acceleration during the 20th century. *J. Comp. Appl. Math.*, 122, 1–21 (2000).

[69] Brezinski C.: Extrapolation algorithms for filtering series of functions, and treating the Gibbs phenomenon. *Numerical Algorithms*, 36, 309–29 (2004).

[70] Brillouin L., Parodi M.: *Waves Propagation in Periodic Structures*. Dover, New York, 1953.

[71] Bruno A.D.: *Power Geometry in Algebraic and Differential Equations*. Elsevier, Amsterdam, 2000.

[72] Bruno A.D.: Self-similar solutions and power geometry. *Russ. Math. Surv.*, 55(1), 1–42 (2000).

[73] Buonomo A.: The perturbation solution of Van der Pol's equation. *SIAM J. Appl. Math.*, 59(1), 156–71 (1998).

[74] Cabay S., Domzy B.: Systems of linear equations with dense univariate polynomial coefficients. *J. Assoc. Comp. Mech.*, 34(3), 46–60 (1987).

[75] Chalbaud E., Martin P.: Two-point quasifractional approximant in physics: method improvement and application to $J_n(x)$. *J. Math. Phys.*, 33, 2483–248 (1992).

[76] Chen G., Coleman M. P., Zhou J.: Analysis of vibration eigenfrequencies of a thin plate by the Keller-Rubinow wave method I: Clamped boundary conditions with rectangular or circular geometry. *SIAM J. Appl. Math.*, 51(4), 967–83 (1991).

[77] Chen G., Coleman M. P., Zhou J.: The equivalence between the wave propagation method and Bolotin's method in the asymptotic estimation of eigenfrequencies of a rectangular plate. *Wave Motion*, 16(3), 285–97 (1992).

[78] Chen G., Zhou J.: *Vibration and Damping in Distributed Systems*, vol. I, II. CRS, Boca Raton, Fl, 1993.

[79] Chen S.H., Cheung Y.K.: An elliptic perturbation method for certain strongly non-linear oscillators. *J. Sound Vibr.*, 192(2), 453–64 (1996).

[80] Chen W., Fish J.: A dispersive model for wave propagation in periodic heterogeneous media based on homogenization with multiple spatial and temporal scales. *J. Appl. Mech.*, 86, 153–61 (2001).

[81] Cherkaev E., Ou M.-J.Y.: Dehomogenization: reconstruction of moments of the spectral measure of the composite. *Inverse Problems*, 24, 065008 (2008).

[82] Cochelin B.: A path-following technique via an asymptotic-numerical method. *Comp. Struct.*, 53(5), 1181–92 (1994).

[83] Cochelin B., Damil N., Potier-Ferry M.: Asymptotic-numerical methods and Padé approximants for non-linear elastic structures. *Int. J. Num. Meth. Eng.*, 37(7), 1187–1213 (1994).

[84] Collins M.A.: A quasi-continuum approximation for solitons in an atomic chain. *Chemical Phys. Lett.*, 77(2), 342–7 (1981).

[85] Common A.K.: Application of Hermite-Padé approximants to water waves and the harmonic oscillator on a lattice. *J. Phys. A*, 15, 3665–77 (1982).

[86] Conca C., Lund F.: Fourier homogenization method and the propagation of acoustic waves through a periodic vortex array. *SIAM J. Appl. Math.*, 59(5), 573–81 (1999).

[87] Corless R.M., Gonnet G.H., Hare D.E.G., Jeffrey D.J., Knuth D.E.: On the Lambert W function. *Advances in Computational Mathematics*, 5, 329–59 (1996).
[88] Crighton D.G.: Asymptotics–an indispensible complement to thought, computation and experiment in applied mathematical modelling. *Seventh European Conf. on Mathematics in Industry*. Eds. Fasano A., Primicerio M.B. Stuttgart: G.Teubner, 3–19 (1994).
[89] Dadfar M.B., Geer J.F.: Power series solution to a simple pendulum with oscillating support. *SIAM J. Appl. Math.*, 47(4), 737–50 (1987).
[90] Damil N., Potier-Ferry M.: A new method to compute perturbed bifurcation: application to the buckling of imperfect elastic structures. *Int. J. Eng. Sc.*, 28, 943–57 (1990).
[91] Del Piero G., Truskinovsky L.: Macro and micro cracking in one-dimensional elasticity. *Int. J. Solids Struct.*, 38, 1135–48 (2001).
[92] Dingle R.B.: *Asymptotic Expansions: Their Derivation and Interpretation*. Academic Press, New York, London, 1973.
[93] Dowell E.H., Tang D.: Multiscale, multiphenomena modelling and simulation at the nanoscale: on constructing reduced-order models for nonlinear dynamical systems with many degrees-of-freedom. *J. Appl. Mech.*, 70, 328–38 (2003).
[94] Drazin P.G., Tourigny Y.: Numerical study of bifurcations by analytic continuation of a function defined by a power series. *SIAM J. Appl. Math.*, 56, 1–18 (1996).
[95] Driscoll T.A., Fornberg B.: A Padé-based algorithm for overcoming the Gibbs phenomenon. *Numerical Algorithms*, 26, 77–92 (2001).
[96] Eckhaus W.: *Matched Asymptotic Expansions and Singular Perturbations*. North-Holland, Amsterdam, 1973.
[97] Eckhaus W.: *Asymptotic Analysis of Singular Perturbations*. North-Holland, Amsterdam, 1979.
[98] Elhage-Hussein A., Potier-Ferry M., Damil N.: A numerical continuation method based on Padé approximants. *Int. J. Solids Struct.*, 37, 6981–7001 (2000).
[99] Emachi E., Vakakis A.F., Andrianov I.V., Mikhlin Yu.V.: Study of two-dimensional axisymmetric breathers using Padé approximants. *Non. Dyn.*, 13, 327–38 (1997).
[100] Estrada R., Kanwal R.P.: *A Distributional Approach to Asymptotics: Theory and Applications*. Birkhäuser, Boston, 2002.
[101] Fedoriyuk M.V.: *Asymptotics: Integrals and Series*. Nauka, Moscow, 1987 (in Russian).
[102] Fidl A.: A new technique of selection between sequence transformations. *Applied Numerical Mathematics*, 25, 21–40 (1997).
[103] Filatov A.N.: *Asymptotic Method in Theory of Differential and Integro-Differential Equations*. FAN, Tashkent, 1974 (in Russian).
[104] Filatov A.N., Sharova L.V.: *Integral Inequalities and Theory of Non-Linear Vibrations*. Nauka, Moscow, 1976 (in Russian).
[105] Filimonov A.M.: Some unexpected results in the classical problem of vibrations of the string with *N* beads. The case of multiple frequencies. *C. R. Acad. Sci.*, 315(8), 957–61 (1991).
[106] Filimonov A.M.: Continuous approximations of difference operators. *J. Difference Eq. Appl.*, 2(4), 411–22 (1998).
[107] Filimonov A.M., Kurchanov P.F., Myshkis A.D.: Some unexpected results in the classical problem of vibrations of the string with n beads when *n* is large. *C. R. Acad. Sci.*, 313, 961–5 (1991).
[108] Floater M.S., Hormann K.: Barycentric rational interpolation with no poles and high rates of approximation. *Numerische Mathematik*, 107, 315–31 (2007).
[109] Friedrichs K.O.: Asymptotic phenomena in mathematical physics. *Bull. Amer. Math. Soc.*, 61, 485–504 (1955).
[110] Friedrichs K.O.: *Perturbation of Spectra in Hilbert Space*. AMS, Providence, RI, 1965.
[111] Friesecke G., Theil F.: Validity and failure of the Cauchy-Born hypothesis in a two-dimensional mass-spring lattice. *J. Nonlin. Sciences*, 12, 445–78 (2002).
[112] Frost P.A., Harper E.Y.: Extended Padé procedure for constructing global approximations from asymptotic expansions: an explication with examples. *SIAM Rev.*, 18(1), 62–91 (1976).
[113] Gaunt D.S., Guttman A.J.: Series expansions: analysis of coefficients. *Phase Trans. Crit. Phen.*, 3, 181–243, (1974).
[114] Gekeler E.: On the solution of system of equations by the epsilon algorithm of Wynn. *Math. Comp.*, 26, 427–36 (1972).
[115] Giacaglia G.E.O.: *Perturbation Methods in Non-Linear Systems*. Springer, New York, 1972.

[116] Gilewicz J., Pindor M.: Padé approximants and noise: A case of geometrical series. *J. Comp. Appl. Math.*, 87, 199–214 (1997).

[117] Gilewicz J., Pindor M.: Padé-type approximants and error of Padé approximants. *J. Comp. Appl. Math.*, 99, 155–65 (1998).

[118] Gilewicz J., Pindor M.: Padé approximants and noise: rational functions. *J. Comp. Appl. Math.*, 105, 285–97 (1999).

[119] Gol'denveizer A.L.: *Theory of Elastic Thin Shells*. Pergamon Press, New York, 1961.

[120] Gonchar A.A.: Some recent convergence results on diagonal Padé approximants. In: *Approximation Theory V*. Academic Press, Boston, 55–70 (1986).

[121] Gontkevich V.S.: *Free Vibrations of Plates and Shells*. Naukova Dumka, Kiev, 1964 (in Russian).

[122] Grigolyuk E.I., Filshtinskiy L.A.: *Perforat Plates and Shells*. Nauka, Moscow, 1970 (in Russian).

[123] Grigolyuk E.I., Filshtinskiy L.A.: *Periodic Piece-Wise Homogeneous Elastic Structures*. Nauka, Moscow, 1992 (in Russian).

[124] Grigolyuk E.I., Kulikov G.M.: *On a Simplified Method of Solution to Non-Linear Problems of Elastic Plates and Shells: Some Problems of Theory of Plates and Shells*. MGU Press, Moscow, 1981 (in Russian).

[125] Grundy R.E.: Laplace transform inversion using two-point rational approximants. *J. Inst. Math. Appl.*, 20, 299–306 (1977).

[126] Grundy R.E.: On the solution of nonlinear Volterra integral equations using two-point Padé approximants. *J. Inst. Math. Appl.*, 22, 317–20 (1978).

[127] Grundy R.E.: The solution of Volterra integral equation of the convolution type using two-point rational approximants. *J. Inst. Math. Appl.*, 22, 147–58 (1978).

[128] Handy C.R.: Harmonic balance method and the theory of generalized Padé approximants. *J. Sound Vibr.*, 102(2), 247–57 (1985).

[129] He Ji-Huan: A new approach to nonlinear partial differential equations. *Com. Non. Sc. Num. Sim.*, 2(4), 230–5 (1997).

[130] He Ji-Huan.: Homotopy perturbation technique. *Comp. Meth. Appl. Mech. Eng.*, 178, 257–256 (1999).

[131] He Ji-Huan.: A coupling method of homotopy technique and a perturbation technique for non-linear problems. *Int. J. Nonlin. Mech.*, 35, 37–43 (2000).

[132] He Ji-Huan: Recent developments of the homotopy perturbation method. *Top. Meth. Nonlin. Anal.*, 31, 205–9 (2008).

[133] Hinch E.J.: *Perturbation Methods*. University Press, Cambridge, 1991.

[134] Holms M.H.: *Introduction to Perturbation Methods*. Springer, New York, 1995.

[135] Jamai R., Damil N.: Quadratic approximants in the asymptotic numerical method. *C.R. Acad. Sci. Paris*, series II b, 329, 809–14 (2001).

[136] Jones W.B., Thron W.J.: *Continued Fractions. Analytic Theory and Its Applications*. Addison-Wesley, New York, 1980.

[137] Kaas-Petersen C.: Continuation methods as the link between perturbation analysis and asymptotic analysis. *SIAM Rev.*, 29(1), 115–20 (1987).

[138] Kalamkarov A.L.: *Composite and Reinforced Elements of Construction*. John Wiley & Sons, Ltd, Chichester, 1992.

[139] Kalamkarov A.L., Kolpakov A.G.: *Analysis, Design and Optimization of Composite Structures*. John Wiley & Sons, Ltd, Chichester, 1997.

[140] Kalamkarov A.L. Kudriavtsev B.A, Parton V.Z.: Asymptotic method of averaging in mechanics of composities of regular structure. *Achievements in Science and Techniques. VINITI, Series Mechanics of Solids*, 19, 78–147 (1987) (in Russian).

[141] Kauderer H.: *Nichtlineare Mechanik*. Springer, Berlin, 1958.

[142] Kantorovich L.V., Krylov V.I.: *Approximate Methods of Higher Analysis*. Noordhoff, Groningen, 1958.

[143] Karpov S.Yu., Stolyarov S.N.: Propagational transformation of electromagnetic waves in one-dimensional periodic structures. *Phys. Usp.*, 36(1), 1–22 (1993).

[144] Kashin A.P., Kvaratskheliya T.M., Maksimov M.Z., Chikovani Z.E.: Higher approximations of the reduced method of matched asymptotic expansions and its convergence. *Theor. Math. Phys.*, 78, 278–84 (1989).

[145] Kayuk Ya.F.: *Some Problems of the Methods Splitting by Parameters*. Naukova Dumka, Kiev, 1991 (in Russian).

[146] Keller J.B., Rubinow S.I.: Asymptotic solution of eigenvalue problems. *Ann. Phys.*, 9(1), 24–75 (1960).

[147] Kevorkian J., Cole J.D.: *Multiple Scale and Singular Perturbation Methods*. Springer, New York, 1996.

[148] Kolpakov A.G.: *Stressed Composite Structures: Homogenized Models for Thin-Walled Nonhomogeneous Structures with Initial Stresses*. Springer-Verlag, Berlin, New York, 2004.
[149] Kosevich A.M., Kovallov A.S.: *Introduction to Non-Linear Physical Mechanics*. Naukova Dumka, Kiev, 1989 (in Russian).
[150] Krivtsov A.M.: *Deformation and Damage of Rigid Bodies with Microstructure*. Fizmat, Moscow, 2007 (in Russian).
[151] Kruskal M.D.: Asymptotology. *Mathematical Models in Physical Sciences. Proceedings of the conference held at Notre Dame*, Indiana, April 1962. Prentice-Hall, Englewood Cliffs, N.J., 17–48, 1963.
[152] Kunin I.A.: *Elastic Media with Microstructure*. vol. 1,2, Springer, Berlin, 1982, 1983.
[153] Kuzmak G.E.: Asymptotic solution of nonlinear second order differential equations with variable coefficients. *J. Appl. Math. Mech.*, 23, 730–44 (1959).
[154] Kurchanov P.F., Myshkis A.D., Filimonov A.M.: Vibrations of rolling stock and a theorem of Kronecker. *J. Appl. Math. Mech.*, 55(6), 870–6 (1991).
[155] Lambert F., Musette M.: Solitary waves, padeons and solitons. *Lect. Not. Math.*, 1071, 197–212 (1984).
[156] Lambert F., Musette M.: Solitons from a direct point of view: padeons. *J. Comp. Appl. Math.*, 15, 235–49 (1986).
[157] Li Jian-Lin: Adomian's decomposition method and homotopy perturbation method in solving nonlinear equations. *J. Comp. Appl. Math.*, 228(1), 168–73 (2009).
[158] Liao S.J.: *Beyond Perturbation: Introduction to the Homotopy Analysis Method*. Chapman and Hall/CRC Press, Boca Raton, 2004.
[159] Liapunov A.M.: An investigation of one of the singular cases of the theory of stability of motion, II. *Mathematicheskiy Sbornik*, 17(2), 253–333 (1893). (Eng. Transl. in 1603, 128–84).
[160] Liapunov A.M.: *Stability of Motion*. Academic Press, New York, 1966.
[161] Lighthill M.J.: A technique for rendering approximate solutions to physical problems uniformly valid. *Phil. Mag.*, 40, 1179–1201 (1949).
[162] Lin S.S., Segel L.A.: *Mathematics Methods Applied to Deterministic Problems in the Natural Sciences*. SIAM, Philadelphia, 1988.
[163] Litvinov G.L.: Approximate construction of rational approximations and the effect of autocorrection error. *Russian J. Math. Phys.*, 1(3), 313–52 (1994).
[164] Litvinov G.L.: Error autocorrection in rational approximation and interval estimation. *Central European J. Math.*, 1(1), 1–33 (2003).
[165] Luke Y.L.: Computations of coefficients in the polynomials of Padé approximation by solving system of linear equations. *J. Comp. Appl. Math.*, 6(3), 213–18 (1980).
[166] Luke Y.L.: A note on evaluation of coefficients in the polynomials of Padé approximation by solving system of linear equations. *J. Comp. Appl. Math.*, 8(2), 93–9 (1982).
[167] Mandelshtam L.I.: *Lectures on the Theory of Oscillations*. Nauka, Moscow, 1972 (in Russian).
[168] Manevitch L.I.: Nonlinear normal modes and solitons. In: Scott E. *Nonlinear Science: the Birth and Development of Coherent Structures*, Fizmatlit, Moscow, 2007, 532–45 (in Russian).
[169] Manevitch L.I., Andrianov I.V., Oshmyan V.O.: *Mechanics of Periodically Heterogeneous Structures*. Springer-Verlag, Berlin, 2002.
[170] Manevitch A.I., Manevitch L.I.: *The Mechanics of Nonlinear Systems with Internal Resonances*. World Scientific, New York, 2005.
[171] Manevitch L.I., Mikhlin Yu.V., Philipchuk V.N.: *Method of Normal Vibrations for Essentialy Non-Linear Systems*. Nauka, Moscow, 1989 (in Russian).
[172] March R., Barone P.: Application of the Padé method to solving the noisy trigonometric moment problem: some initial results. *SIAM J. Appl. Math.*, 58(1), 324–43 (1998).
[173] Marchuk G.I., Agoshkov V.I.: Conjugate operators and algorithms of perturbation in non-linear problems: 1. Principles of construction of conjugate operators. *Soviet J. Num. Math. Modeling*, 1, 21–46 (1988).
[174] Marchuk G.I., Agoshkov V.I.: Conjugate operators and algorithms of perturbation in non-linear problems: 2. Perturbation algorithms. *Soviet J. Num. Math. Modeling*, 2, 115–36 (1988).
[175] Marchuk G.I., Agoshkov V.I., Shutyaev V.P.: *Adjoint Equations and Perturbation Algorithms in Nonlinear Problems*. CRC Press, Boca Raton, RI, 1996.
[176] Martin P., Baker G.A. Jr.,: Two-point quasifractional approximant in physics. Truncation error. *J. Math. Phys.*, 32(6), 313–28 (1991).
[177] Mason J.C.: Some applications and drawbacks of Padé approximants. *Appr. Theory Appl.*, 207–23 (1981).
[178] Maslov V.P.: *Operational Methods*. Mir, Moscow, 1976 (in Russian).

[179] Maslov V.P.: *The Complex WKB Method for Nonlinear Equations I: Linear Theory*. Birkhäuser, Basel, Boston, 1994.
[180] Maslov V.P.: *Asymptotic Methods and Theory of Perturbations*. Nauka, Moscow, 1988 (in Russian).
[181] Maslov V.P., Omel'yanov G.A.: *Geometric Asymptotic for Nonlinear PDE*. I. Providence, RI: AMS, 2001.
[182] McKenzie D.R., McPhedran R.C.: Exact modelling of cubic lattice permittivity and conductivity. *Nature*, 265, 128–9 (1977).
[183] McKenzie D.R., McPhedran R.C., Derrick G.H.: The conductivity of lattices of spheres II. The body-centered and face-centered lattices. *Proc. R. Soc. A*, 362, 211–32 (1978).
[184] McPhedran R.C., McKenzie D.R.: The conductivity of lattices of spheres. I. The simple cubic lattice. *Proc. R. Soc. A*, 359, 45–63 (1978).
[185] McPhedran R.C., Milton G.W.: Bounds and exact theories for the transport properties of inhomogeneous media. *Appl. Phys. A: Mater. Sci. Proc.*, 26, 207–20 (1981).
[186] Meredith R.E., Tobias G.W.: Resistance to potential flow through a cubical array of spheres. *J. Appl. Phys.*, 31, 1270–3 (1960).
[187] Mickens R.E.: A generalization of the method of harmonic balance. *J. Sound Vibr.*, 111, 515–18 (1986).
[188] Miller P.D.: *Applied Asymptotic Analysis*. AMS, Rhode Island, Providence, 2006.
[189] Moiseev N.N.: *Asymptotic Methods in Non-Linear Mechanics*. Nauka, Moscow, 1981 (in Russian).
[190] Mokhtari R.E., Cadou J.-M., Potier-Ferry M.: A two grid algorithm based on perturbation and homotopy method. *Mechanique*, 330, 825–30 (2002).
[191] Molotkov I.A., Vakulenko S.A.: *Localized Nonlinear Waves*. LGU, Leningrad, 1988 (in Russian).
[192] Movchan A.B., Movchan N.V., Poulton C.G.: *Asymptotic Models of Fields in Dilute and Densely Packed Composites*. Imperial College Press, London, 2002.
[193] Myshkis A.D.: Mixed functional differential equations. *J. Math. Sc.*, 129(5), 4111–26 (2005).
[194] Nayfeh A.H.: *Introduction to Perturbation Techniques*. John Wiley and Sons, Inc., New York, 1981.
[195] Nayfeh A.H.: *Perturbation Methods*. John Wiley and Sons, Inc., New York, 2000.
[196] Nemeth G., Paris G.: The Gibbs phenomenon in generalized Padé approximation. *J. Math. Phys.*, 26(6), 1175–8 (1985).
[197] Olver F.W.J.: *Asymptotic and Special Functions*. AK Peters, Wellesley, 1997.
[198] O'Malley R.E.: *Singular Perturbation Methods for Ordinary Differential Equations*. Springer-Verlag, New York, 1991.
[199] Pagano S., Paroni R.: A simple model for phase transition: from the discrete to the continuum problem. *Quart. Appl. Math.*, 61, 89–109 (2003).
[200] Panovko Ya.G.: *Mechanics of Deformable Rigid Body: Modern Concepts, Errors and Paradoxes*. Nauka, Moscow, 1985 (in Russian).
[201] Paroni R.: From discrete to continuum: a Young measure approach. *ZAMP*, 54(2), 328–48 (2003).
[202] Peierls R.E.: *Surprises in Theoretical Physics*. Princeton University Press, Princeton, 1979.
[203] Pichugin A.V., Askes H., Tyas A.: Asymptotic equivalence of homogenization procedures and fine-tuning of continuum theories. *J. Sound Vibr.*, 313, 858–74 (2008).
[204] Pilipchuk V.N.: *Nonlinear Dynamics: Between Linear and Impact Limits*. Springer, New York, 2010.
[205] Pinney E.: *Ordinary Difference-Differential Equations*. Univ. California Press, Berkeley and Los Angeles, 1958.
[206] Pobedrya B.E.: *Mechanics of Composite Material*. MGU, Moscow, 1984 (in Russian).
[207] Poincaré H.: *New Methods in Celestial Mechanics*. Ed. and introd. D.L. Goroff. Woodbury, AIP, 1993.
[208] Poston T., Stewart I.: *Catastrophe Theory and its Applications*. Pitman, London, 1978.
[209] Prudnikov A.P., Brychkov O.A., Marichev O.I.: *Integrals and Series*. Gordon and Breach, New York, 1992.
[210] Reuss A.: Berechnung der Fließgrenze on Mischkristallen auf Grund der Plastizitätsbedingungen für Einkristall. *ZAMM*, 9, 49–58 (1929).
[211] Rosenau Ph.: Dynamics of nonlinear mass-spring chains near the continuum limit. *Physics Letters A*, 118(5), 222–7 (1986).
[212] Rosenau Ph.: Hamiltonian dynamics of dense chains near and lattices: or how to correct the continuum. *Physics Letters A*, 311(5), 39–52 (2003).
[213] Rosenberg R.M.: The normal modes of nonlinear n-degree-of-freedom systems. *J. Appl. Mech.*, 29, 7–14 (1962).
[214] Rosenberg R.M.: The Ateb(h)-functions and their properties. *Quart. Appl. Math.*, 21, 37–47 (1963).
[215] Sangani A.S., Acrivos A.: The effective conductivity of a periodic array of spheres. *Proc. R. Soc. A*, 386, 263–75 (1983).

[216] Scorobogat'ko V. Ya.: *Theory of the Branching Continued Fractions and its Application in the Numerical Mathematics*. Nauka, Moscow, 1983 (in Russian).

[217] Segel L.A.: The importance of asymptotic analysis in applied mathematics. *American Mathematical Monthly*, 73, 7–14 (1966).

[218] Semerdjiev Kh.: Trigonometric Padé Approximants and Gibbs phenomenon. *Rep. United Inst. Nuclear Research* NP5-12484, Dubna, 1979 (in Russian).

[219] Senik P.M.: Inversion of the incomplete Beta function. *Ukrainian Math. J.*, 21, 271–8 (1969).

[220] Sergeev A.V.: A recursive algorithm for Padé-Hermite approximations. *USSR Comp. Math. Math. Phys.*, 26(2), 17–22 (1986).

[221] Shafer R.E.: On quadratic approximation. *SIAM J. Numer. Anal.*, 11, 447–60 (1974).

[222] Shamrovskii A.D.: Asymptotic integration of static equation of the theory of elasticity in Cartesian coordinates with automated search of integration parameters. *J. Appl. Math. Mech.*, 43(5), 925–34 (1979).

[223] Shamrovskii A.D., Andrianov I.V., Awrejcewicz J.: Asymptotic-group analysis of algebraic equations. *Math. Probl. Eng.*, 5, 411–51 (2004).

[224] Shanks D.: Non-linear transformation of divergent and slowly convergent sequence. *J. Math. Phys.*, 34(1), 1–42 (1955).

[225] Shaw S.W., Pierre C.: Normal modes of vibration for non-linear continuous systems. *J. Sound Vibr.*, 169(3), 319–47 (1994).

[226] Shawyer B., Watson B.: *Borel's Method of Summability. Theory and Applications*. Clarendon Press, Oxford, 1994.

[227] Shokin Y.I.: *The Method of Differential Approximation*. Springer, Berlin, 1983.

[228] Slavyanov S.Yu.: *Asymptotic Solutions of the One-dimensional Schrödinger Equation*. AMS, Providence, R.I., 1996.

[229] Slepyan L.I., Yakovlev Yu.S.: *Integral Transforms in Non-steady-state Problems of Mechanics*. Sudostroyenie, Leningrad, 1980 (in Russian).

[230] Stermin B, Shatalov V.: *Borel-Laplace Tranform and Asymptotic Theory. Introduction to Resurgent Analysis*. CRC Press, N.Y., 1995

[231] Suetin S.P.: Padé approximants and efficient analytic continuation of a power series. *Russ. Math. Surv.*, 57(1), 43–141 (2002).

[232] Suetin S.P.: On interpolation properties of diagonal Padé approximants of elliptic functions. *Russ. Math. Surv.*, 59, 201–2 (2004).

[233] Timoshenko S.P., Woinowsky-Krieger S.: *Theory of Plates and Shells*. 2nd ed., McGraw-Hill, New York, 1987.

[234] Tokarzewski S., Telega J.J.: S-continued fraction method for the investigation of a complex dielectric constant of two-phase composite. *Acta Appl. Math.*, 49, 55–83 (1997).

[235] Tovstik P.E., Bauer S.M., Smirnov A.L., Filippov S.B.: *Asymptotic Methods in Mechanics of Thin-Walled Structures*. SP University Press, Sankt-Petersburg, 1995 (in Russian).

[236] Ulam S.: *Collection of Mathematical Problems*. Interscience, New York, 1960.

[237] Vainberg M.M., Trenogin V.A.: *Theory of Branching of Solutions of Non-linear Equations*. Noordhoff International Publishing, Leyden, 1974.

[238] Vakman D.E.: *Signals, Oscillations, and Waves: A Modern Approach*. Artech House, Boston, 1998.

[239] Vakakis A.F., Manevitch L.I., Mikhlin Yu.V., Pilipchuk V.N., Zevin A.A.: *Normal Modes and Localization in Nonlinear Systems*. Wiley Interscience, New York, 1996.

[240] Vakhromeyev Yu.M., Kornev V.M.: Dynamical boundary effect in rods. Formulation truncated problems. *Izv. AN of USSR, Mekhanika Tverdogo Tela*, 4, 95–103 (1972) (in Russian).

[241] Van der Pol B., Bremmer H.: *Operational Calculus Based on the Two-Sided Laplace Integral*. Chelsea Publishing Company, London, 1987.

[242] Van Dyke M.: Analysis and improvement of perturbation series. *Quart. J. Mech. Appl. Math.*, 27, 423–50 (1974).

[243] Van Dyke M.: Computer extension of perturbation series in fluid mechanics. *SIAM J. Appl. Math.*, 28, 720–34 (1975).

[244] Van Dyke M.: *Perturbation Methods in Fluid Mechanics*. The Parabolic Press, Stanford, 1975.

[245] Van Dyke M.: From zero to infinite Reynolds by computing extension of Stokes series. *Lec. Not. Math.*, 594, 506–17 (1977).

[246] Van Dyke M.: *Mathematical Approaches in Hydrodynamics*. SIAM, Philadelphia, 1991.

[247] Vavilov V.V., Tchobanou M.K., Tchobanou P.M.: Design of multi-dimensional recursive systems through Padé type rational approximation. *Non. Anal.: Modell. Contr.*, 7(1), 105–25 (2002).

[248] Verhulst F.: *Methods and Applications of Singular Perturbations, Boundary Layers and Multiple Timescale Dynamics*. Springer-Verlag, Berlin, 2005.

[249] Vinogradov V.N., Gay E.V., Rabotnov N.C.: *Analytical Approximation of Data in Nuclear and Neutron Physics*. Energoatomizdat, Moscow, 1987 (in Russian).

[250] Vishik M.I., Lyusternik L.A.: Solution of some problems about perturbation for matrix and selfadjoint and nonselfadjoint differential equations. *Russ. Math. Surv.*, 15(3), 3–80 (1960).

[251] Vishik M.I., Lyusternik L.A.: The asymptotic behaviour of solutions of linear differential equations with large or quickly changing coefficients and boundary conditions. *Russ. Math. Surv.*, 15(4), 23–91 (1960).

[252] Voigt W. Über die Beziehung zwischen den beiden Elasticitätsconstanten isotroper Körper. *Annalen der Physik und Chemie*, 38, 573–87 (1889).

[253] Volmir A.S.: *Stability of Elastic Systems*. Fizmatgiz, Moscow, 1963 (in Russian).

[254] Vyatchin A.V.: On the convergence of Padé approximants. *Moscow Univ. Math. Bull.*, 37(4), 1–4 (1982).

[255] Wasov W.: *Asymptotic Expansions for Ordinary Differential Equations*. John Wiley and Sons, Inc., New York, 1965.

[256] Whitham G.B.: *Linear and Nonlinear Waves*. John Wiley and Sons, Inc., New York, 1974.

[257] Yudovich V.I.: Vibrodynamics and vibrogeometry of mechanical systems with couplings. *Advances in Mechanics*, 3(1), 26–74 (2006) (in Russian).

[258] Weniger E.I.: Nonlinear sequence transformations for the acceleration of convergence and the summation of divergent series. *Comp. Phys. Rep.*, 10, 189–371 (1989).

[259] Wiener O.: Die Theorie des Mischkörpers für das Feld der stationären Strömung. Erste Abhandlung die Mittelwertsätze für Kraft, Polarisation und Energie. *Der Abhandlungen der Mathematisch-Physischen Klasse der Königlich Sächsischen Gesellschaft der Wissenschaften*, 32, 509–604 (1912).

[260] Zaslavskiy G.M., Usikov D. A., Sagdeev R.Z.: *Nonlinear Physics: From the Pendulum to Turbulence and Chaos*. Harwood Ac. Publ., New York, 1988.

[261] Joukowsky N.E.: The work of continuous and non-continuous traction devices in pulling a train from its position and at the beginning of its motion. In *Complete Collected Works, 8: Theory of Elasticity. Railways. Automobiles*, 221–55, ONTI, Moscow (1937) (in Russian).

2

Computational Methods for Plates and Beams with Mixed Boundary Conditions

2.1 Introduction

2.1.1 Computational Methods of Plates with Mixed Boundary Conditions

A computation of static and dynamic behavior of plates can be reduced to that of integration of biharmonic equations with various BCs [5], [39], [72], [78]–[82], [85]. It is also clear that an exact solution to that problem may be obtained only in rare cases, when the occurred BCs allow separating the spatial variables [75]–[77]. Otherwise, a problem under consideration should be solved approximately, i.e. numerically in majority of the studied cases. Mostly, the Rayleigh-Ritz, Kantorovich, Bubnov-Galerkin, and Trefftz approximation methods (or their various modifications) are used. The mentioned methods have been successively applied for many years to solve various practical problems. However, efficiency of variational approaches goes down when one studies mixed BCs, since a proper choice of functions satisfying different BCs on different parts of a supporting contour does not belong to easy tasks.

The Finite Element Method (FEM) is one of the most popular and efficient variants of the Rayleigh-Ritz approach. It has been widely applied to real world problems in mechanical and civil engineering and beyond. Nowadays we have a wide spectrum of various FEMs allowing us to solve any practical problem. However, even this so deeply developed method possesses its own drawbacks. Namely, it is rather difficult to estimate the error associated with FEM application, in many cases a computational instability occurs in BCs; also the computational process is usually time costly and is not directly oriented to the particular problem that is being studied.

FEM allows us to solve some problems devoted to static and dynamic behavior of plates with mixed BCs. However, the problem associated with the error estimation that has occurred belongs to a separate one and requires an additional and separate study.

Asymptotic Methods in the Theory of Plates with Mixed Boundary Conditions, First Edition.
Igor V. Andrianov, Jan Awrejcewicz, Vladislav V. Danishevs'kyy, Andrey O. Ivankov.

It should be emphasized that an interesting approximation method, devoted to computation of plates with mixed BCs, has been proposed by V.A. Smirnov [72]. In this case a plate is approximated via a set of coupled crossing beams. Moments on the simply supported beam ends are equal to zero, whereas in the clamped beam parts the moment is unknown and is defined by the method of forces. The problem has been studied in a matrix form. Vertical displacements of beam on the contour equal zero. In the case of BCs change, a linear distribution of the boundary moment is assumed starting with zero (simple support) and ending on a given value (clamping). This method has been also successively applied to solve the SSS and eigenvalue problems of rectangular plates having mixed BCs of the form "clamping-simple support." However, in the case of another mixed BCs, a direct application of this method is rather complicated. In addition, a direct application of this method to estimate the critical load values regarding stability problems is still open.

There also exists the V.A. Rvachev R-function method [53]. The R-function method allows us to calculate eigenfrequencies and eigenforms of either natural or forced plate vibrations, the SSS, as well as buckling of a plate with arbitrary BCs. In spite of that, this method seems not to be widely known to the western community.

Analysis of plates with complex boundary conditions can also be carried out by the method proposed by V.V. Bolotin [11]–[15], [21]–[22], [23], [24], [34], [35], [50]. This method relies on the introduction of two components of a solution being sought: one is associated with the plate internal domain, whereas the second one plays the role of a correction term of a boundary layer type being localized in the small plate contour neighborhood. However, an application of this method is rather limited and can be only applied to dynamic problems. In the case of stability problems this method yields rather high inaccuracy.

Mixed BVPs of plates can also be attacked through methods associated with the use of integral transformations. For instance, in the case of half-infinite plates with various variants of mixed BCs Shvabyuk applied the method of infinite integral transformations to solve problems related to SSS of plates; whereas problems of SSS and vibrations of half-infinite plates with various variants of mixed BCs have been studied by Zorski [93], [94].

The method of integral equation also has a wide spectrum of application [83]. In this case a solution is sought in the integral form being equivalent to the initial-boundary value problem. The obtained integral equation is solved approximately. Then, the method has been developed to that of boundary integral equations. The obtained integral equation is reduced to an integral equation regarding plate contour, which allows us to reduce the dimension of the problem, and it can be easily solved numerically. If one approximates the contour via finite elements, then the problem reduces to one that can easily be solved by FEM.

Most reliable results devoted to solutions of mixed BVPs of plates have been obtained through the method of multiple series (or multiple equations). The basic idea of the method follows. A general solution of the governing equations is sought, including a set of constants. On each separated part of the associated BVPs the constants are chosen in a way to satisfy the BCs. Besides, if a solution is presented in the form of a Fourier series, then a number of obtained different series overlaps with a number of BCs changes/variations. When a solution is sought in the form of a Fourier integral, then a set of multiple integral equations is obtained. Next, the obtained set of equations undergoes an action of a finite integral transformation yielding eventually an infinite system of linear algebraic (integral) equations, which are then solved through known procedures. The method described so far has been successfully applied to problems of SSS and vibrations of plates with mixed BCs ([37], [78]–[82]), to stability problems of three-layered plates with mixed BCs ([6]), and also to contact problems of theory of elasticity and bending of plates ([4], [92]).

For periodically mixed BCs the homogenization approach can be used [56].

An important role in application to various real structures with complicated BCs is played by plates with free edges and lying on an elastic foundation. B.G. Korenev proposed the method of compensating loads for their computation [51]. Instead of a domain occupied by a plate the more wide (for instance, a whole plane) with two introduced loads is determined. First one (fundamental) corresponds to that of the really acting load, whereas the second (compensating) is chosen in a way to satisfy BCs on the plate contour. A solution governing infinite plate behavior under the action of the first load is called the fundamental solution, whereas the second is referred to as the compensating solution. A sum of solutions should satisfy the differential equation and all BCs. In the general case, in order to define the compensating load, a Fredholm integral equation is obtained. For many problems they can be reduced to a system of LAEs. Other solution methods of plates with mixed BCs are analysed in [19], [33], [36], [42], [45], [54], [57], [58], [59], [61], [70], [71], [73], [88], [91].

2.1.2 *Method of Boundary Conditions Perturbation*

The numerical and semi-analytical methods presented so far are oriented on application of nowadays computational tools. However, their direct application to optimal design of structures is costly computation. Therefore, an important role in application is still played by the development of approximate analytical methods devoted to computations of plates with complicated BCs, allowing for getting simple and clear formulas, necessary for practical engineering computations with an emphasis on investigation of various factors (geometric and stiffness characteristics, support conditions, etc.) on the structure behavior.

One of the possible approaches, directed to the solution of this problem, is an application of the of perturbation of BCs. This method proposed by Dorodnitsin ([32]) may be treated as a novel variant of the perturbation method. In what follows we briefly describe this approach.

Consider a BVP, governed by the DE

$$L(U) = 0 \tag{2.1}$$

and BCs

$$D(U) = 0 \quad \text{on} \quad \partial G, \tag{2.2}$$

where $L(U)$ and $D(U)$ are certain differential operators, and

$$D(U) = D_0(U) + \varepsilon D_1(U).$$

Parameter ε is introduced initially or in a way to simplify the operator $D_0(U)$. If the BVP

$$L(U) = 0, \qquad D_0(U) = 0 \quad \text{on} \quad \partial G \tag{2.3}$$

is easily solved, then a solution for perturbed BCs can be sought in the form of a PS with respect to ε. In the final formula $\varepsilon = 1$ should be taken. Let us note that the method so far described and applied has been revised recently and is referred to as the HPM (see Chapter 1.2.2).

So, it is necessary to consider PS regarding $\varepsilon = 1$. As a rule they are divergent, which is manifested by the occurrence of poles in the circle of $r \leq |1|$. In order to overcome this drawback A.A. Dorodnitsyn proposed to use analytical continuation ([32]). The analytical continuation ([52], [66]) allows us to solve two fundamental problems: first, singular points of the function using their approximation in powers of ε are detected; second, computation of function values in an arbitrary point of the interval $0 \leq \varepsilon \leq 1$ is carried out. Unfortunately, no progress has been made in that direction so far. A reason is mainly motivated by the fact that in order to detect singular points it is necessary to get higher order terms of PS. In real problems it is impossible. On the other hand, without the exact poles determining standard methods of the parameter change during an analytical continuation procedure will fail.

Sometimes, the information regarding location of poles of a function being sought can be achieved via coefficients of the differential operator $L(U)$ associated with a studied BVP. However, it should be underlined that the quoted goals in the case of civil engineering have not been achieved.

In what follows we demonstrate how to increase efficiency of the continuation procedure. Let us consider BCs (2.2) in the following modified form:

$$D_0(U) + [D(U) - D_0(U)] = 0 \quad \text{on} \quad \partial G. \tag{2.4}$$

Let us introduce the following new parameter ε_1 :

$$D_0(U) + \varepsilon_1[\alpha D(U) - D_0(U)] = 0 \quad \text{on} \quad \partial G, \tag{2.5}$$

where α is a constant different from zero. BCs (2.4) and (2.5) allow us to obtain a link between parameters ε and ε_1:

$$\varepsilon = \frac{\alpha\varepsilon_1}{1 - (1-\alpha)\varepsilon_1}. \tag{2.6}$$

Function (2.6) transforms a circle of unit radius of the plane ε_1 into a circle of radius $1/(2 - \varepsilon)$ with its center at $((1-\alpha)/(2-\alpha); 0)$ of the plane ε. Therefore, if a solution to the BVP (2.1), (2.4) does not possess singularities inside the circle in point $(0.5; 0)$ of the plane ε, and the radius larger then $0.5 + \delta$, where δ is an arbitrarily small positive number, then for sufficient small $\alpha > 0$, a solution to the BVP with BCs (2.4) will be analytical regarding ε_1 for $\varepsilon \leq 1$. This is particularly the case for the well known Euler summation method (see Section 1.3.1).

The discussed variant of the BCs perturbation has been applied in [66] for solutions devoted to the dynamics of viscous fluids.

A more general form can be obtained assuming α depending on ε_1:

$$\varepsilon = \frac{\alpha(\varepsilon_1)\varepsilon_1}{1 - \varepsilon_1[1 - \alpha(\varepsilon_1)]} \quad \text{or} \quad \varepsilon = \frac{\alpha\sigma(P)\varepsilon_1}{1 - \varepsilon_1[1 - \alpha\sigma(P)]},$$

where $\sigma(P)$ is an arbitrary strongly positive function defined on the boundary ∂G.

Unfortunately, a question regarding the proper choice of a coefficient α or the function $\alpha(\varepsilon_1)$ and $\sigma(P)$ remains open for many real problems. This is why an analytical continuation does not guarantee an increase of the velocity convergence of a PS and does not remove the problem related to the construction of the high order approximations.

2.2 Natural Vibrations of Beams and Plates

2.2.1 Natural Vibrations of a Clamped Beam

Consider the problem having the exact solution. Namely, we are going to determine eigenfrequencies and the eigenforms of a clamped beam of the length l. The basic DE is

$$EIz_{\overline{xxxx}} + \rho z_{tt} = 0, \tag{2.7}$$

where EI denotes beam bending stiffness, ρ is the mass per unit beam length, $\overline{x}$ is the spatial coordinate, E is the Young modulus, I is the moment of inertia, $-l/2 \leq \overline{x} \leq l/2$.

Let us introduce the following nondimensional coordinate:

$$x = \overline{x}/l. \tag{2.8}$$

Equation (2.7) takes the form:

$$z_{xxxx} + \frac{\rho l^4}{EI} z_{tt} = 0. \tag{2.9}$$

A solution to Equation (2.9) is sought through the variable separation of the form:

$$z = W(x)T(t). \tag{2.10}$$

After substitution Ansatz (2.10) into Equation (2.9) the following equations are obtained:

$$\dot{T} + \theta^2 T = 0, \tag{2.11}$$

$$W^{IV} - \lambda W = 0, \tag{2.12}$$

where θ^2 is the circular frequency of the beam transversal vibrations, whereas $\lambda = \rho\theta^2 l^4 (EI)^{-1}$ is the corresponding eigenvalue.

Solution to Equation (2.11) follows:

$$T(t) = C_1 \sin\theta t + C_2 \cos\theta t, \tag{2.13}$$

where $C_1, C_2 = const.$

In order to obtain an eigenavalue problem we add to the Equation (2.12) the following BCs:

$$W = 0, \quad W' = 0 \quad \text{for} \quad x = \pm 1/2. \tag{2.14}$$

Let us introduce the parameter ε to the BCs (2.14) in such a way that for $\varepsilon = 0$ one gets a simple support, and for $\varepsilon = 1$ the clamped beam:

$$W = 0, \quad (1-\varepsilon)W'' \pm \varepsilon W' = 0 \quad \text{for} \quad x + \pm 1/2. \tag{2.15}$$

For the remaining values of ε $(0 < \varepsilon < 1)$ we apply conditions of elastic support with the elasticity coefficient $\varepsilon/(1-\varepsilon)$.

The mode W and the associated eigenvalue λ can be cast in the PS:

$$W = \sum_{i=0}^{\infty} W_i \varepsilon^i, \qquad \lambda = \sum_{i=0}^{\infty} \lambda_i \varepsilon^i. \tag{2.16}$$

We substitute Ansatzes (2.16) into Equation (2.12) and BCs (2.15). The following recurrent sequence of successive BVPs is obtained after splitting with respect to powers of ε

$$W_0^{IV} - \lambda_0 W_0 = 0,$$

$$W_0 = 0, \quad W_0'' \quad \text{for} \quad x = \pm 1/2,$$

$$W_i^{IV} - \lambda_0 W_i = \sum_{j=1}^{i} \lambda_i W_{i-j},$$

$$W_i = 0, \quad W_i'' = \pm \sum_{j=0}^{i-1} W_j' \quad \text{for} \quad x = \pm 1/2, i = 1, 2, 3, \ldots.$$

Solution to the zeroth order approximation is as follows:

$$\lambda_0 = \pi^4 n^4, \qquad n = 1, 2, 3, \ldots, \tag{2.17}$$

$$W_0 = C \left\{ \begin{array}{ll} \cos \pi n x, & n = 1, 3, 5, \ldots \\ \sin \pi n x, & n = 2, 4, 6, \ldots \end{array} \right\}. \tag{2.18}$$

Let us consider the BVP of the first order approximation:

$$W_1^{IV} - \lambda_0 W_1 = \lambda_1 C \left\{ \begin{array}{ll} \cos \pi n x, & n = 1, 3, 5, \ldots \\ \sin \pi n x, & n = 2, 4, 6, \ldots \end{array} \right\}, \tag{2.19}$$

$$W_1 = 0, \quad W_1'' = \pm C \pi n \left\{ \begin{array}{c} -(-1)^{(n-1)/2} \\ (-1)^{n/2} \end{array} \right\} \quad \text{for} \quad x = \pm 1/2. \tag{2.20}$$

We rely on the observation that we cannot satisfy all BCs for arbitrary values of λ_1. This drawback can be removed by adding certain conditions, which can be constructed following the way reported in monograph [27].

Let us multiply Equation (2.19) by the further defined function $u(x)$, called the adjoint solution. Integration by parts from $-1/2$ to $1/2$ yields

$$\int_{-1/2}^{1/2} W_1 (u^{IV} - \lambda_0 u) dx + (u W_1''' - u' W_1'' + u'' W_1' - u''' W_1)|_{-1/2}^{1/2} = \tag{2.21}$$

$$\lambda_1 \int_{-1/2}^{1/2} W_0 u dx.$$

We require that the integrand expression standing on the l.h.s. of Equation (2.21) will be equal to zero,

$$u^{IV} - \lambda_0 u = 0. \tag{2.22}$$

Since both w_1''' and w_1' are unknown, we get

$$u W_1''' \Big|_{-1/2}^{1/2} + u'' W_1' \Big|_{-1/2}^{1/2} = 0. \tag{2.23}$$

Relation (2.23) can be satisfied if the coefficients standing by W_1''' and W_1' are equal to zero. BCs for u are as follows:

$$u = 0, \quad u'' = 0 \quad \text{for} \quad x = \pm 1/2. \tag{2.24}$$

Therefore, the function u represents the solution of the BVP for Equation (2.22) and BCs (2.24). The given BVP with the accuracy of the assumed order coincides with the problem of zeroth order approximation. Therefore, $u = W_0$, and Equation (2.21) takes the following form:

$$-W_0' W_1'' \Big|_{-1/2}^{1/2} = \lambda_1 \int_{-1/2}^{1/2} W_0^2 dx. \tag{2.25}$$

This yields the first correction term to the eigenvalue:

$$\lambda_1 = -\frac{W_0' W_1'' \Big|_{-1/2}^{1/2}}{\int_{-1/2}^{1/2} W_0^2 dx} = 4\pi^2 n^2. \tag{2.26}$$

Let us define the first order correcting term to the vibration mode:

$$W_1 = \frac{C}{\pi n} \begin{Bmatrix} \frac{(-1)^{(n-1)/2}}{2\mathrm{ch}(\pi n/2)} \cosh \pi n x - x \sin \pi n x, & n = 1, 3, 5, \ldots \\ \frac{-(-1)^{n/2}}{2\mathrm{sh}(\pi n/2)} \sinh \pi n x + x \cos \pi n x, & n = 2, 4, 6, \ldots \end{Bmatrix}. \tag{2.27}$$

Function W_1 is defined with accuracy up to functions (2.18), but this additional term can be removed because it has already been included in the zeroth order approximation.

Analogously are defined λ_2 and W_2. The eigenvalue approximation truncated to three first terms has the following form:

$$\lambda = \pi^4 n^4 + 4\pi^2 n^2 \varepsilon + 4\pi n \left[\pi n - \frac{1}{2} \left(\coth \frac{\pi n}{2} \right)^{(-1)^n} - \frac{1}{2\pi n} \right] \varepsilon^2 + \ldots. \tag{2.28}$$

Eigenmode W can be rewritten in the following form:

$$W = C \begin{Bmatrix} \cos \pi n x \\ \sin \pi n x \end{Bmatrix} + \frac{C}{\pi n} \begin{Bmatrix} \frac{(-1)^{(n-1)/2}}{2\cosh(\pi n/2)} \cosh \pi n x - x \sin \pi n x \\ \frac{-(-1)^{n/2}}{2\sinh(\pi n/2)} \sinh \pi n x + x \cos \pi n x \end{Bmatrix} \varepsilon +$$

$$\left[\frac{(-1)^{n+1} C}{2\pi^2 n^2} \begin{Bmatrix} \frac{(-1)^{(n-1)/2}}{\cosh(\pi n/2)} \\ \frac{(-1)^{n/2}}{\sinh(\pi n/2)} \end{Bmatrix} \left(\left[\pi n - \left(\coth \frac{\pi n}{2} \right)^{(-1)^n} - \frac{1}{\pi n} \right] \begin{Bmatrix} \cosh \pi n x \\ \sinh \pi n x \end{Bmatrix} + \right. \right.$$

$$\left. x \begin{Bmatrix} \sinh \pi n x \\ \cosh \pi n x \end{Bmatrix} \right) - \frac{(-1)^{n+1} C}{\pi^2 n^2} \left[\pi n - \frac{1}{2} \left(\coth \frac{\pi n}{2} \right)^{(-1)^n} - \frac{1}{\pi n} \right] x \begin{Bmatrix} \sin \pi n x \\ \cos \pi n x \end{Bmatrix} -$$

$$\left. \frac{C}{\pi^2 n^2} x \begin{Bmatrix} \cos \pi n x \\ \sin \pi n x \end{Bmatrix} \right] \varepsilon^2 + \ldots, \qquad \begin{Bmatrix} n = 1, 3, 5, \ldots \\ n = 2, 4, 6, \ldots \end{Bmatrix}. \tag{2.29}$$

The obtained truncated PS (2.28) can be transformed to the following diagonal PA:

$$\lambda_{[1/1]}(\varepsilon) = \frac{a_0 + a_1\varepsilon}{1 + b_1\varepsilon}, \tag{2.30}$$

where $a_0 = \lambda_0$, $a_1 = \lambda_1 + b_1\lambda_0$, $b_1 = -\lambda_2/\lambda_1$.

In what follows we are going to estimate an error associated with definition of eigenvalues λ. Since it is mainly influenced by BCs its main part is represented by the first eigenvalue of the BVP (2.12)-(2.14). Solution to Equation (2.12) for symmetric modes is as follows

$$W = C_1 \cos 2\alpha x + C_2 \cosh 2\alpha x, \quad \alpha = (1/2)(\lambda)^{1/4}. \tag{2.31}$$

Satisfaction of BCs (2.15) yields the following transcendental equation with respect to α:

$$4(1-\varepsilon)\alpha \cosh\alpha \cos\alpha + \varepsilon(\cosh\alpha \sin\alpha + \cos\alpha \sinh\alpha) = 0. \tag{2.32}$$

Numerical solution to Equation (2.32) for $\varepsilon = 1$ gives the value of $\lambda = (1.5056\pi)^4$, truncated PS (2.28) - $\lambda = (1.1542\pi)^4$ (error of 23%), PA (2.30) - $\lambda = (1.5139\pi)^4$ (error of 0.58%).

Dependence of the first eigenvalue versus ε is reported in Figure 2.1. Curves 1, 2, 3 are obtained with the help of the truncated PS (2.28), PA (2.30) and numerical solution to transcendental Equation (2.32), respectively.

It is evident that results associated with getting the first eigenvalue of the BVP (2.12), (2.15) obtained through PA practically coincide with the exact solution for all values of the parameter ε, whereas the truncated series reliable results only up to $\varepsilon = 0.4$.

In Figure 2.2 the change of the relative error in estimation of the first five 15 eigenvalues is reported. Curve 1 corresponds to results obtained via the truncated PS part, whereas curve 2 is associated with results obtained by the PA.

Since for low frequencies a difference between the truncated PS and the PA is sufficiently large, then it is advised to focus on the results obtained via the PA. For higher frequencies ($n > 10$) the error associated with frequencies definition is less than 5% and it decreases with

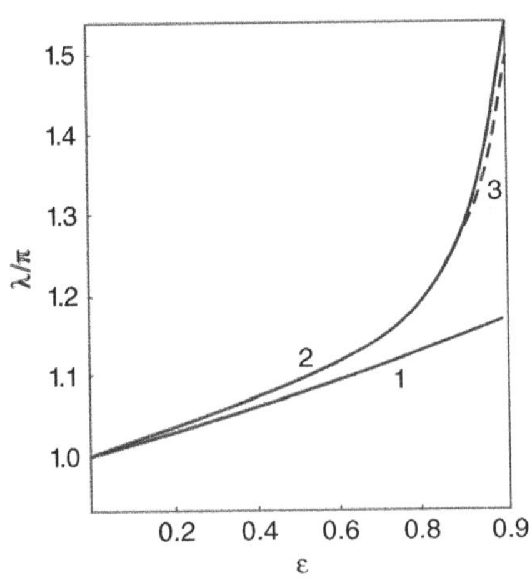

Figure 2.1 Comparison of efficiency of different methods of determination of eigenvalues

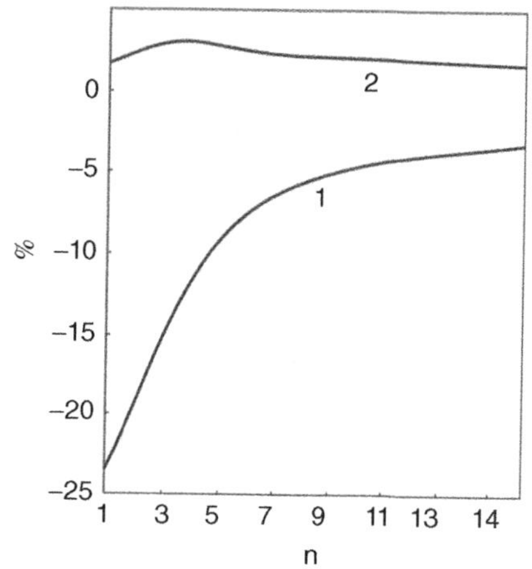

Figure 2.2 Comparison of efficiency of PA and truncated PS for first 15 modes of natural beam vibrations

increase of the wave number. Besides, in the case of high valued frequencies estimation one may get the upper (PA) and lower (truncated PS) values. Those estimations are of higher accuracy for higher frequencies.

Besides the numerical solution there exists a possibility of introduction of asymptotic simplification of the transcendental equation for determination of the eigenvalues. Let us represent α in the form PS:

$$\alpha = \frac{1}{2}\sum_{i=0}^{\infty} \alpha_i \varepsilon^i. \tag{2.33}$$

Substituting Ansatz (2.33) into Equation (2.32) and splitting with respect to powers of ε, the following system of transcendental equations is obtained

$$4\alpha_0 \cos\alpha_0 \cosh\alpha_0 = 0, \tag{2.34}$$

$$\begin{aligned} &4\alpha_1 \cos\alpha_0 \cosh\alpha_0 + 4\alpha_1\alpha_0(\cos\alpha_0 \sinh\alpha_1 - \sin\alpha_0 \cosh\alpha_0) - \\ &4\alpha_0 \cos\alpha_0 \cosh\alpha_0 + \sin\alpha_0 \cosh\alpha_0 + \cos\alpha_0 \sinh\alpha_0 = 0, \end{aligned} \tag{2.35}$$

$$\begin{aligned} &2\left\{4\alpha_2 \cos\alpha_0 + \alpha_1^2(\cos\alpha_0 \sinh\alpha_0 - \sin\alpha_0 \cosh\alpha_0) + 2\alpha_0\left[\left(\alpha_2 \sinh\frac{\alpha_0}{2} + \right.\right.\right. \\ &\left.\left.\left.\frac{\alpha_1^2}{2}\cosh\alpha_0\right)\cos\alpha_0 - \alpha_1^2 \sin\alpha_0 \sinh\alpha_0 - \left(\alpha_2 \sinh\alpha_0 + \frac{\alpha_1^2}{2}\cosh\alpha_0\right)\cosh\alpha_0\right]\right\} - \\ &4[\alpha_1 \cos\alpha_0 \cosh\alpha_0 + \alpha_0\alpha_1(\cos\alpha_0 \sinh\alpha_0 - \sin\alpha_0 \cosh\alpha_0)] = 0. \end{aligned} \tag{2.36}$$

Equation (2.34) yields

$$\cos\alpha_0 = 0, \qquad \alpha_0 = 2\pi n, \qquad n = 1, 3, 5, \ldots.$$

Furthermore, Equations (2.35), (2.36) give

$$\alpha_1 = \frac{2}{\pi n}, \tag{2.37}$$

$$\alpha_2 = \frac{2}{\pi n}\left(1 - \frac{1}{2\pi n}\tanh\frac{\pi n}{2} - \frac{1}{\pi^2 n^2}\right). \tag{2.38}$$

Analogously, are constructed transcendental equations regarding antisymmetric modes, and finally we get

$$\alpha = \pi n + \frac{1}{\pi n}\varepsilon + \frac{1}{\pi n}\left[1 - \frac{1}{2\pi n}\left(\coth\frac{\pi n}{2}\right)^{(-1)^n} - \frac{1}{\pi^2 n^2}\right]\varepsilon^2 + \ldots, \quad n = 1, 3, 5, \ldots. \tag{2.39}$$

Note that the truncated PS in Equation (2.39) is a fourth root from expression (2.28).

While constructing vibration modes we find α for $\varepsilon = 1$; in addition we carry out the reconstruction of the truncated PS into PA according to formula (2.30). Computational results are given in Figures 2.3 and 2.4. Curve 1 corresponds to W_0, curve 2 – W_1, curve 3 – W_2, curve 4 – $W_0 + W_1 + W_2$, curve 5 – PA, curve 6 – exact solution.

2.2.2 *Natural Vibration of a Beam with Free Ends*

In what follows we study natural vibrations of a beam with free ends governed by the following DE:

$$W^{IV} - \lambda W = 0.$$

BCs follow

$$W'' = 0, \qquad W''' = 0 \qquad \text{for} \qquad x = \pm 1/2. \tag{2.40}$$

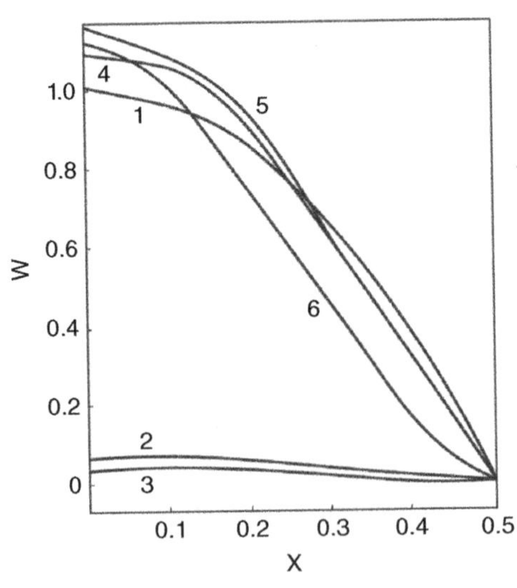

Figure 2.3 First symmetric vibration mode

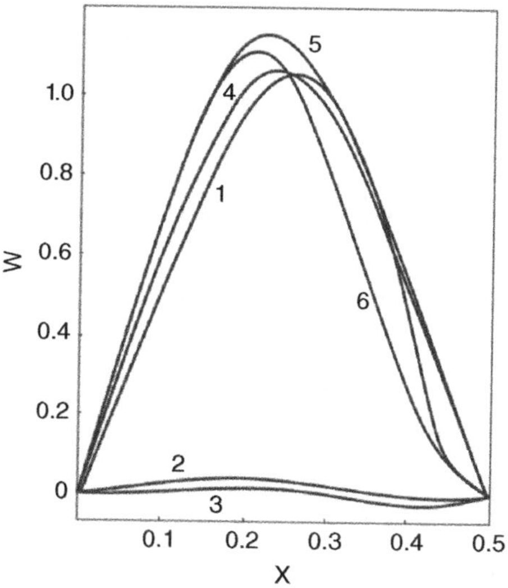

Figure 2.4 First antisymmetric vibration mode

In the rewritten form BCs are

$$W''' = 0, \quad (1-\varepsilon)W' \pm \varepsilon W'' = 0 \quad \text{for} \quad x = \pm 1/2. \tag{2.41}$$

Eigenmodes W and eigenvalue λ can be presented as PS (2.16). Substituting PS (2.16) into Equation (2.12) and BCs (2.41), after splitting with regard to ε, the following recurrent sequence of BVPs is obtained:

$$W_0^{IV} - \lambda_0 W_0 = 0,$$

$$W_0''' = 0, \qquad W_0' = 0 \qquad \text{for} \qquad x = \pm 1/2,$$

$$W_i^{IV} - \lambda_0 W_i = \sum_{j=1}^{i} \lambda_i W_{i-j},$$

$$W_i''' = 0, \quad W_i' = \pm \sum_{j=0}^{i-1} W_j'' \quad \text{for} \quad x = \pm 1/2, \quad i = 1, 2, 3, \ldots.$$

Solution of the zeroth order approximation follows:

$$\lambda_0 = \pi^4 n^4, \qquad n = 2, 3, 4, \ldots, \tag{2.42}$$

$$W_0 = C \begin{Bmatrix} \cos \pi n x, & n = 2, 4, 6, \ldots \\ \sin \pi n x, & n = 3, 5, 7, .. \end{Bmatrix}. \tag{2.43}$$

Eigenform with number $n = 1$ should be rejected, since this case corresponds to a kinematically modified structure.

After defining of W_0, the BVP of the first approximation has the following form:

$$W_1^{IV} - \lambda_0 W_1 = \lambda_1 C \begin{Bmatrix} \cos \pi n x, & n = 2, 4, 6, \ldots \\ \sin \pi n x, & n = 3, 5, 7, .. \end{Bmatrix}, \tag{2.44}$$

$$W_1''' = 0, \quad W_1' = \pm C\pi^2 n^2 \left\{ \begin{array}{c} (-1)^{n/2} \\ (-1)^{(n-1)/2} \end{array} \right\} \quad \text{for} \quad x = \pm 1/2. \tag{2.45}$$

The solvability conditions yield

$$-W_0' W_1'' \Big|_{-1/2}^{1/2} = \lambda_1 \int_{-1/2}^{1/2} W_0^2 dx. \tag{2.46}$$

The first correction term of the eigenavalue follows:

$$\lambda_1 = -4\pi^4 n^4, \tag{2.47}$$

whereas the first correction form of the vibration mode is

$$W_1 = C\pi n \left\{ \begin{array}{ll} \frac{(-1)^{n/2}}{2\sinh\frac{\pi n}{2}} \cosh \pi n x + x \sin \pi n x, & n = 2, 4, 6, \ldots \\ \frac{(-1)^{(n-1)/2}}{2\cosh\frac{\pi n}{2}} \sinh \pi n x - x \cos \pi n x, & n = 3, 5, 7, \ldots \end{array} \right\}. \tag{2.48}$$

Here the zeroth order component in the expression for w_1 should be rejected.

In an analogous way a second order correction term of the eigenvalue λ_2 is constructed, and finally we get

$$\lambda = \pi^4 n^4 - 4\pi^4 n^4 \varepsilon + 2\pi^4 n^4 (3 + \pi n C \tanh^{(-1)^n} 0.5\pi n)\varepsilon^2 + \ldots. \tag{2.49}$$

Note that for the truncated series (2.49) the PA takes the form (2.30). In order to compare the results obtained via the truncated series and PA we use a transcendental equation. We take a symmetric mode like that governed by Equation (2.31). Satisfaction of the BCs (2.41) reduces the problem to the following transcendental equation:

$$2(1-\varepsilon)\sinh\alpha \sin\alpha + \varepsilon\overline{\lambda}(\cosh\alpha \sin\alpha + \cos\alpha \sinh\alpha) = 0. \tag{2.50}$$

The first root of the transcendental Equation (2.50) for $\varepsilon = 1$ gives the eigenvalue $\lambda = (1.5056\pi)^4$. Truncated PS (2.49) for $n = 2$ gives the value $\lambda = (3.9696\pi)^4$ (error – 163.54%), PA (2.30) for $n = 2$ gives $\lambda = (1.4670\pi)^4$ (error – 2.56%).

In Figure 2.5 the dependence of the first eigenvalue versus ε is reported.

The following notation is applied: curve 1 – truncated PS (2.49), curve 2 – PA (2.30), curve 3 – numerical solution. It is clear, that the first eigenvalue of the BVP (2.12), (2.41) that is obtained with a help of PA coincides with the exact solution for all values of the parameter $\varepsilon \leq 1$. At the same time threshold for truncated series $\varepsilon < 0.05$.

In Figure 2.6 is shown the change of relative error regarding the first fifteenth eigenvalues. Curve 1 – truncated series (2.49), curve 2 – zero approximation (2.42), curve 3 – PA (2.30). One may get an upper (truncated PS) and low (PA) error estimations for the eigenvalues associated with this problem. However, it should be emphasized that the first and second term of the truncated PS (2.49) exceed zero term with respect to their module, therefore numerical results obtained with a help of the truncated PS essentially differ from exact values of the eigenvalues. This difference may increase with the increase of eigenvalue number; therefore only the zeroth

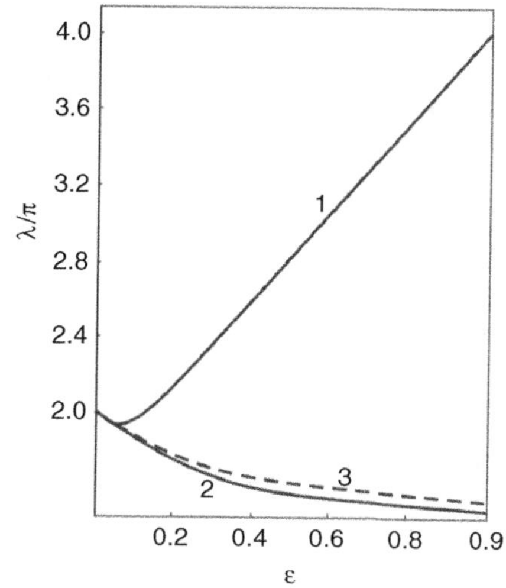

Figure 2.5 Dependence of first eigenvalue of the symmetric vibration mode on ε

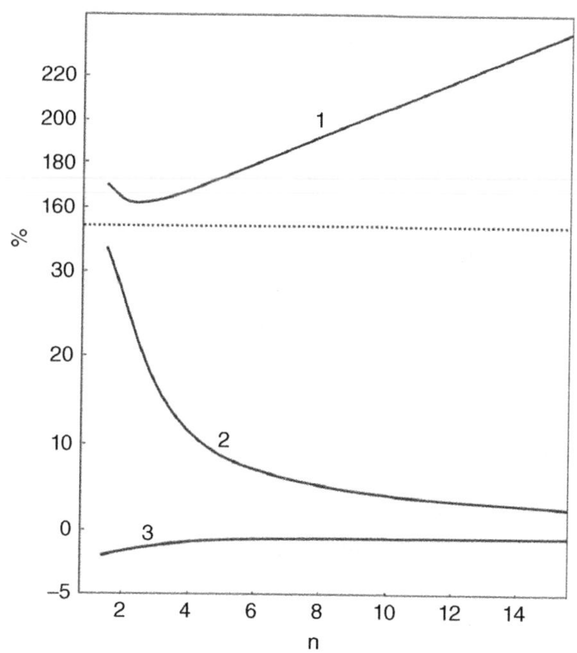

Figure 2.6 Relative errors in definition of first fifteenth eigenvalues for a symmetric vibration form

order term should be kept in order to get the efficient upper estimation in the series part for λ in formula (2.49).

Let us construct an analytical solution of the transcendental equation (2.50). We assume for α the PS (2.33). Substituting Ansatz (2.33) to Equation (2.50) and splitting it regarding ε, the following recurrent system of transcendental equations is obtained:

$$2\sin\alpha_0 \sinh\alpha_0 = 0, \tag{2.51}$$

$$\begin{aligned} &2\alpha_1(\cosh\alpha_0 \sin\alpha_0 + \sinh\alpha_0 \cos\alpha_0) - 2\sinh\alpha_0 \sin\alpha_0 + \\ &2\lambda_0(\cos\alpha_0 \sinh\alpha_0 + \sin\alpha_0 \cosh\alpha_0) = 0, \end{aligned} \tag{2.52}$$

$$\begin{aligned} &2[\alpha_2(\cosh\alpha_0 \sin\alpha_0 + \sinh\alpha_0 \cos\alpha_0) + \alpha_1^2 \cosh\alpha_0 \cos\alpha_0] - \\ &2\alpha_1(\cosh\alpha_0 \sin\alpha_0 + \sinh\alpha_0 \cos\alpha_0) + 4\alpha_0\alpha_1 \cosh\alpha_0 \cos\alpha_0 + \\ &2\alpha_1(\sinh\alpha_0 \cos\alpha_0 + \cosh\alpha_0 \sin\alpha_0) = 0. \end{aligned} \tag{2.53}$$

Equation (2.51) yields

$$\alpha_0 = 2\pi n, \qquad n = 2, 4, 6, \ldots. \tag{2.54}$$

Furthermore, Equations (2.52) and (2.53) allow us finally to get

$$\alpha_1 = -2\pi n, \tag{2.55}$$

$$\alpha_2 = \pi^2 n^2 \coth 0.5\pi n. \tag{2.56}$$

The final α form follows:

$$\alpha = \pi n - \pi n\varepsilon + \frac{\pi^2 n^2}{2} \coth^{(-1)^n} \frac{\pi n}{2} \varepsilon^2 + \ldots. \tag{2.57}$$

2.2.3 *Natural Vibrations of a Clamped Rectangular Plate*

In this section we are aiming at generalization of the method so far proposed for the two-dimensional BVPs. As an example we consider the natural vibrations of a rectangular plate clamped on its contour ($-0.5a \le \overline{x} \le 0.55a, -0.5b \le \overline{y} \le b$). The governing equation is

$$D\nabla^4 W + \rho W_{tt} = 0, \tag{2.58}$$

where $D = Eh^3/[12(1-\nu^2)]$; ρ is the plate mass per its unit; $\overline{x}, \overline{y}$ are the spatial variables; ν is the Poissons ratio; $\nabla^2 = \partial^2/\partial\overline{x}^2 + \partial^2/\partial\overline{y}^2$.

Let us introduce the following notations:

$$y = \overline{y}/b, \qquad x = \overline{x}/b, \qquad k = a/b. \tag{2.59}$$

After substitution of variables (2.59) into Equation (2.58) we get

$$\nabla^4 W + \frac{\rho b^4}{D} W_{tt} = 0. \tag{2.60}$$

A solution to Equation (2.60) is sought in the following form:

$$W = W(x, y) \cdot T(t). \tag{2.61}$$

After substitution of Ansatz (2.61) into Equation (2.58) we get

$$T''(t) + \theta^2 T = 0, \tag{2.62}$$

$$\nabla^4 W - \lambda W = 0. \tag{2.63}$$

where θ is the circular frequency of transversal plate vibrations, and $\lambda = \rho\theta^2 b^4 D^{-1}$ is its eigenvalue.

In order to get the eigenvalue problem, we attach the following BCs to Equation (2.63):

$$W = 0, \quad W_x = 0 \quad \text{for} \quad x = \pm 0.5k, \tag{2.64}$$

$$W = 0, \quad W_y = 0 \quad \text{for} \quad y = \pm 0.5. \tag{2.65}$$

BCs (2.64), (2.65) are introduced via ε in the following form:

$$W = 0, \quad (1-\varepsilon)W_{xx} \pm \varepsilon k W_x = 0 \quad \text{for} \quad x = \pm 0.5k, \tag{2.66}$$

$$W = 0, \quad (1-\varepsilon)W_{yy} \pm \varepsilon W_y = 0 \quad \text{for} \quad y = \pm 0.5. \tag{2.67}$$

For $\varepsilon = 0$ one obtains a simply supported plate, whereas for $\varepsilon = 1$ one obtains BCs (2.66), (2.67).

Further, we represent an eigenvalue λ and eigenform W in the form of PS (2.16). The following recurrent system of the BVPs is obtained by substituting PS (2.16) into Equation (2.63) and into BCs (2.66), (2.67), after splitting regarding ε:

$$\nabla^4 W_0 - \lambda_0 W_0 = 0,$$

$$W_0 = 0, \quad W_{0xx} = 0 \quad \text{for} \quad x = \pm 0.5k,$$

$$W_0 = 0, \quad W_{0yy} = 0 \quad \text{for} \quad y = \pm 0.5,$$

$$\nabla^4 W_j - \lambda_0 W_j = \sum_{i=1}^{j} \lambda_i W_{j-i},$$

$$W_j = 0, \quad W_{jxx} = \mp k \sum_{i=0}^{j-1} W_{ix} \quad \text{for} \quad x = \pm 0.5k,$$

$$W_j = 0, \quad W_{jyy} = \mp \sum_{i=0}^{j-1} W_{iy} \quad \text{for} \quad y = \pm 0.5.$$

In what follows we present the construction of eigenvalues and modes for the case, where a symmetry in directions of x and y is exhibited by eigenforms.

In zeroth order approximation we have

$$\lambda_0 = \pi^4 \left(\frac{m^2}{k^2} + n^2\right)^2, \qquad n, m = 1, 3, 5, \ldots, \tag{2.68}$$

$$W_0 = c \cos \frac{nm}{k} x \cos \pi n y. \tag{2.69}$$

The problem regarding first order approximation can be rewritten in the form:

$$\nabla^4 W_1 - \lambda_0 W_1 = \lambda_1 \cos\frac{nm}{k}x\cos\pi ny, \tag{2.70}$$

$$W_1 = 0, \quad W_{1xx} = \pm\pi m(-1)^{\frac{m-1}{2}}\cos\pi ny \quad \text{for} \quad x = \pm 0.5k, \tag{2.71}$$

$$W_1 = 0, \quad W_{1yy} = \pm\pi n(-1)^{\frac{n-1}{2}}\cos\frac{\pi m}{k}x \quad \text{for} \quad y = \pm 0.5. \tag{2.72}$$

A solution is sought by the method of variable separation assuming the function W_1 in the form:

$$W_1 = Y_1 y\cos\frac{\pi m}{k}x + X_1 x\cos\pi ny. \tag{2.73}$$

Eigenvalue λ_1 can be presented as the following sum:

$$\lambda_1 = \lambda_{1x} + \lambda_{1y}. \tag{2.74}$$

After substitution of Ansatzes (2.73) and (2.74) into Equation (2.70) and BCs (2.71)–(2.72), the two following BVPs are obtained:

$$Y_1^{IV} - 2\pi^2\frac{m^2}{k^2}Y_1^{II} - \pi^4 n^2\left(2\frac{m^2}{k^2} + n^2\right)Y_1 = \lambda_{1y}\cos\pi ny, \tag{2.75}$$

$$Y_1 = 0, \quad Y_1^{II} = \pm\pi n(-1)^{\frac{n-1}{2}} \quad \text{for} \quad y = \pm 0.5, \tag{2.76}$$

$$X_1^{IV} - 2\pi^2 n^2 X_1^{II} - \pi^4\frac{m^2}{k^2}\left(\frac{m^2}{k^2} + 2n^2\right)X_1 = \lambda_{1x}\cos\frac{\pi m}{k}x, \tag{2.77}$$

$$X_1 = 0, \quad X_1^{II} = \pm\pi m(-1)^{\frac{m-1}{2}} \quad \text{for} \quad x = \pm 0.5k. \tag{2.78}$$

We present a construction of solvability conditions regarding the BVP (2.75), (2.76). Multiplying Equation (2.75) by a function $u(y)$, to be defined later, and integration with respect to y from -0.5 to 0.5 yields

$$\begin{aligned}&\int_{-0.5}^{0.5} u(y)\left[Y_1^{IV} - 2\frac{\pi^2 m^2}{k^2}Y_1^{II} - \pi^4 n^2\left(2\frac{m^2}{k^2} + n^2\right)Y_1\right]dy = \\ &\lambda_{1y}\int_{-0.5}^{0.5} u(y)\cos\pi ny dy.\end{aligned} \tag{2.79}$$

Integration of Equation (2.79) by parts gives

$$\begin{aligned}&\int_{-0.5}^{0.5} Y_1\left[u^{IV}(y) - 2\frac{\pi^2 m^2}{k^2}u^{II}(y) - \pi^4 n^2\left(2\frac{m^2}{k^2} + n^2\right)u(y)\right]dy + \\ &u(y)Y_1^{III}|_{-0.5}^{0.5} - u^{I}(y)Y_1^{II}|_{-0.5}^{0.5} + u^{II}(y)Y_1^{I}|_{-0.5}^{0.5} - u^{III}(y)Y_1|_{-0.5}^{0.5} - \\ &2\frac{\pi^2 m^2}{k^2}\left(u(y)Y_1^{I}|_{-0.5}^{0.5} - u^{I}(y)Y_1|_{-0.5}^{0.5}\right) = \lambda_{1y}\int_{-0.5}^{0.5} u(y)\cos\pi ny dy.\end{aligned} \tag{2.80}$$

Comparison to zero of the integrand expression in the l.h.s. of Equation (2.80) gives the following equation with respect to $u(y)$:

$$u^{IV}(y) - 2\frac{\pi^2 m^2}{k^2}u^{II}(y) - \pi^4 n^2\left(2\frac{m^2}{k^2} + n^2\right)u(y) = 0. \tag{2.81}$$

The following condition should be satisfied:

$$u(y)\left[Y_1^{III}-2\frac{\pi^2 m^2}{k^2}Y_1^I\right]\Big|_{-0.5}^{0.5}+u^{II}(y)Y_1^I\Big|_{-0.5}^{0.5}=0. \tag{2.82}$$

Equation (2.82) is satisfied when the coefficients standing by Y_1^I and Y_1^{II} are equal to zero, i.e.

$$u(y)=0,\quad u^{II}(y)=0\quad \text{for}\quad y=\pm 0.5. \tag{2.83}$$

The general solution to Equation (2.82) follows:

$$u(y)=c_1\cosh\pi\sqrt{2\frac{m^2}{k^2}+n^2}y+c_2\cos\pi n y. \tag{2.84}$$

BCs (2.83) are satisfied by the second term, and hence:

$$u(y)=c_2\cos\pi n y. \tag{2.85}$$

Equation (2.80) yields the following solvability condition:

$$\lambda_{1y}=4\pi^2 n^2. \tag{2.86}$$

We define also Y_1, namely:

$$Y_1=\frac{n}{\pi\alpha}\left[\frac{(-1)^{\frac{n-1}{2}}}{2\cosh\frac{\pi}{2}\beta_1}\cosh\pi\beta_1 y-y\sin\pi n y\right], \tag{2.87}$$

where $\alpha=n^2+\frac{m^2}{k^2}$; $\beta_1=\sqrt{2\frac{m^2}{k^2}+n^2}$, $n=1,3,5,\ldots$.

BVP (2.77), (2.78) is solved in the analogous way, and we get

$$\lambda_{1x}=4\pi^2\frac{m^2}{k^2}, \tag{2.88}$$

$$X_1=\frac{m/k}{\pi\alpha}\left[\frac{k(-1)^{\frac{m-1}{2}}}{2\cosh\frac{\pi}{2}\beta_2}\cosh\pi\beta_2 x-x\sin\frac{\pi m}{k}x\right]. \tag{2.89}$$

where $\beta_2=\sqrt{\frac{m^2}{k^2}+2n^2}$, $m=1,3,5,\ldots$.

The first correction term of the eigenvalue λ_1 and of the eigenmode W_1 regarding the symmetric form is

$$\lambda_1=4\pi^2\alpha, \tag{2.90}$$

$$W_1=\frac{1}{\pi\alpha}\left\{n\left[\frac{(-1)^{\frac{n-1}{2}}}{2\cosh\frac{\pi}{2}\beta_1}\cosh\pi\beta_1 y-y\sin\pi n y\right]\cos\frac{\pi m}{k}+\right.$$
$$\left.\frac{m}{k}\left[\frac{k(-1)^{\frac{m-1}{2}}}{2\cosh\frac{\pi}{2k}\beta_2}\cosh\pi\beta_2 x-x\sin\frac{\pi m}{k}x\right]\cos\pi n y\right\}, \tag{2.91}$$

$n,m=1,3,5,\ldots$

We define in a similar way also λ_2. Truncated PS for the egenvalue reads

$$\lambda = \pi^4\alpha^2 + 4\pi^2\alpha\varepsilon + 4\pi\left\{\pi\alpha + 2\frac{n^2\frac{m^2}{k^2}}{\pi\alpha^2} - \frac{1}{2\pi} - \right.$$

$$\left.\frac{1}{2\alpha}\left(k\frac{m^2}{k^2}\beta_1\coth^{(-1)^n}\frac{\pi}{2k}\beta_1 + n^2\beta_2\coth^{(-1)^n}\frac{\pi}{2}\beta_2\right)\right\}\varepsilon^2 + \ldots, \qquad (2.92)$$

$$n, m = 1, 3, 5, \ldots.$$

Then, the truncated PS is transformed into the PA (2.30).

Let us now compare results obtained using the truncated PS (2.92) and PA (2.30) with the numerical solution [39] for the first eigenvalue λ_1. For the rectangular plate (for $n = m = 1$) the numerical method yields $\lambda^{(1)} = (1.9093\pi)^4$, whereas the truncated PS (2.92) gives $\lambda^{(1)} = (1.5351\pi)^4$ (difference of 19.61%), the PA (2.30) gives the value $\lambda^{(1)} = (1.9142\pi)^4$ (difference of 0.26%).

In Figure 2.7 the dependence of the first eigenvalue of the studied problem $\lambda^{(1)}$ versus ε for the truncated PS (2.92) (curve 1) and PA (2.30) (curve 2) are reported. It is clear that a threshold value of ε, for which a difference between results obtained via the PA and truncated PS will be within 5% ($\varepsilon = 0.4$). For $\varepsilon = 1$ results obtained with a help of truncated PS (2.92) are far from the numerically obtained values, and they can be used for the eigenvalue estimation from below. Although results obtained through the PA are located higher than those of numerical solution, they can be used for all values of the parameter $0 \leq \varepsilon \leq 1$.

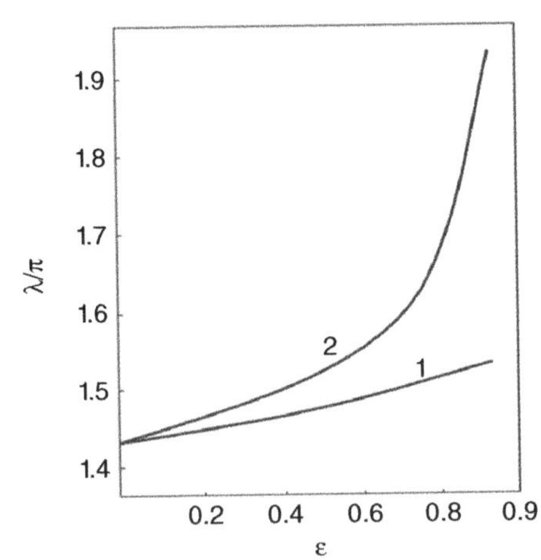

Figure 2.7 First eigenvalue of the square clamped plate

2.2.4 Natural Vibrations of the Orthotropic Plate with Free Edges Lying on an Elastic Foundation

We consider the problem of free vibrations of the lying on the elastic Winkler-Fuss foundation orthotropic rectangular plate ($-a/2 \le \overline{x} \le a/2; -b/2 \le \overline{y} \le b/2$).

PDE governing behavior of free vibrations of the orthotropic plate follows:

$$D_1 \overline{W}_{\overline{xxxx}} + 2D_3 \overline{W}_{\overline{xxyy}} + D_2 \overline{W}_{\overline{yyyy}} + \rho \overline{W}_{tt} + C\overline{W} = 0, \tag{2.93}$$

where $D_1 = E_1h^3/(12(1 - \nu_1\nu_2))$, $D_2 = E_2h^3/(12(1 - \nu_1\nu_2))$, $D_3 = D_1\nu_2 + Gh^3/6$, C is the elastic foundation coefficient; D_1, D_2 are the cylindrical stiffness in directions $\overline{x}$ and $\overline{y}$, respectively; D_3 is the rotational stiffness; E_1, E_2 are the elasticity modulus in directions $\overline{x}$ and $\overline{y}$, respectively; G - shear modulus; ν_1, ν_2 are the Poisson's ratios in directions of $\overline{x}$ and $\overline{y}$, respectively.

After separation of time and spatial variables (2.61), and introduction of the non-dimensional variables (2.59), Equation (2.93) can be rewritten in the following form:

$$W_{xxxx} + 2\alpha_3 W_{xxyy} + \alpha_2 W_{yyyy} - \lambda W = 0, \tag{2.94}$$

where $\alpha_3 = D_3/D_1$, $\alpha_2 = D_2/D_1$, $\lambda_1 = (b^4(\overline{m}\theta^2 - c))/D_1$.

Let us add the following BCs to the Equation (2.94):

$$\begin{aligned} &W_{xxx} + (2 - \nu_1)W_{yyx} = 0, \\ &(1 - \varepsilon)W_x \mp k\varepsilon(W_{xx} + \nu_1 W_{yy}) = 0 \quad \text{for} \quad x = \pm 0.5k, \end{aligned} \tag{2.95}$$

$$\begin{aligned} &W_{yyy} + (2 - \nu_2)W_{xxy} = 0, \\ &(1 - \varepsilon)W_y \mp \varepsilon(W_{yy} + \nu_2 W_{xx}) = 0 \quad \text{for} \quad y = \pm 0.5. \end{aligned} \tag{2.96}$$

Eigenvalues and eigenforms are represented in the form of PS (2.16). Substituting these series into Equation (2.94) and BCs (2.95)–(2.96), and splitting with respect to ε, the following recurrent sequence of BVPs is obtained:

$$W_{0xxxx} + 2\alpha_3 W_{0xxyy} + \alpha_2 W_{0yyyy} - \lambda_0 W_0 = 0,$$

$$W_{0x} = 0, \quad W_{0xxx} = 0 \quad \text{for} \quad x = \pm 0.5k,$$

$$W_{0y} = 0, \quad W_{0yyy} = 0 \quad \text{for} \quad y = \pm 0.5,$$

$$W_{jxxxx} + 2\alpha_3 W_{jxxyy} + \alpha_2 W_{jyyyy} - \lambda_0 W_j = \sum_{i=1}^{j} \lambda_i W_{j-i},$$

$$\left.\begin{aligned} &W_{jxxx} + (2 - \nu_1)W_{jyyx} = 0, \\ &W_{jx} = \varepsilon k \textstyle\sum_{i=0}^{j-1}(W_{ixx} + \nu_1 W_{iyy}), \end{aligned}\right\} \quad \text{for} \quad x = \pm 0.5k,$$

$$\left.\begin{aligned} &W_{jyyy} + (2 - \nu_2)W_{jxxy} = 0, \\ &W_{jy} = \varepsilon \textstyle\sum_{i=0}^{j-1}(W_{iyy} + \nu_2 W_{ixx}), \end{aligned}\right\} \quad \text{for} \quad y = \pm 0.5.$$

In zeroth order approximation we have:

$$\lambda_0 = \pi^4 \left[\frac{m^4}{k^4} + 2\alpha_3 n^2 \frac{m^2}{k^2} + \alpha_2 n^4 \right], \tag{2.97}$$

$$W_0 = X_0 Y_0 = \begin{Bmatrix} \sin \pi n y \\ \cos \pi n y \end{Bmatrix} \begin{Bmatrix} \sin \frac{\pi m}{k} x, & n, m = 3, 5, 7, \dots \\ \cos \frac{\pi m}{k} x, & n, m = 2, 4, 6, \dots \end{Bmatrix}. \tag{2.98}$$

Construction of further approximations will be illustrated and discussed for the symmetric modes $n, m = 3, 5, 7, \dots$. We have the following BVP for the first order approximation:

$$W_{1xxxx} + 2\alpha_3 W_{1xxyy} + \alpha_2 W_{1yyyy} - \lambda_0 W_1 = \lambda_2 \sin \pi n y \sin \frac{\pi m}{k} x, \tag{2.99}$$

$$\left. \begin{aligned} & W_{1xxx} + (2 - \nu_1) W_{yyx} = 0, \\ & W_{1x} = \pm k^2 \pi^2 \left(\tfrac{m^2}{k^2} + \nu_1 n^2 \right) (-1)^{\frac{m-1}{2}} \sin \pi n y \end{aligned} \right\} \quad \text{for} \quad x = \pm 0.5k, \tag{2.100}$$

$$\left. \begin{aligned} & W_{1yyy} + (2 - \nu_2) W_{xxy} = 0, \\ & W_{1y} = \pm \pi^2 \left(n^2 + \nu_2 \tfrac{m^2}{k^2} \right) (-1)^{\frac{n-1}{2}} \sin \tfrac{\pi m}{k} x \end{aligned} \right\} \quad \text{for} \quad y = \pm 0.5. \tag{2.101}$$

Solution to the BVP (2.99)–(2.101) is sought in the form

$$W_1 = X_1(x) \sin \pi n y + Y_1(y) \sin \pi \frac{m}{k} x, \tag{2.102}$$

$$\lambda_1 = \lambda_{1x} + \lambda_{1y}. \tag{2.103}$$

Substituting Ansatzes (2.102), (2.103) into Equation (2.99) and BCs (2.100) and (2.101), the problem is reduced for two one-dimensional problems:

$$X_1^{IV}(x) + 2\alpha_3 \pi^2 n^2 X_1^{II}(x) - \pi^4 \left[\frac{m^4}{k^4} + 2\alpha_3 n^2 \frac{m^2}{k^2} \right] X_1(x) = \lambda_{1x} \sin \frac{\pi m}{k}, \tag{2.104}$$

$$\left. \begin{aligned} & X_1^{I}(x) = \pm k n^2 \left(\tfrac{m^2}{k^2} + \nu_1 n^2 \right) (-1)^{\frac{m-1}{2}}, \\ & X_1^{III}(x) - \pi^2 n^2 (2 - \nu_1) X_1^{I}(x) = 0, \end{aligned} \right\} \quad \text{for} \quad x = \pm 0.5k, \tag{2.105}$$

$$Y_1^{IV}(y) + 2\alpha_3 \pi^2 \frac{m^2}{k^2} Y_1^{II}(y) - \pi^4 \left[2\alpha_3 n^2 \frac{m^2}{k^2} + n^4 \right] Y_1(y) = \lambda_{1y} \sin \pi n y, \tag{2.106}$$

$$\left. \begin{aligned} & Y_1^{I}(y) = \pm \pi^2 \left(n^2 + \nu_2 \tfrac{m^2}{k^2} \right) (-1)^{\frac{n-1}{2}}, \\ & Y_1^{III}(y) - \pi^2 \tfrac{m^2}{k^2} (2 - \nu_2) Y_1^{I}(y) = 0, \end{aligned} \right\} \quad \text{for} \quad y = \pm 0.5. \tag{2.107}$$

Construction of solvability conditions is carried out by means of the previously described algorithm. In what follows we give only final results:

$$\lambda_{1x} = 4\pi^4 \left(\frac{m^2}{k^2} + \nu_1 n^2\right) \left[n^2(2 - \nu_1 - 2\alpha_3) - \frac{m^2}{k^2}\right], \tag{2.108}$$

$$\lambda_{1y} = 4\pi^4 \left(n^2 + \nu_2 \frac{m^2}{k^2}\right) \left[\frac{m^2}{k^2}(2 - \nu_2 - 2\alpha_3) - n^2\right]. \tag{2.109}$$

First correction to the eigenform W_1 follows as

$$W_1 = \frac{\pi\left(\frac{m^2}{k^2} + \nu_1 n^2\right)}{\left(\frac{m^2}{k^2} + \alpha_3 n^2\right)} \left[\frac{k \begin{Bmatrix} (-1)^{\frac{m-1}{2}} \\ (-1)^{\frac{m}{2}} \end{Bmatrix} \left(n^2(2-\nu_1) + \frac{m^2}{k^2}\right)}{2\beta_2 \begin{Bmatrix} \cosh \frac{\pi}{2}\beta_2 k \\ \sinh \frac{\pi}{2}\beta k \end{Bmatrix}} \begin{Bmatrix} \sinh \pi\beta_2 x \\ \cosh \pi\beta k_2 x \end{Bmatrix} - \right.$$

$$\left. (-1)^m \frac{k}{m}\left(n^2(2 - \nu_1 - 2\alpha_3) - \frac{m^2}{k^2}\right) x \begin{Bmatrix} \cos \frac{\pi m}{k} x \\ \sin \frac{\pi m}{k} x \end{Bmatrix} \right] \begin{Bmatrix} \sin \pi n y \\ \cos \pi n y \end{Bmatrix} +$$

$$\frac{\pi\left(n^2 + \nu_2 \frac{m^2}{k^2}\right)}{\left(n^2 + \alpha_3 \frac{m^2}{k^2}\right)} \left[\frac{\begin{Bmatrix} (-1)^{\frac{n-1}{2}} \\ (-1)^{\frac{n}{2}} \end{Bmatrix} \left(\frac{m^2}{k^2}(2-\nu_2) + n^2\right)}{2\beta_1 \begin{Bmatrix} \cosh \frac{\pi}{2}\beta_1 \\ \sinh \frac{\pi}{2}\beta_1 \end{Bmatrix}} \begin{Bmatrix} \sinh \pi\beta_1 y \\ \cosh \pi\beta_1 y \end{Bmatrix} - \right. \tag{2.110}$$

$$\left. (-1)^n \frac{1}{n}\left(\frac{m^2}{k^2}(2 - \nu_2 - 2\alpha_3) - n^2\right) y \begin{Bmatrix} \cos \pi n y \\ \sin \pi n y \end{Bmatrix} \right] \begin{Bmatrix} \sin \frac{\pi m}{k} x \\ \cos \frac{\pi m}{k} x \end{Bmatrix},$$

$$\begin{Bmatrix} n, m = 1, 3, 5, \ldots \\ n, m = 2, 4, 6, \ldots \end{Bmatrix}.$$

In the case of the second order approximation we get

$$\lambda_{2x} = 4\pi^4\left[n^2(2-\nu_1-2\alpha_3)-\frac{m^2}{k^2}\right]\left\{\left(\frac{m^2}{k^2}+\nu_1 n^2\right)-\left\{\frac{\left(\frac{m^2}{k^2}+\nu_1 n^2\right)}{\left(\frac{m^2}{k^2}+\alpha_3 n^2\right)}\right.\right. \times$$

$$\left[\frac{\pi k}{2}\left(n^2(2-\nu_1)+\frac{m^2}{k^2}\right)^2 \coth^{(-1)^m}\frac{\pi}{2}\beta_2 - 2\left(n^2(2-\nu_1-2\alpha_3)-\frac{m^2}{k^2}\right)\right] -$$

$$2\nu_1\frac{\left(n^2+\nu_2\frac{m^2}{k^2}\right)}{\left(n^2+\alpha_3\frac{m^2}{k^2}\right)}\times\left(\frac{m^2}{k^2}(2-\nu_2-2\alpha_3)-n^2\right)-\frac{\left(n^2+\nu_2\frac{m^2}{k^2}\right)}{\left(n^2+\alpha_3\frac{m^2}{k^2}\right)}\times \tag{2.111}$$

$$\left[\frac{\left(\frac{m^2}{k^2}(2-\nu_2)+n^2\right)\left(\frac{m^2}{k^2}(1-2\nu_1\alpha_3)-n^2\right)}{\left(n^2+\alpha_3\frac{m^2}{k^2}\right)}+\frac{1}{2n^2}\left(\frac{m^2}{k^2}(2-\nu_2-2\alpha_3)-n^2\right)\times\right.$$

$$\left.\left.\left(\frac{m^2}{k^2}+\nu_1 n^2\right)\right]\right\}-\frac{\lambda_1}{2}\frac{\left(\frac{m^2}{k^2}+\nu_1 n^2\right)}{\left(\frac{m^2}{k^2}+\alpha_3 n^2\right)}\left\{k\frac{\left(n^2(2-\nu_1)+\frac{m^2}{k^2}\right)}{\left(\frac{m^2}{k^2}+\alpha_3 n^2\right)}+\right.$$

$$\left.\frac{1}{n^2}\left(\frac{m^2}{k^2}(2-\nu_2-2\alpha_3)-n^2\right)\right\},$$

$$\left(\lambda_{2y},\frac{m^2}{k^2},n^2,\nu_1,\nu_2,\beta_1\right)\to\left(\lambda_{2x},n^2,\frac{m^2}{k^2},\nu_2,\nu_1,\beta_2\right).$$

In the orthotropic case ($D_1 = D_2 = D_3 = D, \nu_1 = \nu_2 = \nu$) formulas for λ and W take the following form

$$\lambda = \pi^4\left(\frac{m^2}{k^2}+n^2\right)^2 - 4\pi^4(1+\nu)\left(\frac{m^2}{k^2}+n^2\right)\varepsilon+(\lambda_{2x}+\lambda_{2y})\varepsilon^2+\ldots, \tag{2.112}$$

$$W = \left\{\begin{matrix}\sin\pi n y\sin\frac{\pi m}{k}x\\ \cos\pi n y\cos\frac{\pi m}{k}x\end{matrix}\right\}+\frac{\pi}{\left(n^2+\frac{m^2}{k^2}\right)}\left\{\frac{k\left\{\begin{matrix}(-1)^{\frac{m-1}{2}}\\ (-1)^{\frac{m}{2}}\end{matrix}\right\}\left(n^2(2-\nu)+\frac{m^2}{k^2}\right)}{2\beta_2\left\{\begin{matrix}\cosh\pi\beta_2 k/2\\ \sinh\pi\beta_2 k/2\end{matrix}\right\}}\right.\times$$

$$\left\{\begin{matrix}\sinh\pi\beta_2 x\\ \cosh\pi\beta_2 x\end{matrix}\right\}+(-1)^m\frac{k}{m}\left(\frac{m^2}{k^2}+\nu n^2\right)x\left\{\begin{matrix}\cos\frac{\pi m}{k}x\\ \sin\frac{\pi m}{k}x\end{matrix}\right\}\Bigg]\left\{\begin{matrix}\sin\pi n y\\ \cos\pi n y\end{matrix}\right\}+ \tag{2.113}$$

$$\left(n^2+\frac{m^2}{k^2}\right)\left[\frac{\begin{Bmatrix}(-1)^{\frac{n-1}{2}}\\(-1)^{\frac{n}{2}}\end{Bmatrix}\left(\frac{m^2}{k^2}(2-\nu)+n^2\right)}{\begin{Bmatrix}\cosh\pi\beta_1/2\\\sinh\pi\beta_1/2\end{Bmatrix}}\begin{Bmatrix}\sinh\pi\beta_1 y\\\cosh\pi\beta_1 y\end{Bmatrix}+\right.$$

$$\left.(-1)^{\frac{1}{n}}\left(n^2+\nu\frac{m^2}{k^2}\right)y\begin{Bmatrix}\cos\pi n y\\\sin\pi n y\end{Bmatrix}\right]\begin{Bmatrix}\sin\frac{\pi m}{k}x\\\cos\frac{\pi m}{k}x\end{Bmatrix}\Bigg\}\varepsilon+\ldots,$$

$$\lambda_{2x}=8\pi^4(1+\nu)\left(\frac{m^2}{k^2}+\nu n^2\right)\left[\frac{k}{2}\frac{\left(n^2(2-\nu)+\frac{m^2}{k^2}\right)}{n^2+\frac{m^2}{k^2}}-\frac{k^2}{m^2}\left(\frac{m^2}{k^2}+\nu n^2\right)\right]-$$

$$4\pi^4\left(\frac{m^2}{k^2}+\nu n^2\right)^2+8\pi^4\frac{\left(\frac{m^2}{k^2}+\nu n^2\right)}{\left(n^2+\frac{m^2}{k^2}\right)}\left\{\left(\frac{m^2}{k^2}+\nu n^2\right)\left[\frac{k\left(n^2(2-\nu)+\frac{m^2}{k^2}\right)}{2\beta_2}\times\right.\right.$$

$$\left.\pi\tanh\frac{\pi\beta_2 k}{2}+2\left(\frac{m^2}{k^2}+\nu n^2\right)\right]+2\nu\left(n^2+\nu\frac{m^2}{k^2}\right)+\nu\left(n^2+\nu\frac{m^2}{k^2}\right)\times$$

$$\left.\left[\frac{\left(\frac{m^2}{k^2}(2-\nu)+n^2\right)\left(\frac{m^2}{k^2}(1-2\nu)+n^2\right)}{\left(\frac{m^2}{k^2}+n^2\right)}+\frac{\left(n^2+\nu\frac{m^2}{k^2}\right)\left(\frac{m^2}{k^2}+\nu n^2\right)}{2n^2}\right]\right\},$$

$$\left(\lambda_{2y};n^2;\frac{m^2}{k^2};\beta_2\right)\to\left(\lambda_{2x};\frac{m^2}{k^2};n^2;\beta_1\right).$$

Results verification is carried out for the first eigenvalue of the free square isotropic plate. The PA of the truncated PS (2.112) has the form (2.30). First eigenvalue of the BVP (2.94)–(2.96) for obtained via the PA for $\nu = 1/6$ is equal to $\lambda = (1.100\pi)^4$, whereas the Bubnov-Galerkin method [39] gives $\lambda = (1.2295\pi)^4$ (error - 10.51%).

For $\nu = 0.3$ the PA gives $\lambda = (1.1198\pi)^4$, the Bubnov-Galerkin method - $\lambda = (1.1683\pi)^4$ (error 4.15%), the Southwell method [39] $\lambda = (1.1424\pi)^4$ (error 1.14%). It should be emphasized that both Bubnov-Galerkin and Southwell methods yields upper estimation of the eigenvalue. Zero order approximation yielded by the truncated PS and PA for $\varepsilon = 1$ gives upper and lower bounds of eigenfrequency, respectively.

2.2.5 *Natural Vibrations of the Plate with Mixed Boundary Conditions "Clamping-Simple Support"*

Lets consider the problem of natural plate vibrations $(0.5k \le x \le 0.5k, 0.5 \le y \le 0.5)$. The plate is simply supported for $x = \pm 0.5k$, and on edges $y = \pm 0.5$ it has mixed BCs of the "clamping-simple support" type. In what follows we study a symmetric problem first (Figure 2.8a).

The original PDE has the form (2.63). We apply the BCs through the parameter ε in such a way that for $\varepsilon = 0$ we have BCs of simply supporting on edges $y = \pm 0.5$, whereas for $\varepsilon = 1$ we apply the given BCs:

$$W = 0, \qquad W_{xx} = 0 \qquad \text{for} \qquad x = \pm 0.5, \tag{2.114}$$

$$W = 0, \qquad W_{yy} = \overline{H}(x)\varepsilon(W_{yy} \pm W_y) \qquad \text{for} \qquad y = \pm 0.5, \tag{2.115}$$

where: $\overline{H}(x) = H(x - \mu k) + H(-x - \mu k)$; $H(x)$ is the Heaviside function.

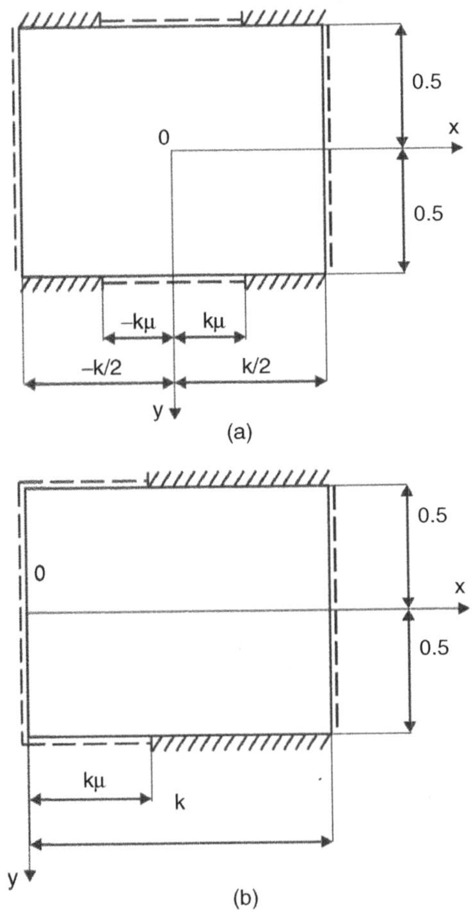

Figure 2.8 Schemes of plates with mixed BCs (– – – simple support, //// clamping)

For this purpose eigenvalues and eigenforms are rewritten in the (2.16) form. Substituting them into Equation (2.63) and BCs (2.114)-(2.115), and carrying out the splitting procedure regarding ε, the following recurrent sequence of the BVPs is obtained:

$$\nabla^4 W_0 - \lambda_0 W_0 = 0,$$

$$W_0 = 0, \quad W_{0xx} = 0 \quad \text{for} \quad x = \pm 0.5k,$$

$$W_0 = 0, \quad W_{0yy} = 0 \quad \text{for} \quad y = \pm 0.5,$$

$$\nabla^4 W_j - \lambda_0 W_j = \sum_{i=1}^{j} \lambda_i W_{j-i},$$

$$W_j = 0, \quad W_{jxx} = 0 \quad \text{for} \quad x = \pm 0.5k,$$

$$W_j = 0, \quad W_{jyy} = \mp \overline{H}(x) \sum_{i=0}^{j-1} W_{iy} \quad \text{for} \quad y = \pm 0.5.$$

Let us study a solution to the problem for the symmetric with respect to eigenform axes x and y case. In zero order approximation we have

$$W_0 = X_0 Y_0 = \cos \frac{\pi m}{k} x \cos \pi n y, \tag{2.116}$$

$$\lambda_0 = \pi^4 \left(n^2 + \frac{m^2}{k^2} \right)^2, \quad n, m = 1, 3, 5, \ldots. \tag{2.117}$$

In the first order approximation, the following BVP is obtained:

$$\nabla^4 W_1 - \lambda_0 W_0 = \lambda_1 W_0, \tag{2.118}$$

$$W_1 = 0, \quad W_{1xx} = 0 \quad \text{for} \quad x = \pm 0.5k, \tag{2.119}$$

$$W_1 = 0, \quad W_{1yy} = \pi n (-1)^{\frac{n-1}{2}} \overline{H}(x) \cos \pi m \frac{x}{k} \quad \text{for} \quad y = \pm 0.5. \tag{2.120}$$

A solution is sought in the following series form:

$$W = \sum_{i=1,3,5,\ldots} Y_{1i} \cos \frac{\pi i}{k} x. \tag{2.121}$$

After substitution of Ansatz (2.121) into Equations (2.118)(2.120), the following two BVPs are obtained:
for $i = m$:

$$Y_{1m}^{IV} - 2\frac{\pi^2 m^2}{k^2} Y_{1m}^{II} - \pi^4 n^2 \left(2\frac{m^2}{k^2} + n^2 \right) Y_{1m} = \lambda_1 \cos \pi n y, \tag{2.122}$$

$$Y_{1m} = 0, \quad Y_{1m}^{II} = \pm \pi n (-1)^{\frac{n-1}{2}} \gamma_{mm} \quad (y = \pm 0.5); \tag{2.123}$$

for $i \neq m$:

$$Y_{1i}^{IV} - 2\frac{\pi^2 i^2}{k^2} Y_{1i}^{II} - \pi^4 \left[\left(n^2 + \frac{m^2}{k^2} \right)^2 - \frac{i^4}{k^4} \right] Y_{1i} = 0, \tag{2.124}$$

$$Y_{1i} = 0, \quad Y_{1i}^{II} = \pm \pi n (-1)^{\frac{n-1}{2}} \gamma_{im} \quad (y = \pm 0.5), \tag{2.125}$$

where

$$\gamma_{im} = \begin{cases} 2(0.5-\mu) - \frac{1}{\pi m}\sin 2\pi m\mu & \text{for} \quad i=m, \\ \frac{4}{\pi}\frac{1}{(m^2-i^2)}[i\sin\pi\mu i\cos\pi\mu m - m\sin\pi\mu m\cos\pi\mu i] & \text{for} \quad i\neq m. \end{cases}$$

Removing secular terms, we define the first correction term:

$$\lambda_1 = 4\pi^2 n^2 \gamma_{mm}. \tag{2.126}$$

After λ_1 determination, we construct Y_{1m}:

$$Y_{1m} = \frac{n}{\pi\alpha}\gamma_{mm}\left[\frac{(-1)^{\frac{n-1}{2}}}{2\text{ch}\frac{\pi}{2}\beta_1}\cosh\pi\beta_1 y - y\sin\pi n y\right], \qquad i=m, \tag{2.127}$$

where $\alpha = n^2 + m^2/k^2$, $\beta_1 = \sqrt{2m^2/k^2+n^2}$.

Solution to the BVP (2.124)–(2.125) does not yield any correction terms to the eigenavalue, but it gives some additional corrections to vibration modes, i.e.

$$Y_{1i} = \frac{n(-1)^{\frac{n-1}{2}}}{2\pi\left(\frac{i^2}{k^2}+n^2\right)}\gamma_{im}\left[\frac{\cosh\alpha_{1i}y}{\cosh\alpha_{1i}/2} - \frac{\begin{Bmatrix}\cosh\gamma_{1i}y\\ \cos\beta_{1i}y\end{Bmatrix}}{\begin{Bmatrix}\cosh\gamma_{1i}/2\\ \cos\beta_{1i}/2\end{Bmatrix}}\right], \quad \begin{Bmatrix} i^2 > m^2+n^2k^2 \\ i^2 < m^2+n^2k^2\end{Bmatrix}, \tag{2.128}$$

where $\alpha_{1i} = \pi\sqrt{\frac{i^2+m^2}{k^2}+n^2}$, $\beta_{1i} = \pi\sqrt{\frac{m^2-i^2}{k^2}+n^2}$, $\gamma_{1i} = \pi\sqrt{\frac{m^2+i^2}{k^2}-n^2}$.

Summing up expressions in (2.126) and (2.127), the following first correction to the eigenform is obtained

$$W_1 = \frac{n}{\pi\alpha}\left\{\gamma_{mm}\left[\frac{(-1)^{\frac{n-1}{2}}}{2\cosh\frac{\pi}{2}\beta_1}\cosh\pi\beta_1 y - y\sin\pi n y\right]\cos\frac{\pi m}{k}x + \right.$$
$$\left.(-1)^{\frac{n-1}{2}}\sum_{i=1,3,5,\ldots}^{\infty}\gamma_{im}\left[\frac{\cosh\alpha_{1i}y}{\cosh\alpha_{1i}/2} - \frac{\begin{Bmatrix}\cosh\gamma_{1i}y\\ \cos\beta_{1i}y\end{Bmatrix}}{\begin{Bmatrix}\cosh\gamma_{1i}/2\\ \cos\beta_{1i}/2\end{Bmatrix}}\right]\cos\frac{\pi m}{k}x\right\}. \tag{2.129}$$

Here operator $\sum_{i=1,3,5,\ldots}$ denotes summation procedure without the term $i=m$.

Proceeding in the analogous way, we get the formula for the second correction term to the eigenvalue:

$$\lambda_2 = 4\pi^2 n^2\gamma_{mm}\left\{1 - \frac{\gamma_{mm}}{\pi^2\alpha}\left[\frac{\pi\beta_1}{2}\tanh\frac{\pi\beta_1}{2} + \frac{n^2}{\alpha} - \frac{3}{2}\right]\right\} -$$
$$\frac{2n^2}{\alpha}\sum_{i=1,3,5,\ldots}\gamma_{im}^2\left[\alpha_i\tanh\frac{\alpha_i}{2} + \begin{Bmatrix}-\varphi_{1i}\tanh\varphi_{1i}/2\\ \beta_{1i}\tan\beta_{1i}/2\end{Bmatrix}\right]. \tag{2.130}$$

In this case the truncated PS has the following form:

$$\lambda = \pi^4\alpha^4 + 4\pi^2 n^2\gamma_{mm}\varepsilon + \left\{4\pi^2 n^2\gamma_{mm}\left(1 - \frac{\gamma_{mm}}{\pi^2\alpha}\left[\frac{\pi\beta_1}{2}\coth^{(-1)^m}\frac{\pi\beta_1}{2} + \frac{n^2}{\alpha} - \frac{3}{2}\right]\right) - \frac{2n^4}{\alpha}\sum_{\left\{\begin{matrix} i=1,3,5,\ldots \\ i=2,4,6,\ldots\end{matrix}\right\}} \gamma_{im}^2\left[\alpha_i\coth^{(-1)^i}\frac{\alpha_{1i}}{2} + \left\{\begin{matrix} -\varphi_{1i}\coth^{(-1)^i}\varphi_{1i}/2 \\ \beta_{1i}\coth^{(-1)^i}\beta_{1i}/2 \end{matrix}\right\}\right]\right\}\varepsilon^2 + \ldots, \tag{2.131}$$

$$W = \left\{\begin{matrix} \cos\frac{\pi m}{k}x\cos\pi n y \\ \sin\frac{\pi m}{k}x\sin\pi n y \end{matrix}\right\} + \left[\frac{n}{\pi\alpha}\gamma_{mm}\left(\frac{\left\{\begin{matrix} (-1)^{\frac{n-1}{2}}\cosh\pi\beta_1 y \\ (-1)^{\frac{n}{2}}\sinh\pi\beta_1 y \end{matrix}\right\}}{\left\{\begin{matrix} \cosh\pi\beta_1/2 \\ \sinh\pi\beta_1/2 \end{matrix}\right\}} + \left\{\begin{matrix} \sin\pi n y \\ \cos\pi n y \end{matrix}\right\}\right)\left\{\begin{matrix} \cos\frac{\pi m}{k}x \\ \sin\frac{\pi m}{k}x \end{matrix}\right\} + \frac{n\left\{\begin{matrix} (-1)^{\frac{n-1}{2}} \\ (-1)^{\frac{n}{2}} \end{matrix}\right\}}{2\pi\alpha}\sum_{\left\{\begin{matrix} i=1,3,5,\ldots \\ i=2,4,6,\ldots\end{matrix}\right\}}^{\infty}\gamma_{im}\left(\frac{\left\{\begin{matrix} \cosh\alpha_{1i}y \\ \sinh\alpha_{1i}y \end{matrix}\right\}}{\left\{\begin{matrix} \cosh\alpha_{1i}/2 \\ \sinh\alpha_{1i}/2 \end{matrix}\right\}} + \frac{\left\{\begin{matrix} A_{1i} \\ A_{2i} \end{matrix}\right\}}{\left\{\begin{matrix} B_{1i} \\ B_{2i} \end{matrix}\right\}}\right)\left\{\begin{matrix} \cos\frac{\pi i}{k}x \\ \sin\frac{\pi i}{k}x \end{matrix}\right\}\right]\varepsilon + \ldots, \quad \left\{\begin{matrix} n,m=1,3,5,\ldots \\ n,m=2,4,6,\ldots \end{matrix}\right\}, \tag{2.132}$$

where

$$\gamma_{im} = \begin{cases} 1 - 2\mu + \frac{(-1)^m}{\pi m}\sin 2\pi m\mu & \text{for} \quad i = m, \\ \frac{4}{\pi}\cdot\frac{1}{(m^2-i^2)}\left[\left\{\begin{matrix} i \\ m \end{matrix}\right\}\sin\pi\mu i\cos\pi\mu m - \left\{\begin{matrix} m \\ i \end{matrix}\right\}\sin\pi\mu m\cos\pi\mu i\right] & \text{for} \quad i \neq m, \end{cases} \tag{2.133}$$

$$\frac{\left\{\begin{matrix} A_{1i} \\ A_{2i} \end{matrix}\right\}}{\left\{\begin{matrix} B_{1i} \\ B_{2i} \end{matrix}\right\}} = \frac{\left\{\begin{matrix} \cosh\varphi_{1i}y \\ \sinh\varphi_{1i}y \end{matrix}\right\}}{\left\{\begin{matrix} \cosh\varphi_{1i}/2 \\ \sinh\varphi_{1i}/2 \end{matrix}\right\}} \quad \text{for} \quad i^2 > m^2 + n^2k^2,$$

$$\frac{\left\{\begin{matrix} A_{1i} \\ A_{2i} \end{matrix}\right\}}{\left\{\begin{matrix} B_{1i} \\ B_{2i} \end{matrix}\right\}} = \frac{\left\{\begin{matrix} \cos\beta_{1i}y \\ \sin\beta_{1i}y \end{matrix}\right\}}{\left\{\begin{matrix} \cos\beta_{1i}/2 \\ \sin\beta_{1i}/2 \end{matrix}\right\}} \quad \text{for} \quad i^2 < m^2 + n^2k^2.$$

We solve the problem regarding computation of the natural frequencies for the plate with nonsymmetric located parts of mixed BCs (Figure 2.8b). Formulas regarding eigenvalues and eigenforms (2.131) and (2.132) remain valid assuming that the following transformations are applied:

$$(-1)^{\frac{m-1}{2}} \to (-1), \quad \left\{\begin{matrix} \cos \pi mx/k \\ \sin \pi mx/k \end{matrix}\right\} \to \sin \frac{\pi m}{k} x, \quad \left\{\begin{matrix} \cos \pi ix/k \\ \sin \pi ix/k \end{matrix}\right\} \to \sin \frac{\pi i}{k} x,$$
$$m = 1, 2, 3, \ldots, \quad i = 1, 2, 3, \ldots,$$

and γ_{im} follows

$$\gamma_{im} = \begin{cases} \mu - \frac{1}{2\pi m} \sin 2\pi m\mu & \text{for} \quad i = m, \\ \frac{2}{\pi} \cdot \frac{1}{(m^2-i^2)} [i \sin \pi\mu m \cos \pi\mu i - m \sin \pi\mu i \cos \pi\mu m] & \text{for} \quad i \neq m. \end{cases} \tag{2.134}$$

Further, we use PA (2.30), and we compute the value of the first eigenvalue associated with the BVP (2.63), (2.114), (2.115) for $\varepsilon = 1$. Computational results are reported in Figure 2.9. A solid curve represents the dependence of the eigenvalue versus the parameter μ for the plate with symmetrically located clamping, whereas dashed curves correspond to the nonsymmetric problem. In the limiting case (edges $y = \pm 0.5$ are completely clamped) the first eigenvalue obtained numerically $\lambda = (1.7050\pi)^4$, whereas the PA (2.30) yields $\lambda = (1.7081\pi)^4$ (error - 0.18%). A dashed curve is associated with results obtained through the method of integral equations. One may see that the difference in results is small.

In the frame of the proposed method, we may include an influence of the support stiffness on the clamped plate parts. The dependence of the first eigenvalue of BVP (2.63), (2.114), (2.115) versus the parameter ε for various values of μ is shown. One may conclude that the influence of clamping stiffness onto the eigenvalue is mainly exhibited for elastic supports being similar to completely developed supports.

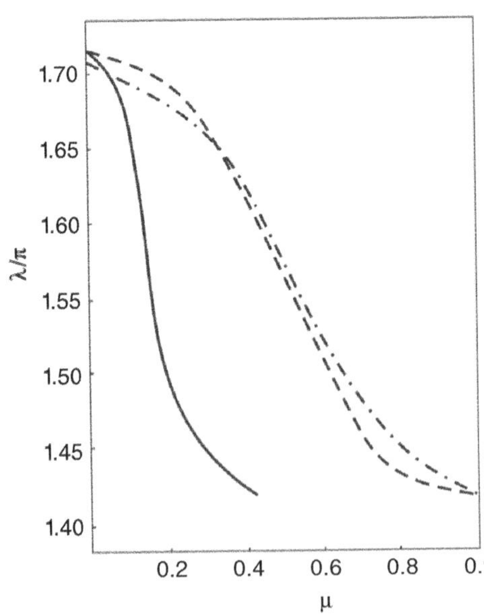

Figure 2.9 Influence of plate clamping parts on the first eigenvalue

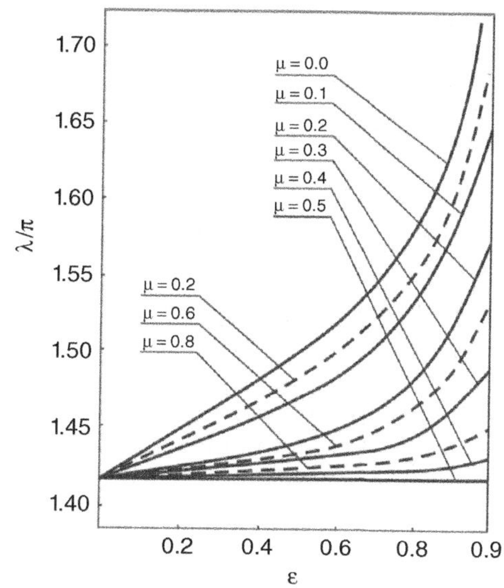

Figure 2.10 Influence of support stiffness on first eigenvalue

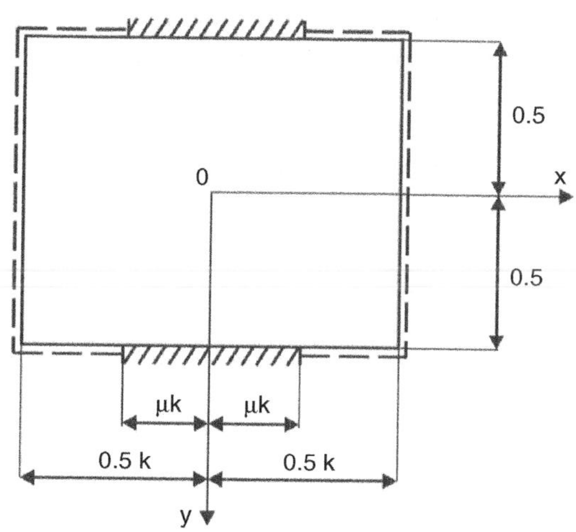

Figure 2.11 Scheme of the plate with mixed BCs

2.2.6 *Comparison of Theoretical and Experimental Results*

In this section we study the problem regarding natural vibrations of the plate ($-0.5k \leq x \leq 0.5k; -0.5 \leq y \leq 0.5$), simply supported for $x = \pm 0.5k$, and having on $y = \pm 0.5$ mixed BCs of the "clamping-simple support" type, where clamping parts are located symmetrically with respect to plate sides $y = \pm 0.5$ (Figure 2.11).

The governing PDE has the form (2.63), wheres in BCs (2.114), (2.115) function $\overline{H}(x)$ should be taken as follows:

$$\overline{H}(x) = H(-x + \mu k) - H(-x - \mu k).$$

Applying this to the problem so far defined in our approach, one gets formulas for eigenvalues and eigenforms of (2.131) and (2.132). Formulas for γ_{im} in this case follow

$$\gamma_{im} = \begin{cases} 2\mu - \frac{(-1)^m}{4\pi m} \sin 2\pi m\mu \quad \text{for} \quad i = m, \\ \frac{4}{\pi} \cdot \frac{1}{(m^2-i^2)} \left[\begin{Bmatrix} i \\ -m \end{Bmatrix} \sin \pi\mu i \cos \pi\mu m + \right. \\ \left. \begin{Bmatrix} -m \\ i \end{Bmatrix} \sin \pi\mu m \cos \pi\mu i \right] \quad \text{for} \quad i \neq m. \end{cases} \tag{2.135}$$

Furthermore, the truncated PS is transformed into the PA (2.30), and the first eigenvalue is computed. Computational results are shown in Figure 2.12. A solid curve corresponds to the obtained result, whereas the dashed and dash-dot curves correspond to theoretical results reported in references [49], [65]. It is evident that all three curves almost coincide with each other for all values of the parameter μ. Experimental data [65] (points) are also close to computed ones.

Observe that the graph of eigenvalue dependence on the parameter μ possesses two zones: $0.0 \leq \mu \leq 0.3$ and $0.3 < \mu \leq 0.5$. On the first part an increase of the parameter μ implies an essential increase of λ. In the second zone the parameter μ variation does not practically influence the eigenvalues positions.

Therefore, a plate having clamping parts of length being larger than $\mu = 0.3$, may be treated as that clamped on edges $y = \pm 0.5$.

In Figure 2.13 the dependence of eigenvalue λ versus parameter ε for various μ is reported. It is clear that the influence of support stiffness is strongly exhibited for $\varepsilon > 0.7$.

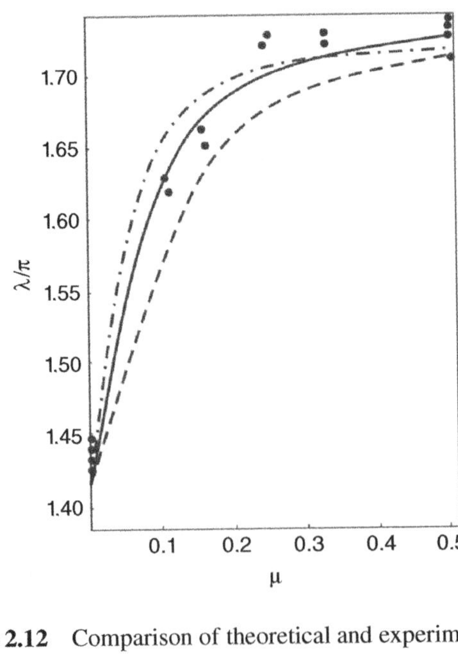

Figure 2.12 Comparison of theoretical and experimental data

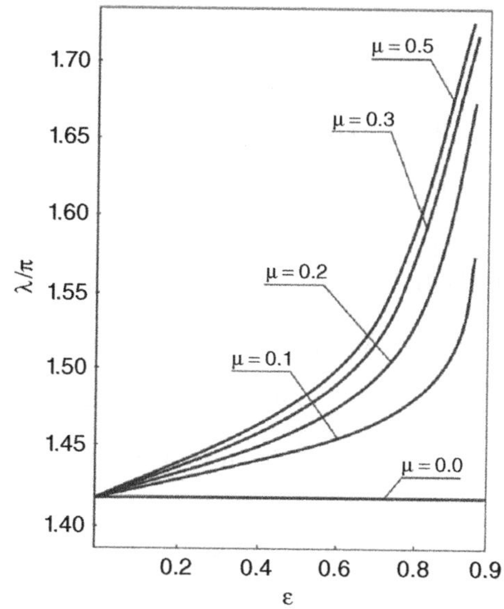

Figure 2.13 Influence of clamping lengths on the first eigenvalue

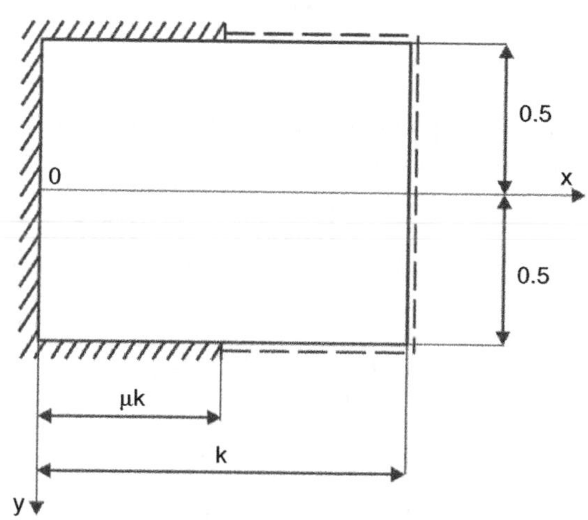

Figure 2.14 Plate with mixed BCs

2.2.7 *Natural Vibrations of a Partially Clamped Plate*

We consider now the more complicated case of the BCs, i.e. partially clamped plate along its rectangular contour ($0 \le x \le k; -0.5 \le y \le 0.5$). It is simply supported on the side $x = k$, clamped on the side $x = 0$, and has mixed BCs of the type "clamping - simple support" on the sides $y = \pm 0.5$ (Figure 2.14). Plate vibrations are governed by the PDE (2.63), and the following BCs are attached:

$$\begin{aligned}
&W = 0, \quad W_{yy} = \varepsilon \overline{H}(x)(W_{yy} \mp W_y) \quad \text{for} \quad y = \pm 0.5,\\
&W = 0, \quad W_{xx} = \varepsilon (W_{xx} + kW_x) \quad \text{for} \quad x = 0, \qquad (2.136)\\
&W = 0, \quad W_{xx} = 0 \quad \text{for} \quad x = k,
\end{aligned}$$

where $\overline{H}(x) = H(x) - H(x - \mu k)$.

Substituting PS (2.16) into the BVP (2.63), (2.136) yields the following system of equations:

$$\nabla^4 W_0 - \lambda_0 W_0 = 0,$$

$$W_0, \quad W_{0yy} = 0 \quad \text{for} \quad y = \pm 0.5,$$

$$W_0, \quad W_{0xx} = 0 \quad \text{for} \quad x = 0, k,$$

$$\nabla^4 W_j - \lambda_0 W_j = \sum_{i=1}^{j} \lambda_i W_{j-i},$$

$$W_j = 0, \quad W_{jyy} = \mp \overline{H}(x) \sum_{i=0}^{j-1} W_{iy} \quad \text{for} \quad y = \pm 0.5,$$

$$W_j = 0, \quad W_{jxx} = k \sum_{i=0}^{j-1} W_{ix} \quad \text{for} \quad x = 0,$$

$$W_j = 0, \quad W_{jxx} = 0 \quad \text{for} \quad x = k.$$

In zero order approximation the problem is reduced to that of the plate simply supported along its contour, and hence its solution follows:

$$W_0 = X_0 Y_0 = \sin \frac{\pi m}{k} x \begin{Bmatrix} \cos \pi n y, & n = 1, 3, 5, \ldots \\ \sin \pi n y, & n = 2, 4, 6, \ldots \end{Bmatrix}, \qquad (2.137)$$

$$\lambda_0 = \pi^4 \left(\frac{m^2}{k^2} + n^2 \right)^2 = \pi^4 \alpha. \qquad (2.138)$$

First we consider the BVP of the first order approximation:

$$\nabla W_1 - \lambda_0 W_1 = \lambda_1 W_0, \qquad (2.139)$$

$$W_1 = 0, \quad W_{1yy} = \mp \begin{Bmatrix} (-1)^{\frac{n-1}{2}} \\ (-1)^{\frac{n}{2}} \end{Bmatrix} \pi n \overline{H}(x) \sin \frac{\pi m}{k} x \quad \text{for} \quad y = \pm 0.5, \qquad (2.140)$$

$$W_1 = 0, \quad W_{1xx} = \pi m \begin{Bmatrix} \cos \pi n y \\ \sin \pi n y \end{Bmatrix} \quad \text{for} \quad x = 0, \qquad (2.141)$$

$$\overline{H}(x) = H(-x + \mu k) - H(-x - \mu k), \qquad (2.142)$$

where $\lambda_1 = \lambda_{1x} + \lambda_{1y}$.

Its solution has the following form:

$$W_1 = X_1 \begin{Bmatrix} \cos \pi n y \\ \sin \pi n y \end{Bmatrix} + \sum_{i=1}^{\infty} Y_{1i} \sin \frac{\pi i}{k} x. \tag{2.143}$$

Substitution of Ansatz (2.143) into Equation (2.139) and BCs (2.140)–(2.141) yields the following three BVPs:

$$X_1^{IV} - 2\pi^2 n^2 X_1^{II} - \pi^4 \frac{m^2}{k^2}\left(2n^2 + \frac{m^2}{k^2}\right) X_1 = \lambda_{1x} \sin \frac{\pi m}{k} x, \tag{2.144}$$

$$X_1 = 0, \quad X_1^{II} = \pi m \quad \text{for} \quad x = 0, \tag{2.145}$$

$$X_1 = 0, \quad X_1^{II} = 0 \quad \text{for} \quad x = k, \tag{2.146}$$

for $i = m$:

$$X_{1m}^{IV} - 2\pi^2 \frac{m^2}{k^2} Y_{1m}^{II} - \pi^4 n^2 \left(2\frac{m^2}{k^2} + n^2\right) Y_{1m} = \lambda_{1y} \begin{Bmatrix} \cos \pi n y \\ \sin \pi n y \end{Bmatrix}, \tag{2.147}$$

$$Y_{1m}^{IV} = 0, \quad Y_{1m}^{II} = \mp \begin{Bmatrix} -(-1)^{\frac{n-1}{2}} \\ (-1)^{\frac{n}{2}} \end{Bmatrix} \pi n \gamma_{mm} \quad \text{for} \quad y = \pm 0.5, \tag{2.148}$$

for $i \neq m$:

$$Y_{1i}^{IV} - 2\pi^2 \frac{i^2}{k^2} Y_{1i}^{II} - \pi^4 \left[\left(\frac{m^2}{k^2} + n^2\right)^2 - \frac{i^4}{k^4}\right] Y_{1i} = 0, \tag{2.149}$$

$$Y_{1i}^{IV} = 0, \quad Y_{1i}^{II} = \mp \begin{Bmatrix} -(-1)^{\frac{n-1}{2}} \\ (-1)^{\frac{n}{2}} \end{Bmatrix} \pi n \gamma_{im} \quad \text{for} \quad y = \pm 0.5. \tag{2.150}$$

where γ_{im} is defined through formula (2.134).

BVPs (2.144)–(2.146) and (2.147)–(2.148) possess secular terms, and their removal allows us to define λ_{1x} and λ_{1y}:

$$\lambda_{1x} = 2\frac{\pi^2 m^2}{k^2}, \tag{2.151}$$

$$\lambda_{1y} = 4\pi^2 n^2 \gamma_{mm}. \tag{2.152}$$

Eigenfunctions of BVPs (2.144)-(2.146) and (2.147)-(2.148) follow:

$$X_1 = \frac{m}{2\pi\alpha}\left[\cosh \beta_2 x - \coth \beta_2 k \sinh \beta_2 x - \cos \frac{\pi m}{k} x + \frac{1}{k} x \cos \frac{\pi m}{k} x\right], \tag{2.153}$$

$$Y_{1m} = \frac{(-1)^{n+1} n \gamma_{mm}}{\pi \alpha} \left[\frac{\begin{Bmatrix} (-1)^{\frac{n-1}{2}} \\ (-1)^{\frac{n}{2}} \end{Bmatrix}}{2\begin{Bmatrix} \cosh \beta_1/2 \\ \sinh \beta_1/2 \end{Bmatrix}} \begin{Bmatrix} \cosh \beta_1 y \\ \sinh \beta_1 y \end{Bmatrix} - \right. \tag{2.154}$$

$$\left. y \begin{Bmatrix} \sin \pi n y \\ \cos \pi n y \end{Bmatrix} \right], \quad \begin{Bmatrix} n = 1, 3, 5, \dots \\ n = 2, 4, 6, \dots \end{Bmatrix}.$$

Solutions to the BVPs (2.149)–(2.150) do not influence the eigenvalues but they bring in corrections to the eigenforms, i.e. we get

$$Y_{1i} = \frac{n}{2\pi\alpha}\gamma_{im}\begin{Bmatrix}(-1)^{\frac{n-1}{2}}\\ (-1)^{\frac{n}{2}}\end{Bmatrix}\left[\frac{\begin{Bmatrix}\cosh\alpha_{1i}y\\ \sinh\alpha_{1i}y\end{Bmatrix}}{\begin{Bmatrix}\cosh\alpha_{1i}/2\\ \sinh\alpha_{1i}/2\end{Bmatrix}} - \frac{\begin{Bmatrix}A_{1i}\\ A_{2i}\end{Bmatrix}}{\begin{Bmatrix}B_{1i}\\ B_{2i}\end{Bmatrix}}\right]. \tag{2.155}$$

Summing up formulas (2.151) and (2.152), the first correction term is found:

$$\lambda_1 = 2\pi^2\left(\frac{m^2}{k^2} + 2n^2\gamma_{mm}\right). \tag{2.156}$$

The first correction term of the eigenform is defined by formulas (2.143), (2.153)–(2.155):

$$W_1 = \frac{m}{2\pi\alpha}\left[\cosh\beta_2 x - \coth\beta_2 k \sinh\beta_2 x - \cos\frac{\pi m}{k}x + \frac{1}{k}x\cos\frac{\pi m}{k}x\right]\begin{Bmatrix}\cos\pi n y\\ \sin\pi n y\end{Bmatrix} +$$

$$(-1)^{n+1}\frac{n}{\pi\alpha}\gamma_{mm}\left[\frac{\begin{Bmatrix}(-1)^{\frac{n-1}{2}}\\ (-1)^{\frac{n}{2}}\end{Bmatrix}}{2\begin{Bmatrix}\cosh\beta_1/2\\ \sinh\beta_1/2\end{Bmatrix}}\begin{Bmatrix}\cosh\beta_1 y\\ \sinh\beta_1 y\end{Bmatrix} - Y\begin{Bmatrix}\sin\pi n y\\ \cos\pi n y\end{Bmatrix}\right]\sin\frac{\pi m}{k}x +$$

$$\frac{n}{2\pi\alpha}\begin{Bmatrix}(-1)^{\frac{n-1}{2}}\\ (-1)^{\frac{n}{2}}\end{Bmatrix}\sum_{i=1}^{\infty}\gamma_{im}\left[\frac{\begin{Bmatrix}\cosh\alpha_{1i}\\ \sinh\alpha_{1i}\end{Bmatrix}}{\begin{Bmatrix}\cosh\alpha_{1i}/2\\ \sinh\alpha_{1i}/2\end{Bmatrix}} - \frac{\begin{Bmatrix}A_{1i}\\ A_{2i}\end{Bmatrix}}{\begin{Bmatrix}B_{1i}\\ B_{1i}\end{Bmatrix}}\right]\sin\frac{\pi i}{k}x. \tag{2.157}$$

In an analogous way we find second correction to the eigenvalue. Finally, the eigenvalue being estimated has the following form:

$$\lambda = \pi^4\alpha^2 + 2\pi^2\left(\frac{m^2}{k^2} + 2n^2\gamma_{mm}\right)\varepsilon + (\lambda_{2x} + \lambda_{2y})\varepsilon^2 + \ldots \tag{2.158}$$

where

$$\lambda_{2x} = 2\frac{\pi^2 m^2}{k^2}\left\{1 + \frac{k}{2\pi^2\alpha}\left(\frac{1}{k} - \beta_2\coth\beta_2 k\right) - \frac{\gamma_{mm}}{2\pi^2\alpha}\left(n^2 - \frac{m^2}{k^2}\right) +\right.$$

$$\left.\frac{4}{\pi^2}\cdot\frac{n^2}{m}\sum_{i=1}^{\infty}\gamma_{im}\frac{1}{\left(\frac{i^2-m^2}{k^2}\right)\left(\frac{i^2+m^2}{k^2} + 2n^2\right)} - \frac{\lambda_1}{2\pi^2\alpha^2}\left(\frac{m^2}{k^2} - n^2\right)\right\},$$

$$\lambda_{2y} = 4\pi^2 n^2\left\{\gamma_{mm} + \frac{m}{2\pi\alpha}\left[\frac{1}{k}\cdot\frac{1}{\pi^2\alpha}\left(\beta_2\sinh\beta_2\mu k\sin\pi\mu k -\right.\right.\right.$$

$$\frac{\pi m}{k}\cosh\beta_2\mu k\cos\pi m\mu + \frac{\pi m}{k} - \coth\beta_1 k(\beta_1\cosh\beta_1\mu k\sin\pi m\mu -$$

$$\frac{\pi m}{k}\sinh\beta_1\mu k\cos\pi m\mu)\Big)\Big) + \frac{1}{2\pi m}\left(\frac{1}{2\pi m}\sin 2\pi m\mu +\right.$$

$$(1-\mu)\cos 2\pi m\mu - 1)] - \frac{\gamma_{mm}^2}{\pi^2\alpha}\left(\frac{\beta_1}{2}\tanh\frac{\beta_1}{2} - 1\right) +$$

$$\frac{1}{2\pi^2\alpha}\sum_{i=1}^{\infty}\gamma_{im}^2\left[\alpha_i \coth^{(-1)^n}\frac{\alpha_{1i}}{2} - \left\{\begin{array}{c}\varphi_{1i}\coth^{(-1)^n}\frac{\varphi_{1i}}{2}\\ \\ (-1)^n\beta_{1i}\coth^{(-1)^n}\frac{\beta_{1i}}{2}\end{array}\right\}\right]\Bigg\} -$$

$$\lambda_1\frac{\gamma_{mm}}{8\pi^2\alpha^2}\left(n^2 - \frac{m^2}{k^2}\right).$$

In the next step we transform the results obtained via truncated PS (2.158) into the PA (2.30).

Let us compare the obtained results with known solutions obtained for some limiting cases. For $\mu = 0$ we get a plate clamped on one side and simply supported on the remaining plate sides. In this case the first eigenvalue $\lambda^{(1)}$, obtained with a help of PA in (2.30) is equal to $(1.5520\pi)^4$. The eigenvalue obtained numerically - $\lambda^{(1)} = (1.5501\pi)^4$, and the error is 0.12%.

In the second limiting case ($\mu = 1$) we have a plate simply supported on one side and clamped on the remaining three sides. First eigenvalue obtained through our method yields $\lambda^{(1)} = (1.7963\pi)^4$. The comparison has been carried out with results reported in [53], where the eigenvalues have been obtained using the method of the series, finite differences, R-function method, and the Bolotin method. Largest error achieves 0.76%.

For a plate clamped on its half of the contour $\mu = 0.5$, the PA gives $\lambda^{(1)} = (1.6076\pi)^4$. In reference [53] eigenvalues obtained via the R-function method are reported, and the largest difference is 6.5%. For the eigenvalue obtained via finite difference method [53], the error is 3%.

In Figure 2.15 the dependence $\lambda^{(1)}$ versus parameter μ is reported. The graph has three characteristic zones: [0, 0.3], [0.3, 0.85], [0.85, 1]. In the first and third zones the eigenvalue practically does not depend on the parameter μ. In the second zone $\lambda^{(1)}$ essentially increases with increase of μ. The discussed results show that the occurrence of the clamped side $x = 0$ influences the eigenvalue of the plate vibrations essentially. Occurrence of symmetrically located plate support on opposite plate sides practically does not influence the fundamental frequency

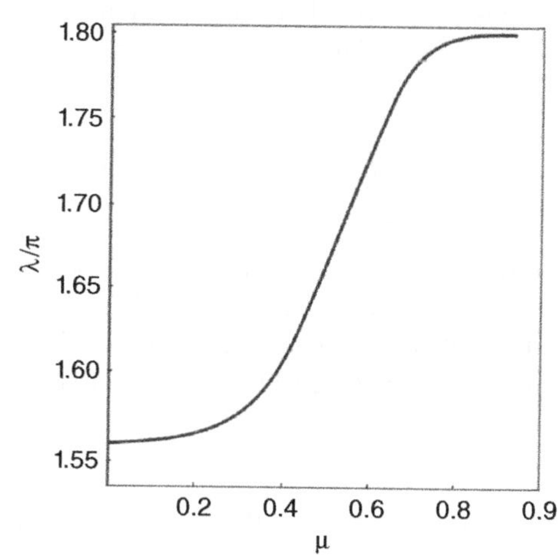

Figure 2.15 Dependence of the first eigenvalue versus the clamping length

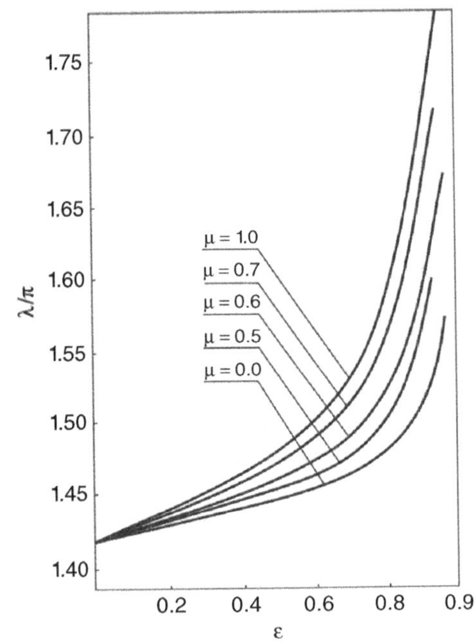

Figure 2.16 Dependence of the eigenvalue $\lambda^{(1)}$ versus parameter ε

up to the value of $\mu \approx 0.3$. In this case, where plate sides $y = \pm 0.5$ are almost completely clamped, occurrence of simply supported parts does not influence the fundamental vibration frequency. In some cases a minor change of the clamping length changes values of the fundamental frequency essentially.

One may also estimate the influence of the clamping stiffness on the fundamental frequency. In Figure 2.16 the dependence of $\lambda^{(1)}$ versus parameter ε for various values of the parameter μ is given. Contrary to the previous problem, influence of the parameter ε on $\lambda^{(1)}$ is essential for the whole interval of the parameter ε variation, although for $0.8 \leq \varepsilon \leq 1.0$ with the increase of ε the eigenvalue increases fast.

2.2.8 *Natural Vibrations of a Plate with Mixed Boundary Conditions "Simple Support-Moving Clamping"*

Let us consider natural vibrations of a rectangular plate ($-0.5k \leq x \leq 0.5k; -0.5 \leq y \leq 0.5$) simply supported on edges $x = \pm 0.5k$, and having mixed BCs "free edge moving clamping." First, we solve the problem for a plate having symmetry in two directions (Figure 2.17a).

Plate vibrations are governed by the PDE (2.63). Let us attach the modified BCs of the form

$$W_{yyy} + (2 - \nu)W_{xxy} = 0,$$

$$W_y = \varepsilon \overline{H}(x)[W_y \mp (W_{yy} + \nu W_{xx})] \quad \text{for} \quad y = \pm 0.5, \tag{2.159}$$

$$W = 0, \quad W_{xx} = 0 \quad \text{for} \quad x = \pm 0.5k, \tag{2.160}$$

where $\overline{H}(x) = H(-x + \mu k) - H(-x - \mu k)$.

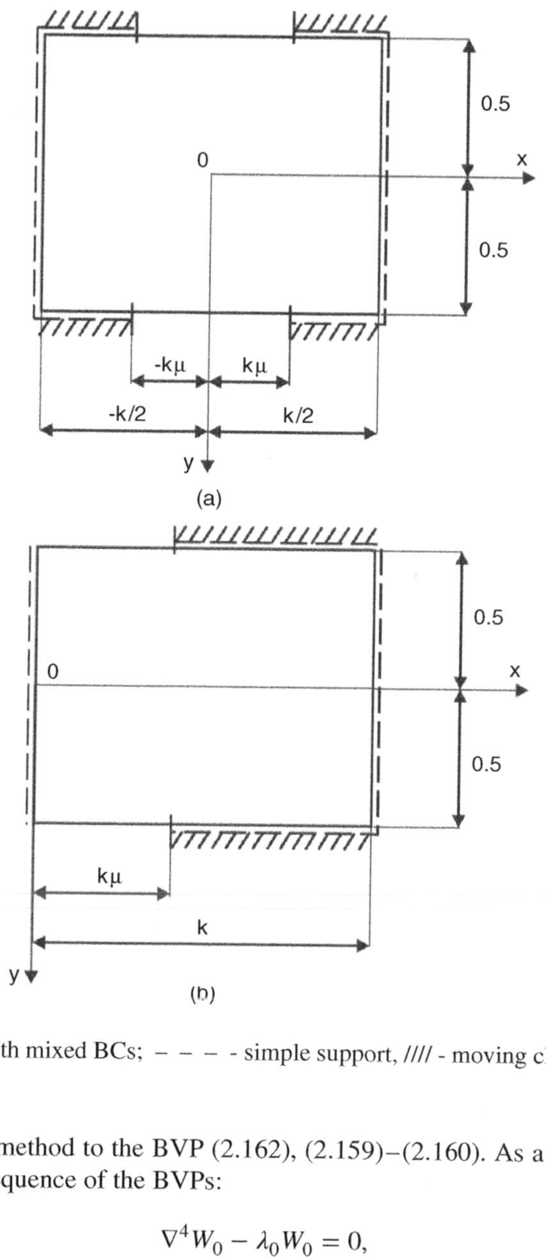

Figure 2.17 Plates with mixed BCs; – – – - simple support, //// - moving clamping, – - free edge

Next we apply our method to the BVP (2.162), (2.159)–(2.160). As a result we obtain the following recurrent sequence of the BVPs:

$$\nabla^4 W_0 - \lambda_0 W_0 = 0,$$

$$W_0 = 0, \quad W_{0xx} = 0 \quad \text{for} \quad x = \pm 0.5k,$$

$$W_{0y} = 0, \quad W_{0yyy} = 0 \quad \text{for} \quad y = \pm 0.5,$$

$$\nabla^4 W_j - \lambda_0 W_j = \sum_{i=1}^{i} \lambda_i W_{j-i},$$

$$W_j = 0, \quad W_{jxx} = 0 \quad \text{for} \quad x = \pm 0.5k,$$

$$W_{jyyy} + (2-\nu)W_{jxxy} = 0,$$

$$W_{jy} = \pm \sum_{i=0}^{j-1}(W_{iyy} + \nu W_{ixx}) \quad \text{for} \quad y = \pm 0.5.$$

Solving the stated problems, the following formula for the eigenvalue is obtained:

$$\lambda = \pi^4\alpha^2 - 4\pi^4\left(n^2 + \nu\frac{m^2}{k^2}\right)\gamma_{mm}\varepsilon + \left\{\lambda_1\left[1 - \frac{\gamma_{mm}}{\alpha}\left(\frac{\pi}{2\beta_1}\left(n^2 - (2-\nu)\frac{m^2}{k^2}\right)\times\right.\right.\right.$$

$$\left.\left.\left.\coth^{(-1)^m}\frac{\pi}{2}\beta_1 - 2\left(n^2 + \nu\frac{m^2}{k^2}\right)\left(\frac{n^2 + (2-\nu)\frac{m^2}{k^2}}{\alpha} - \frac{n^2 + \nu\frac{m^2}{k^2}}{2n^2}\right)\right)\right]\right. +$$

$$4\pi^2\left(n^2 + \nu\frac{m^2}{k^2}\right)\times \sum_{\left\{\begin{matrix} i = 1,3,5,\dots \\ i = 2,4,6,\dots \end{matrix}\right\}} \gamma_{im}^2\left[(1-\nu)\frac{i^2}{k^2} + \alpha\right]\coth^{(-1)^i}\frac{\alpha_{1i}}{2}\cdot\frac{1}{\alpha_{1i}}[(1-$$

$$\left.\nu)\frac{i^2}{k^2} - \alpha\right]^2\left\{\begin{matrix}\coth^{(-1)^i}\frac{\varphi_{1i}}{2}\cdot\frac{1}{\varphi_{1i}} \\ \\ \coth^{(-1)^i}\frac{\beta_{1i}}{2}\cdot\frac{1}{\beta_{1i}}\end{matrix}\right\}\Bigg]\Bigg]\varepsilon^2 + \dots, \quad \left\{\begin{matrix} n,m = 1,3,5,\dots \\ n,m = 2,4,6,\dots \end{matrix}\right\}, \tag{2.161}$$

and the eigenmode W are obtained

$$W = \left\{\begin{matrix}\cos\pi ny\cos\frac{\pi m}{k}x \\ \sin\pi ny\sin\frac{\pi m}{k}x \end{matrix}\right\} +$$

$$\frac{\pi\left(n^2 + \nu\frac{m^2}{k^2}\right)}{\alpha}\left\{\gamma_{mm}\left[\frac{\left(n^2 + (2+\nu)\frac{m^2}{k^2}\right)\left\{\begin{matrix}(-1)^{\frac{n-1}{2}} \\ (-1)^{\frac{n}{2}}\end{matrix}\right\}}{\left\{\begin{matrix}\sinh\pi\beta_1/2 \\ \cosh\pi\beta_1/2\end{matrix}\right\}}\left\{\begin{matrix}\cosh\pi\beta_1 y \\ \sinh\pi\beta_1 y\end{matrix}\right\} +\right.\right.$$

$$\left.(-1)^n\left(n^2 + \nu\frac{m^2}{k^2}\right)Y\left\{\begin{matrix}\sin\pi ny \\ \cos\pi ny\end{matrix}\right\}\right]\left\{\begin{matrix}\sin\frac{\pi m}{k}x \\ \cos\frac{\pi m}{k}x\end{matrix}\right\} +$$

$$\pi\alpha(-1)^{\frac{m-1}{2}}\sum_{\left\{\begin{matrix} i = 1,3,5,\dots \\ i = 2,4,6,\dots \end{matrix}\right\}}^{\infty}\gamma_{im}\times\frac{\left[\frac{(1-\nu)i^2+m^2}{k^2} + n^2\right]}{\alpha_{1i}\left\{\begin{matrix}\sinh\alpha_{1i}/2 \\ \cosh\alpha_{1i}/2\end{matrix}\right\}}\left\{\begin{matrix}\cosh\alpha_{1i}Y \\ \sinh\alpha_{1i}Y\end{matrix}\right\} +$$

$$(-1)^i \frac{\left[\frac{(1-\nu)i^2-m^2}{k^2} - n^2\right]}{\begin{Bmatrix} C_{1i} \\ C_{2i} \end{Bmatrix}} \begin{Bmatrix} D_{1i} \\ D_{2i} \end{Bmatrix} \begin{Bmatrix} \cos \frac{\pi i}{k} x \\ \sin \frac{\pi i}{k} x \end{Bmatrix} \Bigg\} \varepsilon + \ \dots, \begin{Bmatrix} n, m = 1, 3, 5, \dots \\ n, m = 2, 4, 6, \dots \end{Bmatrix}. \quad (2.162)$$

where

$$\gamma_{im} = \begin{cases} 2\mu - \frac{(-1)^m}{4\pi m} \sin 2\pi m\mu \quad \text{for} \quad i = m, \\ \frac{4}{\pi} \cdot \frac{1}{(m^2-i^2)} \left[\begin{Bmatrix} i \\ -m \end{Bmatrix} \sin \pi \mu i \cos \pi \mu m + \right. \\ \left. \begin{Bmatrix} -m \\ i \end{Bmatrix} \sin \pi \mu m \cos \pi m i \right] \quad \text{for} \quad i \neq m, \end{cases}$$

$$\begin{Bmatrix} C_{1i} \\ C_{2i} \end{Bmatrix} = \begin{Bmatrix} \sinh \varphi_{1i}/2 \\ \cosh \varphi_{1i}/2 \end{Bmatrix} \quad \text{for} \quad i^2 > m^2 + k^2 n^2,$$

$$\begin{Bmatrix} C_{1i} \\ C_{2i} \end{Bmatrix} = \begin{Bmatrix} \sin \beta_{1i}/2 \\ \cos \beta_{1i}/2 \end{Bmatrix} \quad \text{for} \quad i^2 < m^2 + k^2 n^2,$$

$$\begin{Bmatrix} D_{1i} \\ D_{2i} \end{Bmatrix} = \begin{Bmatrix} \frac{1}{\varphi_{1i}} \cosh \varphi_{1i} y \\ \frac{1}{\varphi_{1i}} \sinh \varphi_{1i} y \end{Bmatrix} \quad \text{for} \quad i^2 > m^2 + k^2 n^2,$$

$$\begin{Bmatrix} D_{1i} \\ D_{2i} \end{Bmatrix} = \begin{Bmatrix} \frac{1}{\beta_{1i}} \cos \beta_{1i} y \\ \frac{1}{\beta_{1i}} \sin \beta_{1i} y \end{Bmatrix} \quad \text{for} \quad i^2 < m^2 + k^2 n^2.$$

The obtained formulas also have application in the case of nonsymmetric position of the free plate part (Figure 2.17b). For this purpose it is necessary to change $\begin{Bmatrix} \cos(\pi m x/k) \\ \sin(\pi m x/k) \end{Bmatrix}$ by $\sin(\pi m x/k)$, and $\begin{Bmatrix} \cos(\pi i x/k) \\ \sin(\pi i x/k) \end{Bmatrix}$ by $\sin(\pi i x/k)$, $m = 1, 2, 3, \dots$, $i = 1, 2, 3, \dots$ in formulas (2.161) and (2.162).

In this case coefficients γ_{im} are defined by formula (2.134). Furthermore, truncated PS (2.161) is converted into PA (2.30), and we compute its value for $\varepsilon = 1$. Computational results are shown in Figure 2.18. A solid (dashed) curve corresponds to the case of symmetric (nonsymmetric) free edge location. In the limiting case, where edges $y = \pm 0.5$ are free, the solution obtained so far can be compared with a numerical one. For $\nu = 0.3$ the numerical solution yields $\lambda = (1.2758\pi)^4$; the PA solution$-\lambda = (1.2766\pi)^4$ (error of 0.15%). For $\nu = 1/6$ the numerical solution gives $\lambda = (1.3132\pi)^4$; PA - $\lambda = (1.3122\pi)^4$ (error 0.08%). Error regarding determination of eigenvalues decreases with a decrease of the Poisson's ratio. Decrease of the Poisson's ratio causes a shift of the BCs to that of the moving clamping, and hence a contribution of zero order approximation increases in the series of the eigenvalue estimation.

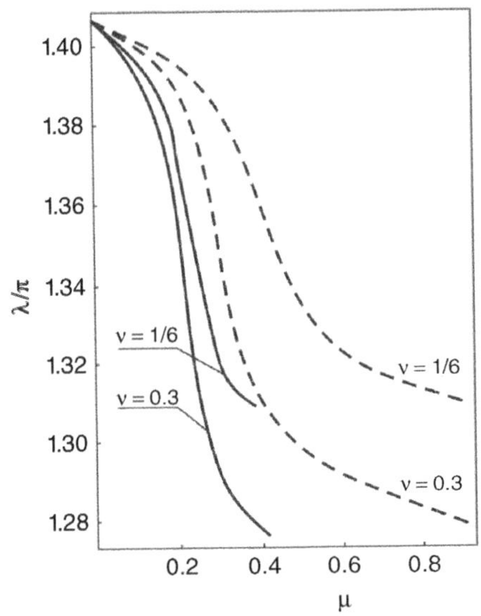

Figure 2.18 Eigenvalue versus length of mixed BCs

Let us consider influence of the geometric dimension of mixed BCs on the eigenfrequency of plate vibrations for $\nu = 0.3$. In Figure 2.18 one may distinguish three parts: [0,0.5], [0.15,0.35], [0.35,0.5] in the case symmetrically located free plate edge and [0,0.2], [0.2,0.6], [0.6,1.0] in the nonsymmetric case. In the first zone the eigenvalue λ depends on geometric dimensions of the plate free edge in essentially nonlinear way, and it decreases negligibly with increase of μ. In the second zone almost linear dependence of λ versus μ is obtained, and small increase of the geometric dimensions of the free plate edge yields remarkable changes of the eigenvalue. In third zone with increase of dimension of the free edge the eigenfrequency decreases insignificantly. This zone is wide one in particular in the case of nonsymmetrically located free edge part. The obtained and discussed so far results show that small dimension BCs do not have significant influence on the eigenavalue.

2.3 Nonlinear Vibrations of Rods, Beams and Plates

2.3.1 *Vibrations of the Rod Embedded in a Nonlinear Elastic Medium*

We begin investigating nonlinear vibrations of spatially finite continuous systems with a study of longitudinal vibrations of the rod embedded into a nonlinear-elastic medium. This will serve as an example for introduction of our asymptotic techniques suitable for solution of similar problems as well as for illustration of some peculiarities of the obtained solution being typical and common for many other problems exhibited by continuous mechanical systems. We study the following equation regarding displacements:

$$a^2 \frac{\partial^2 u}{\partial x^2} - \frac{\partial^2 u}{\partial t^2} - \beta_1 u - \varepsilon \beta_2 u^3 = 0. \qquad (2.163)$$

Rod ends are clamped

$$u|_{x=0,l} = 0, \tag{2.164}$$

where: u is the longitudinal displacement; l is the rod length; $a = \sqrt{E/\rho}$; β_1, β_2 are the coefficients characterizing influence of the external medium; ε is the nondimensional small parameter.

We are aimed on founding a periodic solution of the following form

$$u(x,t) = u(x,t+T), \tag{2.165}$$

where: $T = 2\pi/\omega$ - period, ω is the frequency of vibrations.

We rescale the time

$$\tau = \omega t, \tag{2.166}$$

and we propose the following PS of the solutions being sought:

$$u = u_0 + \varepsilon u_1 + \varepsilon^2 u_2 + \dots, \tag{2.167}$$

$$\omega = \omega_0 + \varepsilon\omega_1 + \varepsilon^2\omega_2 + \dots, \tag{2.168}$$

where $\omega_0 = \sqrt{(a\pi/l)^2 + \beta_1}$ is the fundamental frequency of the associated linear system for $\varepsilon = 0$.

Substituting Ansatzes (2.166)–(2.168) into the BVPs (2.487)–(2.165) and comparing terms standing by the same power of ε, the following recurrent system of linear equations is obtained

$$a^2\frac{\partial^2 u_0}{\partial x^2} - \omega_0^2\frac{\partial^2 u_0}{\partial \tau^2} - \beta_1 u_0 = 0, \tag{2.169}$$

$$a^2\frac{\partial^2 u_1}{\partial x^2} - \omega_0^2\frac{\partial^2 u_1}{\partial \tau^2} - \beta_1 u_1 = 2\omega_0\omega_1\frac{\partial^2 u_0}{\partial \tau^2} + \beta_2 u_0^3, \tag{2.170}$$

....

BCs (2.164) and periodicity conditions (2.165) take the following form

$$u_i|_{x=0,l} = 0, \tag{2.171}$$

$$u_i(x,\tau) = u_i(x,\tau+2\pi), \quad i = 0,1,2,\dots \tag{2.172}$$

Solution of the BVP (2.169), (2.171), (2.172) corresponds to the following zero order approximation:

$$u_0 = \sum_{i=1}^{\infty} A_i \sin\left(\frac{\omega_i^{\text{lin}}}{\omega_0}\tau\right)\sin\left(\frac{\pi i}{l}x\right), \tag{2.173}$$

where: A_1 is the amplitude of the fundamental mode defined by initial conditions; A_j, $j = 2,3,4,\dots$ are the amplitudes of successive harmonics; $\omega_i^{\text{lin}} = \sqrt{(a\pi i/l)^2 + \beta_1}$, $i = 1,2,3,\dots$ are the frequencies of harmonics associated with the corresponding linear case, $\omega_0 = \omega_1^{\text{lin}}$.

Next approximation is found via solution to the BVP (2.170)–(2.172). Note that in order to avoid secular terms, coefficients standing by $\sin\left(\frac{\omega_i^{\text{lin}}}{\omega_0}\tau\right)\sin\left(\frac{\pi i}{l}x\right)$, $i = 1,2,3,\dots$ are set to zero in the r.h.s. of Equation (2.170). Taking into account Ansatz (2.173), this condition yields the infinite system of nonlinear algebraic equations of the form

$$\frac{2A_i\omega_1}{\beta_2\omega_0}\left(\omega_i^{\text{lin}}\right)^2 = \frac{9}{16}A_i^2 + \frac{3}{4}A_i\left(\sum_{k=1}^{i-1}A_k^2 + \sum_{k=i+1}^{\infty}A_k^2\right), \quad i = 1, 2, 3, \ldots. \tag{2.174}$$

Solution to system (2.174) allows defining the second term of PS (2.168), i.e. correction term ω_1 of the frequency being sought generated by the nonlinearity of the problem. In what follows we analyze solutions of system (2.174). Vibrations modes are defined as follows:

$$u = \sum_{i=1}^{\infty} A_i \sin(\Omega_i t)\sin\left(\frac{\pi i}{l}x\right) + O(\varepsilon), \tag{2.175}$$

where $\Omega_i = \frac{\omega_i^{\text{lin}}}{\omega_1^{\text{lin}}}$, ω are the frequencies.

In a general case one deals with only one i-th harmonics, and therefore

$$A_j = 0, \quad \omega_1 = \frac{9A_i^2\beta_2\omega_0}{32\left(\omega_i^{\text{lin}}\right)^2}, \quad j \in N, \quad j \neq i. \tag{2.176}$$

Wanted amplitude-frequency relation has the following form

$$\Omega_i = \omega_i^{\text{lin}} + 0.28125\frac{A_i^2\beta_2}{\omega_i^{\text{lin}}}\varepsilon + O(\varepsilon^2), \tag{2.177}$$

where $i = 1, 2, 3, \ldots$.

Positive (negative) value of ε corresponds to the stiff (weak) characteristic of the restoring force.

Occurrence of internal resonances between harmonics belongs to peculiarities of a continuous system vibrations [8], [9]. In the case considered the phenomenon mentioned so far appears for $\beta_1 = 0$. In this case, the form of nonlinear system (2.174) is changed qualitatively to yield

$$\begin{aligned}
&\frac{2A_1\omega_1}{\beta_2\omega_0}(\omega_1^{\text{lin}})^2 = \frac{9}{16}A_1^3 + \frac{3}{4}A_1(A_2^2 + A_3^2 + A_4^2 + A_5^2) + \\
&\frac{3}{8}(A_1A_2A_4 + A_2A_3A_4 + A_1A_3A_5 + A_2A_4A_5) + \frac{3}{16}(A_1^2A_3 + A_2^2A_3) + \ldots, \\
&\frac{2A_2\omega_1}{\beta_2\omega_0}(\omega_2^{\text{lin}})^2 = \frac{9}{16}A_2^3 + \frac{3}{4}A_2(A_1^2 + A_3^2 + A_4^2 + A_5^2) + \\
&\qquad \frac{3}{8}(A_1A_2A_3 + A_1A_3A_4 + A_1A_2A_5 + A_1A_4A_5 + A_3A_4A_5) + \\
&\qquad \frac{3}{16}(A_1^2A_4 + A_3^2A_4) + \ldots, \\
&\frac{2A_3\omega_1}{\beta_2\omega_0}(\omega_3^{\text{lin}})^2 = \frac{1}{16}A_1^3 + \frac{9}{16}A_3^3 + \frac{3}{4}A_3(A_1^2 + A_2^2 + A_4^2 + A_5^2) + \\
&\frac{3}{8}(A_1A_2A_4 + A_2A_3A_4 + A_1A_3A_5 + A_2A_4A_5) + \\
&\frac{3}{16}(A_1A_2^2 + A_1^2A_5 + A_4^2A_5) + \ldots,
\end{aligned} \tag{2.178}$$

$$\frac{2A_4\omega_1}{\beta_2\omega_0}(\omega_4^{\mathrm{lin}})^2 = \frac{9}{16}A_4^3 + \frac{3}{4}A_4(A_1^2 + A_2^2 + A_3^2 + A_5^2) +$$
$$\frac{3}{8}(A_1A_2A_3 + A_1A_2A_5 + A_2A_3A_5 + A_3A_4A_5) +$$
$$\frac{3}{16}(A_2A_1^2 + A_2A_3^2) + \ldots,$$
$$\frac{2A_5\omega_1}{\beta_2\omega_0}(\omega_5^{\mathrm{lin}})^2 = \frac{9}{16}A_5^3 + \frac{3}{4}A_5(A_1^2 + A_2^2 + A_3^2 + A_4^2) +$$
$$\frac{3}{8}(A_1A_2A_4 + A_2A_3A_4) + \frac{3}{16}(A_1A_2^2 + A_1^2A_3 + A_3^2A_1 + A_3A_4^2) + \ldots,$$

....

An HPM can be applied in order to solve the infinite system of nonlinear algebraic equations (2.178). On the r.h.s. of each i-th equation of the system (2.178), where the following condition is satisfied $(k > i) \cup (l > i) \cup (m > i)$ the μ parameter is introduced before each of the following terms $A_kA_lA_m$, $k, l, m = 1, 2, 3, \ldots$:

$$\frac{2A_1\omega_1}{\beta_2\omega_0}(\omega_1^{\mathrm{lin}})^2 = \frac{9}{16}A_1^3 + \mu\left(\frac{3}{4}A_1(A_2^2 + A_3^2 + A_4^2 + A_5^2) +\right.$$
$$\left.\frac{3}{8}(A_1A_2A_4 + A_2A_3A_4 + A_1A_3A_5 + A_2A_4A_5) + \frac{3}{16}(A_1^2A_3 + A_2^2A_3) + \ldots\right),$$
$$\frac{2A_2\omega_1}{\beta_2\omega_0}(\omega_2^{\mathrm{lin}})^2 = \frac{9}{16}A_2^3 + \frac{3}{4}A_1^2A_2 + \mu\left(\frac{3}{4}A_2(A_3^2 + A_4^2 + A_5^2) +\right.$$
$$\frac{3}{8}(A_1A_2A_3 + A_1A_3A_4 + A_1A_2A_5 + A_1A_4A_5 + A_3A_4A_5) + \tag{2.179}$$
$$\left.\frac{3}{16}(A_1^2A_4 + A_3^2A_4) + \ldots\right),$$
$$\frac{2A_3\omega_1}{\beta_2\omega_0}(\omega_3^{\mathrm{lin}})^2 = \frac{1}{16}A_1^3 + \frac{9}{16}A_3^3 + \frac{3}{4}A_3(A_1^2 + A_2^2) + \frac{3}{16}A_1A_2^2 +$$
$$\mu\left(\frac{3}{4}A_3(A_4^2 + A_5^2) + \frac{3}{8}(A_1A_2A_4 + A_2A_3A_4 + A_1A_3A_5 + A_2A_4A_5) +\right.$$
$$\left.\frac{3}{16}(A_1^2A_5 + A_4^2A_5) + \ldots\right),$$
$$\frac{2A_4\omega_1}{\beta_2\omega_0}(\omega_4^{\mathrm{lin}})^2 = \frac{9}{16}A_4^3 + \frac{3}{4}A_4(A_1^2 + A_2^2 + A_3^2) + \frac{3}{8}A_1A_2A_3 +$$
$$\frac{3}{16}(A_2A_1^2 + A_2A_3^2) + \mu\left(\frac{3}{4}A_4A_5^2 + \frac{3}{8}(A_1A_2A_5 + A_2A_3A_5 + A_3A_4A_5) + \ldots\right),$$
$$\frac{2A_5\omega_1}{\beta_2\omega_0}(\omega_5^{\mathrm{lin}})^2 = \frac{9}{16}A_5^3 + \frac{3}{4}A_5(A_1^2 + A_2^2 + A_3^2 + A_4^2) +$$
$$\frac{3}{8}(A_1A_2A_4 + A_2A_3A_4) + \frac{3}{16}(A_1A_2^2 + A_1^2A_3 + A_3^2A_1 + A_3A_4^2) + \ldots,$$

....

In what follows system (2.179) for $\mu = 0$ takes a triangular form and it can be reduced to the recurrent sequence of equations, whereas for $\mu = 1$ it transits into the input form (2.178).

Unknown quantities are sought in the forms of PS:

$$\omega_1 = \omega_1^{(0)} + \mu\omega_1^{(1)} + \mu^2\omega_1^{(2)} + \dots, \tag{2.180}$$

$$A_j = A_j^{(0)} + \mu A_j^{(1)} + \mu^2 A_j^{(2)} + \dots, \quad j = 2, 3, 4, \dots. \tag{2.181}$$

The first equation of system (2.179) is used for defining the first term of PS (2.180), i.e. $\omega_1^{(0)}$ through the condition of a lack of the secular terms in solution (2.167) being produced by the fundamental mode. Remaining terms $\omega_1^{(j)}, j = 2, 3, 4, \dots$ are defined via lack of secular terms generated by resonance harmonics. Furthermore, our considerations in series are restricted (2.180) to only the first two terms.

A solution to system (2.179) corresponds to the case where all odd harmonics appear simultaneously:

$$\begin{aligned} &A_{2i} = 0, \quad i = 1, 2, 3, \dots, \\ &A_3 = 0.014493151 A_1, \\ &A_5 = 0.000207090 A_1, \\ &\dots\dots\dots\dots\dots\dots, \\ &\omega_1 = 0.282688 A_1^2 \beta_2 / \omega_0. \end{aligned} \tag{2.182}$$

The amplitude – frequency relation follows:

$$\Omega_i = i\omega_0 \left(1 + 0.282688 \frac{A_1^2 \beta_2}{\omega_0^2} \varepsilon \right) + O(\varepsilon^2), \quad i = 1, 3, 5, \dots, \tag{2.183}$$

where $\omega_0 = a\pi/l$.

For $\beta_1 = 0$ solutions (2.182) and (2.183) correspond to the internal resonance between harmonics. Further, we investigate the case, when our system is in a neighborhood of the resonance, i.e. the so called detuning occurs (β_1 in (2.487) tends to zero). The governing Equation (2.487) is cast to the following form

$$a^2 \frac{\partial^2 u}{\partial x^2} - \frac{\partial^2 u}{\partial t^2} - \delta\beta_1^* u - \varepsilon\beta_2 u^3 = 0, \tag{2.184}$$

where β_1^*, β_2 are the some coefficients; $\delta = \beta_1/\beta_1^*$ is the nondimensional detuning small parameter; $\delta \to 0$.

We rescale time in Equation (2.166), and a solution to the BVP (2.164), (2.165), (2.184) is sought in the form of the PS:

$$u = u_0 + \delta u_1 + \delta^2 u_2 + \dots, \tag{2.185}$$

$$\omega = \omega_0 + \delta\omega_1 + \delta^2\omega_2 + \dots, \tag{2.186}$$

terms of which are sought in the form of the following PS

$$u_0 = u_{00} + \varepsilon u_{01} + \varepsilon^2 u_{02} + \dots, \tag{2.187}$$

$$\omega_0 = \omega_{00} + \varepsilon\omega_{01} + \varepsilon^2\omega_{02} + \dots, \tag{2.188}$$

$$u_1 = u_{10} + \varepsilon u_{11} + \varepsilon^2 u_{12} + \ldots, \tag{2.189}$$

$$\omega_1 = \omega_{10} + \varepsilon\omega_{11} + \varepsilon^2\omega_{12} + \ldots, \tag{2.190}$$

where $\omega_{00} = a\pi/l$ is the fundamental frequency for $\varepsilon = 0$ and $\delta = 0$.

Splitting the BVP (2.164), (2.165), (2.184) with respect to powers of δ and ε, results in the following recurrent sequence of linear PDEs:

$$a^2\frac{\partial^2 u_{00}}{\partial x^2} - \omega_{00}^2\frac{\partial^2 u_{00}}{\partial \tau^2} = 0, \tag{2.191}$$

$$a^2\frac{\partial^2 u_{01}}{\partial x^2} - \omega_{00}^2\frac{\partial^2 u_{01}}{\partial \tau^2} = 2\omega_{00}\omega_{01}\frac{\partial^2 u_{00}}{\partial \tau^2} + \beta_2 u_{00}^3, \tag{2.192}$$

$$a^2\frac{\partial^2 u_{10}}{\partial x^2} - \omega_{00}^2\frac{\partial^2 u_{10}}{\partial \tau^2} = 2\omega_{00}\omega_{10}\frac{\partial^2 u_{00}}{\partial \tau^2} + \beta_1^* u_{00}, \tag{2.193}$$

$$a^2\frac{\partial^2 u_{11}}{\partial x^2} - \omega_{00}^2\frac{\partial^2 u_{11}}{\partial \tau^2} = 2\omega_{00}\omega_{10}\frac{\partial^2 u_{01}}{\partial \tau^2} + \beta_1^* u_{01} +$$

$$+ 2(\omega_{01}\omega_{10} + \omega_{00}\omega_{11})\frac{\partial^2 u_{00}}{\partial \tau^2} + 2\omega_{00}\omega_{01}\frac{\partial^2 u_{10}}{\partial \tau^2} + 3\beta_2 u_{00}^2 u_{10}, \tag{2.194}$$

....

BCs (2.164) and periodicity conditions (2.165) are given in the form

$$u_{ij}|_{x=0,l} = 0, \tag{2.195}$$

$$u_{ij}(x,\tau) = u_{ij}(x, \tau + 2\pi), \quad i,j = 0, 1, 2, \ldots. \tag{2.196}$$

The first equation of sequence (2.191) together with conditions (2.195), (2.196) allow defining

$$u_{00} = \sum_{i=1}^{\infty} A_i \sin\left(\frac{\pi i}{l}x\right)\sin(i\tau). \tag{2.197}$$

Second approximation u_{01} can be found from the BVP (2.192), (2.195), (2.196). In order to avoid secular terms in Equation (2.187), set to zero the coefficients standing by terms $\sin\left(\frac{\pi i}{l}x\right)\sin(i\tau)$, $i = 1, 2, 3, \ldots$ in the r.h.s. of Equation (2.192):

$$\frac{2A_1}{\beta_2}\omega_{00}\omega_{01} = \frac{9}{16}A_1^3 + \frac{3}{4}A_1(A_2^2 + A_3^2 + A_4^2 + A_5^2) +$$

$$\frac{3}{8}(A_1A_2A_4 + A_2A_3A_4 + A_1A_3A_5 + A_2A_4A_5) + \frac{3}{16}(A_1^2A_3 + A_2^2A_3) + \ldots,$$

$$\frac{8A_2}{\beta_2}\omega_{00}\omega_{01} = \frac{9}{16}A_2^3 + \frac{3}{4}A_2(A_1^2 + A_3^2 + A_4^2 + A_5^2) +$$

$$\frac{3}{8}(A_1A_2A_3 + A_1A_3A_4 + A_1A_2A_5 + A_1A_4A_5 + A_3A_4A_5) + \tag{2.198}$$

$$\frac{3}{16}(A_1^2A_4 + A_3^2A_4) + \ldots,$$

$$\frac{18A_3}{\beta_2}\omega_{00}\omega_{01} = \frac{1}{16}A_1^3 + \frac{9}{16}A_3^3 + \frac{3}{4}A_3(A_1^2 + A_2^2 + A_4^2 + A_5^2) +$$
$$\frac{3}{8}(A_1A_2A_4 + A_2A_3A_4 + A_1A_3A_5 + A_2A_4A_5) +$$
$$\frac{3}{16}(A_1A_2^2 + A_1^2A_5 + A_4^2A_5) + \ldots,$$
$$\frac{32A_4}{\beta_2}\omega_{00}\omega_{01} = \frac{9}{16}A_4^3 + \frac{3}{4}A_4(A_1^2 + A_2^2 + A_3^2 + A_5^2) +$$
$$\frac{3}{8}(A_1A_2A_3 + A_1A_2A_5 + A_2A_3A_5 + A_3A_4A_5) +$$
$$\frac{3}{16}(A_2A_1^2 + A_2A_3^2) + \ldots,$$
$$\frac{50A_5}{\beta_2}\omega_{00}\omega_{01} = \frac{9}{16}A_5^3 + \frac{3}{4}A_5(A_1^2 + A_2^2 + A_3^2 + A_4^2) +$$
$$\frac{3}{8}(A_1A_2A_4 + A_2A_3A_4) + \frac{3}{16}(A_1A_2^2 + A_1^2A_3 + A_3^2A_1 + A_3A_4^2) + \ldots,$$
$$\ldots\ldots\ldots\ldots .$$

Solving system (2.198) with the help of the HPM, we obtain

$$A_{2i} = 0, \quad i = 1, 2, 3, \ldots,$$
$$A_3 = 0.014493151A_1,$$
$$A_5 = 0.000207090A_1, \tag{2.199}$$
$$\ldots\ldots\ldots,$$
$$\omega_{01} = 0.282688A_1^2\beta_2/\omega_{00}.$$

Function u_{01} is approximated by the series

$$u_{01} = \sum_{i=1}^{\infty} f_i(x)\big(C_i^{(1)}\sin(i\pi) + C_i^{(2)}\cos(i\pi)\big), \tag{2.200}$$

where $C_i^{(1)}$, $C_i^{(2)}$, $f_i(x)$ are some coefficients and functions.

Solution to the BVP (2.193), (2.195), (2.196) allows us to define the term u_{10}. Lack of secular terms in Equation (2.185) requires that coefficients standing by $\sin\frac{\pi i}{l}x\sin(i\tau)$, $i = 1, 2, 3, \ldots$ in the r.h.s. of Equation (2.193) should be equal to zero. Finally, we get

$$\omega_{10} = \frac{\beta_1^*}{2\omega_{00}i^2}, \tag{2.201}$$

$$u_{10} = \sum_{i=1}^{\infty} B_i \sin\left(\frac{i\pi}{l}x\right)\sin(i\pi). \tag{2.202}$$

Observe that here correction term ω_{10} to the frequency of the i-th harmonics depends on the harmonics number.

Term u_{11} is defined through the BVP (2.194)–(2.196). In this case, taking into account the earlier introduced relations for u_{01} in Ansatz (2.200), a condition of lack of secular terms in the PS (2.189) yields the infinite system of equations linear with respect to ω_{11} and B_i, $i = 1, 2, 3, \ldots$ of the form

$$
\begin{aligned}
&\frac{2}{3\beta_2}(\omega_{01}\omega_{10}A_1 + \omega_{00}\omega_{11}A_1 + \omega_{00}\omega_{01}B_1) = \\
&\left(\frac{9}{16}A_1^2 + \frac{1}{4}(A_2^2 + A_3^2 + A_4^2 + A_5^2) + \frac{1}{8}(A_1A_3 + A_2A_4 + A_3A_5)\right)B_1 + \\
&\left(\frac{1}{2}A_1A_2 + \frac{1}{8}(A_2A_3 + A_1A_4 + A_3A_4 + A_2A_5 + A_4A_5)\right)B_2 + \\
&\left(\frac{1}{16}(A_1^2 + A_2^2) + \frac{1}{2}A_1A_3 + \frac{1}{8}(A_2A_4 + A_1A_5 + A_3A_5)\right)B_3 + \\
&\quad\left(\frac{1}{2}A_1A_4 + \frac{1}{8}(A_1A_2 + A_2A_3 + A_2A_5)\right)B_4 + \\
&\quad\left(\frac{1}{16}(A_2^2 + A_3^2) + \frac{1}{2}A_1A_5 + \frac{1}{8}(A_1A_3 + A_2A_4)\right)B_5 + \ldots, \\[2ex]
&\frac{8}{3\beta_2}(\omega_{01}\omega_{10}A_2 + \omega_{00}\omega_{11}A_2 + \omega_{00}\omega_{01}B_2) = \\
&\left(\frac{1}{2}A_1A_2 + \frac{1}{8}(A_2A_3 + A_1A_4 + A_3A_4 + A_2A_5 + A_4A_5)\right)B_1 + \\
&\left(\frac{9}{16}A_2^2 + \frac{1}{4}(A_1^2 + A_3^2 + A_4^2 + A_5^2) + \frac{1}{8}(A_1A_3 + A_1A_5)\right)B_2 + \\
&\left(\frac{1}{2}A_2A_3 + \frac{1}{8}(A_1A_2 + A_1A_4 + A_3A_4 + A_4A_5)\right)B_3 + \\
&\left(\frac{1}{16}(A_1^2 + A_3^2) + \frac{1}{2}A_2A_4 + \frac{1}{8}(A_1A_3 + A_1A_5 + A_3A_5)\right)B_4 + \\
&\left(\frac{1}{2}A_2A_5 + \frac{1}{8}(A_1A_2 + A_1A_4 + A_3A_4)\right)B_5 + \ldots, \\[2ex]
&\frac{6}{\beta_2}(\omega_{01}\omega_{10}A_3 + \omega_{00}\omega_{11}A_3 + \omega_{00}\omega_{01}B_3) = \\
&\left(\frac{1}{16}(A_1^2 + A_2^2) + \frac{1}{2}A_1A_3 + \frac{1}{8}(A_2A_4 + A_1A_5 + A_3A_5)\right)B_1 + \\
&\left(\frac{1}{2}A_2A_3 + \frac{1}{8}(A_1A_2 + A_1A_4 + A_3A_4 + A_4A_5)\right)B_2 + \\
&\left(\frac{9}{16}A_3^2 + \frac{1}{4}(A_1^2 + A_2^2 + A_4^2 + A_5^2) + \frac{1}{8}(A_2A_4 + A_1A_5)\right)B_3 + \\
&\left(\frac{1}{2}A_3A_4 + \frac{1}{8}(A_1A_2 + A_2A_3 + A_2A_5 + A_4A_5)\right)B_4 + \\
&\left(\frac{1}{16}(A_1^2 + A_4^2) + \frac{1}{2}A_3A_5 + \frac{1}{8}(A_1A_3 + A_2A_4)\right)B_5 + \ldots,
\end{aligned}
\tag{2.203}
$$

$$\frac{32}{3\beta_2}(\omega_{01}\omega_{10}A_4+\omega_{00}\omega_{11}A_4+\omega_{00}\omega_{01}B_4)=$$
$$\left(\frac{1}{2}A_1A_4+\frac{1}{8}(A_1A_2+A_2A_3+A_2A_5)\right)B_1+$$
$$\left(\frac{1}{16}(A_1^2+A_3^2)+\frac{1}{2}A_2A_4+\frac{1}{8}(A_1A_3+A_1A_5+A_3A_5)\right)B_2+$$
$$\left(\frac{1}{2}A_3A_4+\frac{1}{8}(A_1A_2+A_2A_3+A_2A_5+A_4A_5)\right)B_3+$$
$$\left(\frac{9}{16}A_4^2+\frac{1}{4}(A_1^2+A_2^2+A_3^2+A_5^2)+\frac{1}{8}A_3A_5\right)B_4+$$
$$\left(\frac{1}{2}A_4A_5+\frac{1}{8}(A_1A_2+A_2A_3+A_3A_4)\right)B_5+\ldots,$$
$$\frac{50}{3\beta_2}(\omega_{01}\omega_{10}A_5+\omega_{00}\omega_{11}A_5+\omega_{00}\omega_{01}B_5)=$$
$$\left(\frac{1}{16}(A_2^2+A_3^2)+\frac{1}{2}A_1A_5+\frac{1}{8}(A_1A_3+A_2A_4)\right)B_1+$$
$$\left(\frac{1}{2}A_2A_5+\frac{1}{8}(A_1A_2+A_1A_4+A_3A_4)\right)B_2+$$
$$\left(\frac{1}{16}(A_1^2+A_4^2)+\frac{1}{2}A_3A_5+\frac{1}{8}(A_1A_3+A_2A_4)\right)B_3+$$
$$\left(\frac{1}{2}A_4A_5+\frac{1}{8}(A_1A_2+A_2A_3+A_3A_4)\right)B_4+$$
$$\left(\frac{9}{16}A_5^2+\frac{1}{4}(A_1^2+A_2^2+A_3^2+A_4^2)\right)B_5+\ldots,$$
$$\ldots\ldots\ldots\ldots\ldots\ldots\ldots\ldots\ldots\ldots\ldots .$$

Finally, we get

$$B_{2i}=0,\quad i=1,2,3,\ldots,$$
$$B_3=0.0144344B_1,$$
$$B_5=0.000200445B_1, \tag{2.204}$$
$$\ldots\ldots\ldots\ldots\ldots\ldots,$$
$$\omega_{11}=0.565352\frac{\beta_2A_1B_1}{\omega_{00}}-0.141344\frac{\beta_1^*\beta_2A_1^2}{i^2\omega_{00}^3},$$

and

$$u=\sum_{i=1}^{\infty}(A_i+\delta B_i)\sin\left(\frac{i\pi}{l}x\right)\sin(\Omega_it)+O(\varepsilon)+O(\varepsilon\delta)+O(\sigma^2), \tag{2.205}$$
$$\Omega_i=\sqrt{i^2\omega_{00}^2+\beta_1^*\delta}+0.282688\frac{i^2\beta_2A_1^2}{\sqrt{i^2\omega_{00}^2+\beta_1^*\delta}}\varepsilon+ \tag{2.206}$$
$$0.565352\frac{i\beta_2A_1B_1}{\omega_{00}}\varepsilon\delta+o(\varepsilon)+o(\delta)+o(\varepsilon\delta),\quad i=1,2,3,\ldots,$$

where $\omega_{00}=a\pi/l$.

The obtained solution coincides with the results obtained in the one above. For instance, for $\delta = 0$ relation (2.206) coincides with formula (2.183), and for $\varepsilon = 0$ with formula (2.177).

2.3.2 *Vibrations of the Beam Lying on a Nonlinear Elastic Foundation*

Let us consider the problem of bending vibrations of the beam lying on a nonlinear-elastic foundation. We assume enough high beam stiffness (in order to neglect occurrence of the longitudinal forces in the beam during its bending). This allows us to study the following PDE:

$$c^2\frac{\partial^4 w}{\partial x^4} + \frac{\partial^2 w}{\partial t^2} + \beta_1 w + \varepsilon\beta_2 w^3 = 0. \tag{2.207}$$

BCs follow

$$\frac{\partial^2 w}{\partial x^2}|_{x=0,l} = w|_{x=0,l} = 0, \tag{2.208}$$

where: w is the transversal beam displacement; $c = \sqrt{EI/\rho S}$, S is the area of the transversal beam cross section.

The solution being sought satisfies the following periodicity condition

$$w(x,t) = w(x,t+T). \tag{2.209}$$

After rescaling of time (2.166), a solution is sought in the form of PS:

$$w = w_0 + \varepsilon w_1 + \varepsilon^2 w_2 + \dots, \tag{2.210}$$

$$\omega = \omega_0 + \varepsilon\omega_1 + \varepsilon^2\omega_2 + \dots, \tag{2.211}$$

where $\omega_0 = \sqrt{\pi^4c^2/l^4 + \beta_1}$ is the eigenfrequency of the fundamental mode of linear system (for $\varepsilon = 0$).

Substituting Ansatzes (2.166), (2.210) and (2.211) into the input BVP (2.207)–(2.209), and after splitting with respect to ε the following recurrent system of linear equations is obtained

$$c^2\frac{\partial^4 w_0}{\partial x^4} + \omega_0^2\frac{\partial^2 w_0}{\partial \tau^2} + \beta_1 w_0 = 0, \tag{2.212}$$

$$c^2\frac{\partial^4 w_1}{\partial x^4} + \omega_0^2\frac{\partial^2 w_1}{\partial \tau^2} + \beta_1 w_1 = -2\omega_0\omega_1\frac{\partial^2 w_0}{\partial \tau^2} - \beta_2 w_0^3, \tag{2.213}$$

....

BCs (2.208) and periodicity conditions (2.209) take the following form

$$\partial^2 w_i/\partial x^2|_{x=o,l} = w_i|_{x=o,l} = 0, \tag{2.214}$$

$$w_i(x,\tau) = w_i(x,\tau+2\pi), \quad i = 0,1,2,\dots. \tag{2.215}$$

Solving the BVP (2.212), (2.214), (2.215) yields zero order approximation

$$w_0 = \sum_{i=1}^{\infty} A_i \sin\left(\frac{\omega_i^{\text{lin}}}{\omega_0}\tau\right)\sin\left(\frac{\pi i}{l}x\right), \quad i = 1,2,3,\dots, \tag{2.216}$$

where $\omega_i^{\text{lin}} = \sqrt{\pi^4c^2i^4/l^4 + \beta_1}$ are eigenfrequencies of the linear system, $\omega_1^{\text{lin}} = \omega_0$.

In order to find the next approximation, the BVP (2.213)–(2.215) should be solved.

We compare coefficients standing by $\sin\left(\frac{\omega_i^{\text{lin}}}{\omega_0}\tau\right)\sin\left(\frac{\pi i}{l}x\right)$, $i = 1, 2, 3, \ldots$ in the r.h.s. of Equation (2.213) to zero to remove secular terms. The conditions mentioned so far yield the infinite system of nonlinear algebraic Equations. (2.174).

Modes of vibrations have the following form

$$w = \sum_{i=1}^{\infty} A_i \sin(\Omega_i t)\sin\left(\frac{\pi i}{l}x\right) + O(\varepsilon), \tag{2.217}$$

where $\Omega_i = \frac{\omega_i^{\text{lin}}}{\omega_1^{\text{lin}}}$.

In a general case, system (2.174) possesses solution (2.176). In this case only one i-th harmonic appears, whose frequency is governed by formula (2.177), where $\omega_i^{\text{lin}} = \sqrt{\pi^4 c^2 i^4/l^4 + \beta_1}$.

For $\beta_1 = 9c^2\pi^4/l^4$ in the system the internal resonance takes place between the first and third harmonics. Owing to the lack of secular terms requirement, the nonlinear system of equations finally takes the following form:

$$\begin{aligned}
\frac{2A_1\omega_1}{\beta_2\omega_0}(\omega_1^{\text{lin}})^2 &= \frac{9}{16}A_1^3 + \frac{3}{16}A_1^2A_3 + \frac{3}{4}A_1\sum_{k=2}^{\infty}A_k^2,\\
\frac{2A_3\omega_1}{\beta_2\omega_0}(\omega_3^{\text{lin}})^2 &= \frac{1}{16}A_1^3 + \frac{9}{16}A_3^3 + \frac{3}{4}A_3\sum_{k=1}^{2}A_k^2 + \frac{3}{4}A_3\sum_{k=4}^{\infty}A_k^2,\\
\frac{2A_i\omega_1}{\beta_2\omega_0}(\omega_i^{\text{lin}})^2 &= \frac{9}{16}A_i^2 + \frac{3}{4}A_i\left(\sum_{k=1}^{i-1}A_k^2 + \sum_{k=i+1}^{\infty}A_k^2\right), \qquad i = 2, 4, 5, 6, \ldots.
\end{aligned} \tag{2.218}$$

In order to solve system (2.218), we again apply the HPM. Unknown quantities are sought through series (2.180), (2.181). Restricting considerations only to the two first terms in PS (2.180) we get

$$\begin{aligned}
A_i &= 0, \quad i = 2, 4, 5, 6, \ldots,\\
A_3 &= 0.0144072A_1,\\
\omega_1 &= 0.282679A_1^2\beta_2/\omega_0.
\end{aligned} \tag{2.219}$$

It should be emphasized that the solution obtained having frequencies (2.220) corresponds to the internal resonance of the first and third harmonics:

$$\Omega_i = i\omega_0\left(1 + 0.282679\frac{A_1^{2\beta_2}}{\omega_0^2}\varepsilon\right) + O(\varepsilon^2), \quad i = 1, 3, \tag{2.220}$$

where $\omega_0 = c\pi^2\sqrt{10}/l$.

Positive (negative) values correspond to stiff (weak) characteristics of the restoring force.

2.3.3 Vibrations of the Membrane on a Nonlinear Elastic Foundation

Now let us consider natural vibrations of the rectangular membrane lying on the nonlinear elastic support. The basic Equation has the following form:

$$a^2\nabla^2 w - \frac{\partial^2 w}{\partial t^2} - \beta_1 w - \varepsilon\beta_2 w^3 = 0. \tag{2.221}$$

Membrane edges are clamped, and hence BCs and periodicity condition have the following form:

$$\begin{aligned} w|_{x=0,l_1} &= w|_{y=0,l_2} = 0, \\ w(x,y,t) &= w(x,y,t+T), \end{aligned} \tag{2.222}$$

where l_1, l_2 are membrane edges length in directions x, y, respectively; $a = \sqrt{N/\rho}$; N is the stretching force.

When in the linear case one half-wave appears on each of membrane sides the eigenfrequencies correspond to a fundamental vibration mode.

We rescale time as in (2.166). A solution to the BVP (2.221), (2.222) is sought in the form of Ansatzes (2.210), (2.211). Now in the PS (2.211) we take $\omega_0 = \sqrt{a^2\left(\frac{\pi^2}{l_1^2} + \frac{\pi^2}{l_2^2}\right) + \beta_1}$ as the eigenfrequency of the fundamental vibration mode of the corresponding linear system. By substitution of the Ansatzes (2.166), (2.210), (2.211) into the input BVP (2.221), (2.222), the following recurrent system of linear equations is obtained:

$$a^2\nabla^2 w_0 - \omega_0^2\frac{\partial^2 w_0}{\partial \tau^2} - \beta_1 w_0 = 0, \tag{2.223}$$

$$a^2\nabla^2 w_1 - \omega_0^2\frac{\partial^2 w_1}{\partial \tau^2} - \beta_1 w_1 = 2\omega_0\omega_1\frac{\partial^2 w_0}{\partial \tau^2} + \beta_2 w_0^3, \tag{2.224}$$

....

BCs and periodicity condition (2.222) take the form

$$\begin{aligned} w_i|_{x=0,l_1} &= w_i|_{y=0,l_2} = 0, \\ w_i(x,y,\tau) &= w_i(x,y,\tau+2\pi), \quad i = 0,1,2,\dots. \end{aligned} \tag{2.225}$$

Zero order approximation is yielded via a solution to the BVP (2.223), (2.225):

$$w_0 = \sum_{m=1}^{\infty}\sum_{n=1}^{\infty} A_{m,n}\sin\left(\frac{\omega_{m,n}^{\text{lin}}}{\omega_0}\tau\right)\sin\left(\frac{\pi m}{l_1}x\right)\sin\left(\frac{\pi n}{l_2}\right), \tag{2.226}$$

where $A_{1,1}$ is the amplitude of the fundamental vibrations mode; $A_{m,n}$, $m,n = 1,2,3,\dots$, $m,n \neq (1,1)$ are the amplitudes of the successive harmonics; $\omega_{m,n}^{\text{lin}} = \sqrt{a^2\left(\frac{\pi^2 m^2}{l_1^2} + \frac{\pi^2 n^2}{l_2^2}\right) + \beta_1}$, $m,n = 1,2,3,\dots$ are the eigenfrequencies of the linear system vibrations, $\omega_{1,1}^{\text{lin}} = \omega_0$.

In order to find the next approximation we need to solve the BVP (2.224), (2.225). The condition of absence of secular terms requires that on the r.h.s. of Equation (2.224) the coefficients near the terms $\sin\left(\frac{\omega^{\text{lin}}_{m,n}}{\omega_0}\tau\right)\sin\left(\frac{\pi m}{l_1}x\right)\sin\left(\frac{\pi n}{l_2}\right)$, $m,n=1,2,3,\dots$ should be equal to zero. As the result the following system of nonlinear algebraic equations is obtained:

$$\begin{aligned}
\frac{2A_{1,1}}{\beta_2}\omega_0\omega_1 &= \frac{27}{64}A^3_{1,1}+\frac{9}{16}A_{1,1}(A^2_{1,2}+A^2_{2,1})+\frac{3}{8}A_{1,1}A^2_{2,2}+\dots,\\
\frac{2A_{1,2}}{\beta_2}\frac{\omega_1}{\omega_0}(\omega^{\text{lin}}_{1,2})^2 &= \frac{27}{64}A^3_{1,2}+\frac{9}{16}A_{1,2}(A^2_{1,1}+A^2_{2,2})+\frac{3}{8}A_{1,2}A^2_{2,1}+\dots,\\
\frac{2A_{2,1}}{\beta_2}\frac{\omega_1}{\omega_0}(\omega^{\text{lin}}_{2,1})^2 &= \frac{27}{64}A^3_{2,1}+\frac{9}{16}A_{2,1}(A^2_{1,1}+A^2_{2,2})+\frac{3}{8}A_{2,1}A^2_{1,2}+\dots,\\
\frac{2A_{2,2}}{\beta_2}\frac{\omega_1}{\omega_0}(\omega^{\text{lin}}_{2,2})^2 &= \frac{27}{64}A^3_{2,2}+\frac{9}{16}A_{2,2}(A^2_{1,2}+A^2_{2,1})+\frac{3}{8}A_{2,2}A^2_{1,1}+\dots,\\
&\dots\dots\dots\dots\dots\dots\dots\dots\dots\dots\dots\dots
\end{aligned}\tag{2.227}$$

Vibrations mode is defined by the series

$$w=\sum_{m=1}^{\infty}\sum_{n=1}^{\infty}A_{m,n}\sin(\Omega_{m,n}t)\sin\left(\frac{\pi m}{l_1}x\right)\sin\left(\frac{\pi n}{l_2}y\right)+O(\varepsilon),\tag{2.228}$$

where $\Omega_{m,n}=\frac{\omega^{\text{lin}}_{m,n}}{\omega^{\text{lin}}_{1,1}}\omega$ are the frequencies of the corresponding harmonics.

A solution to the system (2.227) corresponds to the case where only one harmonic with the numbers m,n appears

$$A_{i,j}=0,\quad i,j=1,2,3,\dots,\quad (i,j)\neq(m,n);\quad \omega_1=\frac{27}{128}\frac{A^2_{m,n}\beta_2\omega_0}{(\omega^{\text{lin}}_{m,n})^2}.\tag{2.229}$$

In the case of frequency $\Omega_{m,n}$, we obtain

$$\Omega_{m,n}=\omega^{\text{lin}}_{m,n}+0.2109375\frac{A^2_{m,n}\beta_2}{\omega^{\text{lin}}_{m,n}}\varepsilon+O(\varepsilon^2),\tag{2.230}$$

where $m,n=1,2,3,\dots$.

Positive (negative) values of ε correspond to stiff (weak) restoring force value. However, a solution changes qualitatively if the linear part of the restoring force is equal to zero ($\beta_1=0$). In this case the internal resonance occurs between the natural vibration harmonics. A condition of the absence of secular terms yields the following set of nonlinear algebraic equations:

$$\begin{aligned}
&\frac{2A_{1,1}}{\beta_2}\omega_0\omega_1=\frac{27}{64}A^3_{1,1}+\frac{3}{8}A_{1,1}(A^2_{2,2}+A^2_{3,3}+A^2_{4,4}+A^2_{5,5})+\\
&\frac{3}{32}(-A_{1,1}A_{2,2}A_{4,4}+A_{2,2}A_{3,3}A_{4,4}-A_{1,1}A_{3,3}A_{5,5}+A_{2,2}A_{4,4}A_{5,5})+\\
&\frac{3}{64}(-A^2_{1,1}A_{3,3}+A^2_{2,2}A_{3,3}-A^2_{2,2}A_{5,5}+A^2_{3,3}A_{5,5})+\dots,\\
&\frac{2A_{2,2}}{\beta_2}\frac{\omega_1}{\omega_0}(\omega^{\text{lin}}_{2,2})^2=\frac{27}{64}A^3_{2,2}+\frac{3}{8}A_{2,2}(A^2_{1,1}+A^2_{3,3}+A^2_{4,4}+A^2_{5,5})+\\
&\frac{3}{32}(A_{1,1}A_{2,2}A_{3,3}+A_{1,1}A_{3,3}A_{4,4}-A_{1,1}A_{2,2}A_{5,5}+A_{1,1}A_{4,4}A_{5,5}+
\end{aligned}$$

$$A_{3,3}A_{4,4}A_{5,5} + \frac{3}{64}(-A_{1,1}^2A_{4,4} + A_{3,3}^2A_{4,4}) + \dots,$$

$$\frac{2A_{3,3}}{\beta_2}\frac{\omega_1}{\omega_0}(\omega_{3,3}^{\text{lin}})^2 = -\frac{1}{64}A_{1,1}^3 + \frac{27}{64}A_{3,3}^3 + \frac{3}{8}A_{3,3}(A_{1,1}^2 + A_{2,2}^2 + A_{4,4}^2 + A_{5,5}^2)+$$

$$\frac{3}{32}(A_{1,1}A_{2,2}A_{4,4} + A_{2,2}A_{3,3}A_{4,4} + A_{1,1}A_{3,3}A_{5,5} + A_{2,2}A_{4,4}A_{5,5})+$$

$$\frac{3}{64}(A_{1,1}A_{2,2}^2 - A_{1,1}^2A_{5,5} + A_{4,4}^2A_{5,5}) + \dots,$$

$$\frac{2A_{4,4}}{\beta_2}\frac{\omega_1}{\omega_0}(\omega_{4,4}^{\text{lin}})^2 = \frac{27}{64}A_{4,4}^3 + \frac{3}{8}A_{4,4}(A_{1,1}^2 + A_{2,2}^2 + A_{3,3}^2 + A_{5,5}^2)+$$

$$\frac{3}{32}(A_{1,1}A_{2,2}A_{3,3} + A_{1,1}A_{2,2}A_{5,5} + A_{2,2}A_{3,3}A_{5,5} + A_{3,3}A_{4,4}A_{5,5})+$$

$$\frac{3}{64}(-A_{1,1}^2A_{2,2} + A_{2,2}A_{3,3}^2) + \dots,$$

$$\frac{2A_{5,5}}{\beta_2}\frac{\omega_1}{\omega_0}(\omega_{5,5}^{\text{lin}})^2 = \frac{27}{64}A_{5,5}^3 + \frac{3}{8}A_{5,5}(A_{1,1}^2 + A_{2,2}^2 + A_{3,3}^2 + A_{4,4}^2)+$$

$$\frac{3}{32}(A_{1,1}A_{2,2}A_{4,4} + A_{2,2}A_{3,3}A_{4,4})+ \tag{2.231}$$

$$\frac{3}{64}(-A_{1,1}A_{2,2}^2 + A_{1,1}^2A_{3,3} + A_{1,1}A_{3,3}^2 + A_{3,3}A_{4,4}^2) + \dots,$$

$$\dots\dots\dots\dots\dots\dots\dots\dots\dots\dots .$$

The solution to system (2.231) is sought with the help of the HPM. Namely, we introduce the homotopy parameter μ into the r.h.s. of system (2.231). Unknown quantities are written in the form of the following PS

$$\omega_1 = \omega_1^{(0)} + \mu\omega_1^{(1)} + \mu^2\omega_1^{(2)} + \dots, \tag{2.232}$$

$$A_{m,n} = A_{m,n}^{(0)} + \mu A_{m,n}^{(1)} + \mu^2 A_{m,n}^{(2)} + \dots, \quad m,n = 1,2,3,\dots \quad (m,n) \neq (1,1). \tag{2.233}$$

System (2.231) gives a solution, where all even "diagonal" harmonics appear simultaneously

$$\begin{aligned} &A_{m,n} = 0, \\ &\quad m,n = 1,2,3,\dots \quad (m,n) \neq (1,1), \\ &\quad (m,n) \neq (2i-1, 2i-1), \quad i = 1,2,3,\dots, \\ &A_{3,3} = -0.004566222A_{1,1}, \\ &A_{5,5} = 0.000021139A_{1,1}, \\ &\dots\dots\dots\dots \\ &\omega_1 = 0.211048A_{1,1}^2\beta_2/\omega_0. \end{aligned} \tag{2.234}$$

The amplitude-frequency relation required possesses the following effective asymptotic formula

$$\Omega_{m,n} = m\omega_0\left(1 + 0.211048\frac{A_{1,1}^2\beta_2}{\omega_0^2}\varepsilon\right) + O(\varepsilon^2), \quad m = n = 1,3,5,\dots, \tag{2.235}$$

where $\omega_0 = \sqrt{\frac{\pi^2}{l_1^2} + \frac{\pi^2}{l_2^2}}$.

2.3.4 Vibrations of the Plate on a Nonlinear Elastic Foundation

Now consider the case of the rectangular simply supported plate lying on a nonlinear elastic foundation. Basic Equation is

$$D\nabla^4 w + \rho h \frac{\partial^2 w}{\partial t^2} + \beta_1 w + \varepsilon \beta_2 w^3 = 0, \tag{2.236}$$

whereas BCs and periodicity conditions are cast in the form

$$\begin{gathered} w|_{x=0,l_1} = w|_{y=0,l_2} = 0, \quad \left|\frac{\partial^2 w}{\partial x^2}\right|_{x=0,l_1} = \left|\frac{\partial^2 w}{\partial y^2}\right|_{y=0,l_2} = 0, \\ w(x,y,t) = w(x,y,t+T). \end{gathered} \tag{2.237}$$

Proceeding in a way similar to that of the previous section, we are looking for frequencies corresponding to such a form of the fundamental vibrations mode, where in the linear case only vibrations exhibiting one half-wave in the direction of each plate sides is realized.

Then, we change the time scale regarding (2.166). The solution to the BVP (2.236), (2.237) is approximated through Ansatzes (2.210), (2.211). In the PS (2.211) we take the following eigenfrequency of the fundamental mode of linear vibrations: $\omega_0 = \sqrt{\frac{D}{\rho h}\left(\frac{\pi^2}{l_1^2} + \frac{\pi^2}{l_2^2}\right) + \frac{\beta_1}{\rho h}}$.
Substituting Ansatzes (2.166), (2.210) and (2.211) into the basic BVP (2.236), (2.237), after splitting with respect to ε, the following recurrent system of equations is obtained:

$$D\nabla^4 w_0 + \rho h \omega_0^2 \frac{\partial^2 w_0}{\partial \tau^2} + \beta_1 w_0 = 0, \tag{2.238}$$

$$D\nabla^4 w_1 + \rho h \omega_0^2 \frac{\partial^2 w_1}{\partial \tau^2} + \beta_1 w_1 = -2\rho h \omega_0 \omega_1 \frac{\partial^2 w_0}{\partial \tau^2} - \beta_2 w_0^3, \tag{2.239}$$

$$\dots\dots\dots\dots\dots\dots\dots\dots .$$

BCs and periodicity conditions (2.237) have the following form:

$$\begin{gathered} w_i|_{x=0,l_1} = w_i|_{y=0,l_2} = 0, \quad \left|\frac{\partial^2 w_i}{\partial x^2}\right|_{x=0,l_1} = \left|\frac{\partial^2 w_i}{\partial y^2}\right|_{y=0,l_2} = 0, \\ w_i(x,y,\tau) = w_i(x,y,\tau+2\pi), \quad i = 0,1,2,\dots. \end{gathered} \tag{2.240}$$

Solution to BVP (2.238), (2.240), corresponding to zero order approximation, is defined via formula (2.226), where $\omega_{m,n}^{\text{lin}} = \sqrt{\frac{D}{\rho h}\left(\frac{\pi^2 m^2}{l_1^2} + \frac{\pi^2 n^2}{l_2^2}\right) + \frac{\beta_1}{\rho h}}$, $m,n = 1,2,3,\dots$ are the eigenfrequencies of vibrations of the associated linear system (for $\varepsilon = 0$), $\omega_{1,1}^{\text{lin}} = \omega_0$.

Next approximation is found solving the BVP (2.239), (2.240). In order to exclude secular terms, coefficients standing by the terms $\sin\left(\frac{\omega_{m,n}^{\text{lin}}}{\omega_0}\tau\right)\sin\left(\frac{\pi m}{l_1}x\right)\sin\left(\frac{\pi n}{l_2}y\right)$, $m,n = 1,2,3,\dots$ on the r.h.s. of Equation (2.75) are set to zero. The latter condition produces the following infinite system of nonlinear algebraic equations:

$$\frac{2\rho h A_{1,1}}{\beta_2}\omega_0\omega_1 = \frac{27}{64}A_{1,1}^3 + \frac{9}{16}A_{1,1}(A_{1,2}^2 + A_{2,1}^2) + \frac{3}{8}A_{1,1}A_{2,2}^2 + \dots,$$
$$\frac{2\rho h A_{1,2}}{\beta_2}\frac{\omega_1}{\omega_0}(\omega_{1,2}^{\text{lin}})^2 = \frac{27}{64}A_{1,2}^3 + \frac{9}{16}A_{1,2}(A_{1,1}^2 + A_{2,2}^2) + \frac{3}{8}A_{1,2}A_{2,1}^2 + \dots,$$
$$\frac{2\rho h A_{2,1}}{\beta_2}\frac{\omega_1}{\omega_0}(\omega_{2,1}^{\text{lin}})^2 = \frac{27}{64}A_{2,1}^3 + \frac{9}{16}A_{2,1}(A_{1,1}^2 + A_{2,2}^2) + \frac{3}{8}A_{1,2}^2A_{2,1} + \dots,$$
$$\frac{2\rho h A_{2,2}}{\beta_2}\frac{\omega_1}{\omega_0}(\omega_{2,2}^{\text{lin}})^2 = \frac{27}{64}A_{2,2}^3 + \frac{9}{16}A_{2,2}(A_{1,2}^2 + A_{2,1}^2) + \frac{3}{8}A_{1,1}^2A_{2,2} + \dots, \tag{2.241}$$
$$\dots\dots\dots\dots\dots\dots .$$

Similarly to the previous case, the vibration form is defined by formula (2.228). Solution to system (2.241) has the following form

$$A_{i,j} = 0, \quad i,j = 1,2,3,\dots, \quad (i,j) \neq (m,n),$$
$$\omega_1 = \frac{27}{128}\frac{A_{m,n}^2\beta_2\omega_0}{\rho h(\omega_{m,n}^{\text{lin}})^2}.$$

Only one harmonic with numbers (m,n) appears, and its frequency is

$$\Omega_{m,n} = \omega_{m,n}^{\text{lin}} + 0.2109375\frac{A_{m,n}^2\beta_2}{\rho h\omega_{m,n}^{\text{lin}}}\varepsilon + O(\varepsilon^2), \quad m,n = 1,2,3,\dots.$$

Positive (negative) values of ε corresponds to stiff (weak) characteristics of the restoring force.

For $\beta_1 = 9\pi^4 D(1/l_1^2 + 1/l_2^2)^2$ a solution to the problem changes qualitatively. Namely, in this case simultaneously appear harmonics $(1,1)$ and $(3,3)$. Absence of secular terms yields the following system of nonlinear equations:

$$\frac{2\rho h A_{1,1}}{\beta_2}\omega_0\omega_1 = \frac{27}{64}A_{1,1}^3 - \frac{3}{64}A_{1,1}^2A_{3,3} +$$
$$\frac{3}{8}A_{1,1}(A_{2,2}^2 + A_{3,3}^2 + A_{4,4}^2 + A_{5,5}^2) + \dots,$$
$$\frac{2\rho h A_{2,2}}{\beta_2}\frac{\omega_1}{\omega_0}(\omega_{2,2}^{\text{lin}})^2 = \frac{27}{64}A_{2,2}^3 +$$
$$\frac{3}{8}A_{2,2}(A_{1,1}^2 + A_{3,3}^2 + A_{4,4}^2 + A_{5,5}^2) + \dots,$$
$$\frac{2\rho h A_{3,3}}{\beta_2}\frac{\omega_1}{\omega_0}(\omega_{3,3}^{\text{lin}})^2 = -\frac{1}{64}A_{1,1}^3 + \frac{27}{64}A_{3,3}^3 +$$
$$\frac{3}{8}A_{3,3}(A_{1,1}^2 + A_{2,2}^2 + A_{4,4}^2 + A_{5,5}^2) + \dots,$$
$$\frac{2\rho h A_{4,4}}{\beta_2}\frac{\omega_1}{\omega_0}(\omega_{4,4}^{\text{lin}})^2 = \frac{27}{64}A_{4,4}^3 + \tag{2.242}$$

$$\frac{3}{8}A_{4,4}(A_{1,1}^2 + A_{2,2}^2 + A_{3,3}^2 + A_{5,5}^2) + \dots,$$

$$\frac{2\rho h A_{5,5}}{\beta_2}\frac{\omega_1}{\omega_0}(\omega_{5,5}^{\text{lin}})^2 = \frac{27}{64}A_{5,5}^3 +$$

$$\frac{3}{8}A_{5,5}(A_{1,1}^2 + A_{2,2}^2 + A_{3,3}^2 + A_{4,4}^2) + \dots,$$

$$\dots\dots\dots\dots\dots\dots\dots\dots\dots\dots\dots\dots .$$

In order to solve system (2.242) we introduce the homotopy parameter μ on its r.h.s. in a way similar to that described previously. Unknown quantities are rewritten in the form of PS (2.232), (2.233). By taking in PS (2.232) only the first two terms, the following solution to system (2.242) is obtained:

$$A_{i,j} = 0, \quad i,j = 1,2,3,\dots, \quad (i,j) \neq (1,1),(3,3),$$

$$A_{3,3} = -0.004566222A_{1,1}, \quad \omega_1 = 0.211048\frac{A_{1,1}^2\beta_2}{\rho h\omega_0}.$$

Frequencies of the harmonics follow:

$$\Omega_{m,n} = m\omega_0\left(1 + 0.211048\frac{A_{1,1}^2\beta_2}{\rho h\omega_0^2}\varepsilon\right) + O(\varepsilon^2), \quad m = n = 1,3,$$

where $\omega_0 = \sqrt{10\frac{\pi^4 D}{\rho h}}\left(\frac{1}{l_1^2} + \frac{1}{l_2^2}\right)$.

2.4 SSS of Beams and Plates

2.4.1 *SSS of Beams with Clamped Ends*

In this section we apply our approach to compute SSS of a beam ($-0.5l \leq \bar{x} \leq 0.5l$) clamped on its ends. The beam is loaded by the uniformly distributed load $\bar{q}$. After nondimensional procedure owing to (2.8), the basic Equation takes the following form

$$W^{IV} = q, \qquad q = \frac{\bar{q}l^3}{EI}. \tag{2.243}$$

In order to close the BVP associated with the given equation, we attach the BCs (2.15). Exact solution to BVP (2.243), (2.15) takes the following form:

$$W = \frac{q}{24}x^4 - \frac{q}{48}\frac{6-5\varepsilon}{2-\varepsilon}x^2 + \frac{q}{384}\frac{10-9\varepsilon}{2-\varepsilon}. \tag{2.244}$$

In what follows we are going to compare how the solution obtained via our approach coincides with the exact one governed by (2.244). First, we present the displacement W in the form of (2.16). Substituting this Ansatz into Equation (2.243) and BCs (2.15), and splitting with respect to powers of ε, the following recurrent sequence of the BVPs is obtained:

$$W_0^{IV} = q;$$
$$W_0 = 0, \quad W_0^{II} = 0 \quad \text{for} \quad x = \pm 0.5,$$
$$W_j^{IV} = 0;$$
$$W_j = 0, \quad W_j^{II} = \mp \sum_{i=0}^{j-1} W_i^{I} \quad \text{for} \quad x = \pm 0.5, \quad j = 1, 2, 3, \ldots \quad .$$

From BVPs one obtains:

$$W = \frac{q}{24} x^4 + C_1 x^2 + C_2, \tag{2.245}$$

where

$$C_1 = \frac{q}{24}\left(-\frac{3}{2} + \sum_{j=1}^{\infty} \frac{1}{2^j}\varepsilon^j\right), \tag{2.246}$$

$$C_2 = \frac{q}{96}\left(\frac{5}{4} - \sum_{j=1}^{\infty} \frac{1}{2^j}\varepsilon^j\right). \tag{2.247}$$

The series in formulas (2.246) and (2.247) is the geometric progression with the denominator 0.5ε, having the radius of convergence $\varepsilon = 2$, and the following sum:

$$\sum_{j=1}^{\infty} \frac{\varepsilon^j}{2^j} = \frac{\varepsilon}{2 - \varepsilon}. \tag{2.248}$$

Due to Equation (2.248), formula (2.245) coincides with formula (2.244). Therefore, in the studied case our approach gives the exact solution to the problem. However, in practice for the majority of the cases only a few of the first terms of the PS can be constructed. This is why we apply PA to the PS.

Let us verify benefits yielded by the PA for the studied case. Let us take three first terms of the PS for coefficients C_1 and C_2, and let us apply the PA:

$$C_{1[1/1]}(\varepsilon) = -\frac{q}{48}\frac{6 - 5\varepsilon}{2 - \varepsilon}, \qquad C_{2[1/1]}(\varepsilon) = \frac{q}{384}\frac{10 - 9\varepsilon}{2 - \varepsilon}.$$

We get the exact solution. Now let us estimate what is the difference between solutions obtained via the PA and PS. For this purpose we estimate the bending moment using the exact solution (2.244)

$$M = -\left[\frac{q}{2}x^2 - \frac{q}{12}\frac{6 - 5\varepsilon}{2 - \varepsilon}\right]. \tag{2.249}$$

Next by taking three first terms we estimate the error in determination of the displacement and bending moment at the beam center and at the beam end via the PS. In Figure 2.19 computational results are shown. One may observe that the largest error is obtained in the beams center (for $\varepsilon = 1$ error achieves 100%). Error regarding the bending moment in the beam center (edge) is 50% (25%) for $\varepsilon = 1$. On the other hand for the value of $\varepsilon < 0.5$, the error associated with the SSS factors estimation is less than 5%.

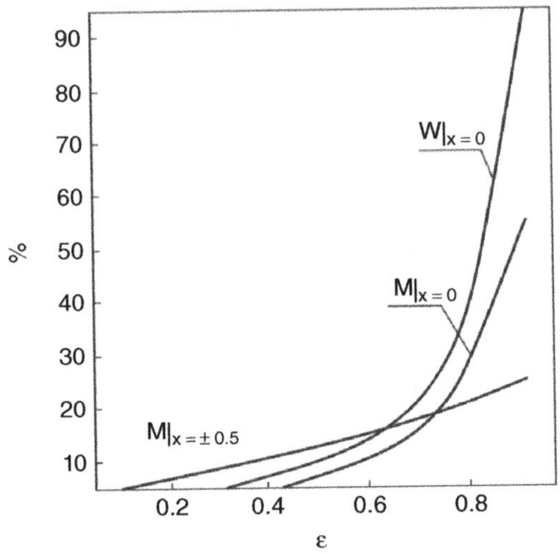

Figure 2.19 Estimation of the PS accuracy

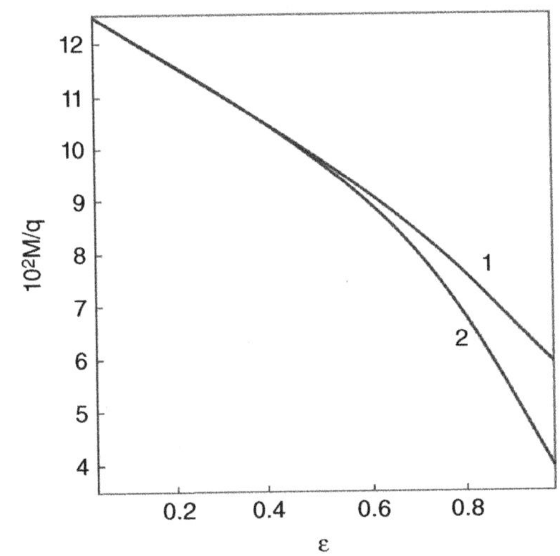

Figure 2.20 Efficiency comparison of PS and PA, $M|_{x=0}$

In Figures 2.20–2.22 various factors of the SSS versus ε are presented for an exact solution (2.244), (2.249) (curve 2) and the solution is obtained with the help of PS (curve 1). Largest error between them is achieved in the interval from $\varepsilon = 0.8$ up to $\varepsilon = 1.0$. In this case the PA allows us to achieve the exact solutions.

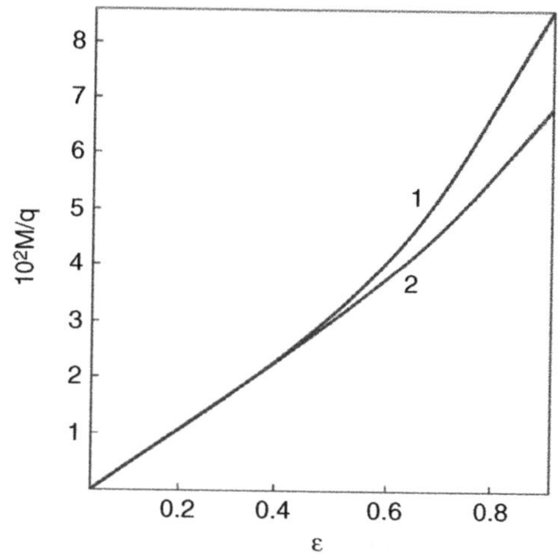

Figure 2.21 Efficiency comparison of PS and PA, $M|_{x=\pm 0.5}$

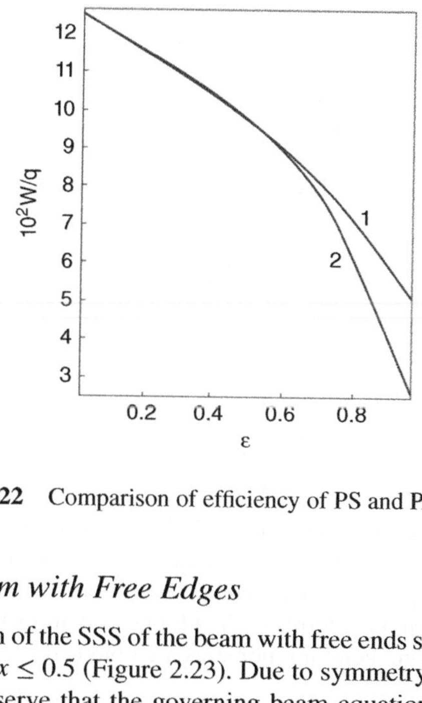

Figure 2.22 Comparison of efficiency of PS and PA, $W|_{x=0}$

2.4.2 SSS of the Beam with Free Edges

Let us focus on computation of the SSS of the beam with free ends subjected to a self-balanced load $q(x) = q(4x - 1)$, $0 \leq x \leq 0.5$ (Figure 2.23). Due to symmetry of the problem, we study only half of the beam. Observe that the governing beam equation coincides with Equation (2.243) assuming that instead of $q = const$ we take $q(x) = q(4x - 1)$ for $0 \leq x \leq 0.5$. BCs take the form (2.41). The exact solution to the BVP (2.243), (2.41) reads:

$$W = \frac{q}{6}x^4\left(\frac{x}{5} - \frac{1}{4}\right) + \frac{q}{96}\frac{1+3\varepsilon}{1+\varepsilon}x^2 \quad \text{for} \quad 0 \leq x \leq 0.5. \tag{2.250}$$

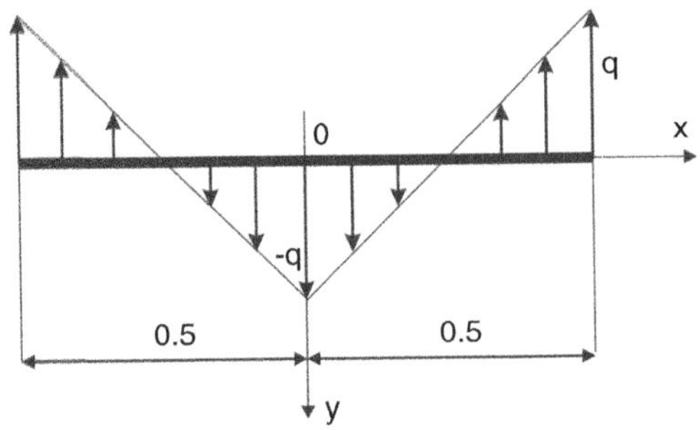

Figure 2.23 Beam subjected to self-balanced load

We assume the displacement W in the form of a PS. After splitting with respect to ε, the following recurrent set of BVPs is obtained:

$$W_0^{IV} = q(x);$$

$$W_0^{III} = 0, \quad W_0^{I} = 0 \quad \text{for} \quad x = \pm 0.5,$$

$$W_j^{IV} = 0;$$

$$W_j^{III} = 0, \quad W_j^{I} = \mp \sum_{i=0}^{j-1} W_i^{I} \quad \text{for} \quad x = \pm 0.5, \quad j = 1, 2, 3, \ldots \quad .$$

Successively solving the obtained BVPs one finally obtains the beam dsplacement as

$$W = \frac{q}{6} x^4 \left(\frac{x}{5} - \frac{1}{4} \right) + \frac{q}{48} \left(\frac{1}{2} + \sum_{j=1}^{\infty} (-1)^{j-1} \varepsilon \right). \tag{2.251}$$

The series appeared in (2.251) for $\varepsilon = 1$ is divergent; however, we are able to find its sum:

$$\sum_{j=1}^{\infty} (-1)^{j-1} \varepsilon = \frac{\varepsilon}{1 + \varepsilon}.$$

Observe that the beam displacement (2.251) coincides with the exact solution (2.512). However, it should be emphasized that the exact solution obtained with the help of PS has been found only due to the obtained summation procedure. An arbitrary finite number of approximations cannot yield the appropriate and validated solution.

Applying the PA to the first three terms of the PS one obtains exact solution. Using the exact solution (2.512), the following bending moments are obtained:

$$M = \frac{q}{6}(4x^3 - 3x^2) + \frac{q}{48} \frac{1 + 3\varepsilon}{1 + \varepsilon}.$$

In what follows we estimate the error in determination of the displacement W and the bending moment M for various values of x (Figure 2.24). It is clear that accuracy of determination of the displacement is better than that of the bending moment estimation.

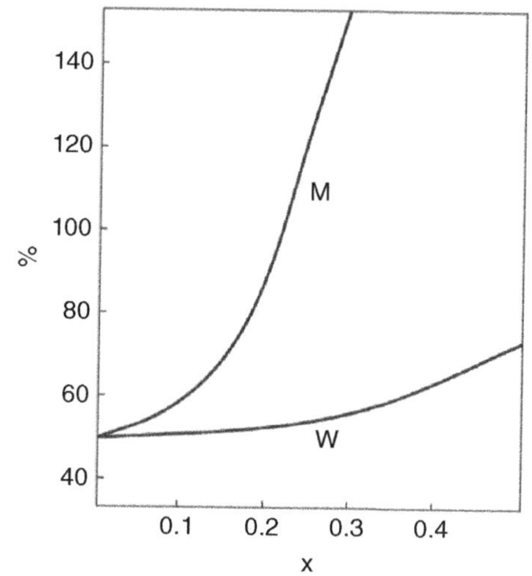

Figure 2.24 Estimation of accuracy of the normal displacement and bending moment of the beam using the PS

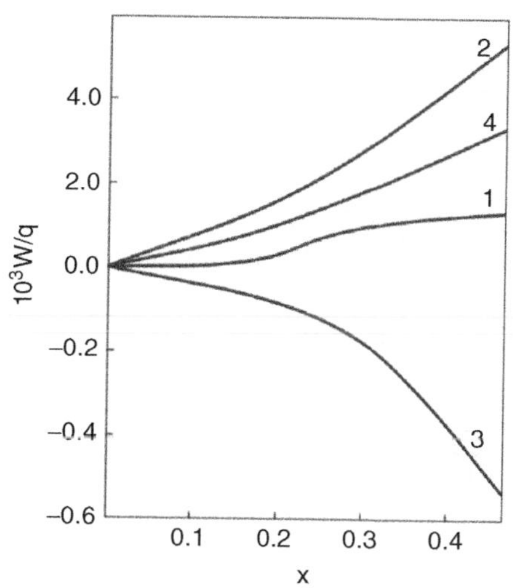

Figure 2.25 Accuracy estimation by partial sums of PS of normal beam displacements: 1 - W_0, 2 - W_1, 3 - W_2, 4 - exact solution

In Figure 2.25 graphs of various approximations of the function W through partial sums of the PS in comparison with the exact solution (2.512) are given. In Figure 2.26 graphs present the first, second and third approximations of the PS for the function M.

Here, similarly to the case of displacement estimation, a sum of three first terms yields the result coinciding with that of the zeroth order approximation, i.e. the accuracy is not improved. At the same time, the PA constructed with the inclusion of only three terms yield the exact solution.

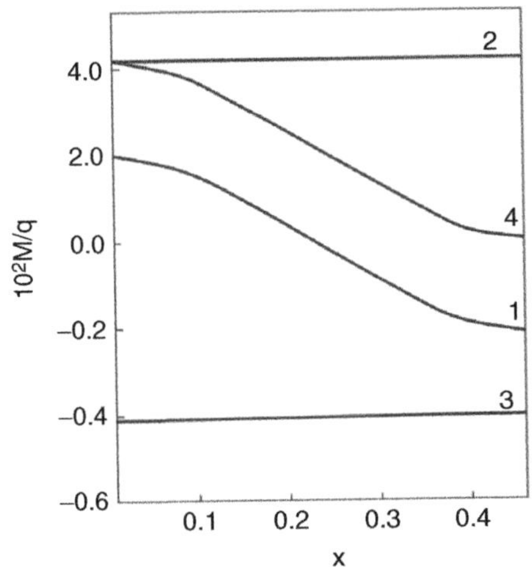

Figure 2.26 Accuracy estimation by partial sums of PS of bending beam moment: 1 - M_0, 2 - M_1, 3 - M_2, 4 - exact solution

2.4.3 SSS of Clamped Plate

In what follows we study the SSS of the clamped rectangular plate ($-0.5a \leq \overline{x} \leq 0.5a, -0.5b \leq \overline{y} \leq 0.5b$) uniformly loaded. Governing PDE can be written in the following form:

$$\nabla^4 W = q, \tag{2.252}$$

where $y = \overline{y}/b$, $x = \overline{x}/b$, $k = a/b$, $q = \overline{q}b^4/D$.

BCs are given in the form (2.66), (2.67). Let us introduce the plate displacement as a sum of three components:

$$W = W_1 + W_2 + W_3, \tag{2.253}$$

$$W_1 = \frac{q}{8k} \sum_{m=1,3,5} \frac{(-1)^{\frac{m-1}{2}}}{\alpha_m^5} \left(1 - \frac{\alpha_m \tanh \alpha_m + 2}{2\cosh \alpha_m} \cosh 2\alpha_m y + \right.$$
$$\left. \frac{\alpha_m}{\cosh \alpha_m} y \sinh 2\alpha_m y \right) \cos 2\alpha_m x, \tag{2.254}$$

$$W_2 = \frac{1}{8} \sum_{m=1,3,5} \frac{(-1)^{\frac{m-1}{2}} A_m}{\alpha_m^2 \cosh \alpha_m} (\alpha_m \tanh \alpha_m \cosh 2\alpha_m y -$$
$$2\alpha_m y \sinh 2\alpha_m y) \cos 2\alpha_m x, \tag{2.255}$$

$$W_3 = \frac{k^2}{8} \sum_{m=1,3,5} \frac{(-1)^{\frac{m-1}{2}} B_m}{\beta_m^2 \cosh \beta_m} \left(\beta_m \tanh \beta_m \cosh \frac{2}{k}\beta_m x - \right.$$
$$\left. \frac{2}{k}\beta_m x \sinh \frac{2}{k}\beta_m x \right) \cos \frac{2}{k}\beta_m y, \tag{2.256}$$

where $\alpha_m = \frac{\pi m}{2k}$, $\beta_m = \frac{\pi m k}{2}$.

Formula (2.254) describes the deflection of a simply supported plate and subjected to the uniformly distributed load. Formula (2.255) describes simply supported plate bent by moments M_y acting on the edges $y = \pm 0.5$, and has the following form:

$$M_y|_{y=\pm 0.5} = \sum_{m=1,3,5,\ldots} (-1)^{\frac{m-1}{2}} A_m \cos \frac{\pi m}{k} x.$$

Finally, formula (2.256) models the deflection of simply supported plate and subjected to bending moment M_x located on its edges $x = \pm 0.5k$, which is given by the series

$$M_x|_{x=\pm 0.5k} = \sum_{m=1,3,5,\ldots} (-1)^{\frac{m-1}{2}} B_m \cos \pi m y.$$

Choice of the plate deformation form (2.253) allows satisfaction to one of the BCs, i.e. the absence of displacements on the external plate contour. Satisfaction to BCs (2.66), (2.67) yields the following equations regarding coefficients A_m and B_m:

$$\begin{aligned}
&-(1-\varepsilon) \sum_{m=1,3,5,\ldots} (-1)^{\frac{m-1}{2}} B_m \cos \pi m y + \\
&\varepsilon k \left\{ -\frac{q}{4} \sum_{m=1,3,5,\ldots} \frac{1}{\alpha_m^4} \underline{\underline{\left(1 - \frac{\alpha_m \tanh \alpha_m + 2}{2\cosh \alpha_m} \cosh 2\alpha_m y + \frac{\alpha_m}{\cosh \alpha_m} y \sinh 2\alpha_m y \right)}} + \right. \\
&\frac{1}{4} \sum_{m=1,3,5,\ldots} \frac{A_m}{\cosh \alpha_m} \underline{\underline{(\tanh \alpha_m \cosh 2\alpha_m y - 2y \sinh 2\alpha_m y)}} + \\
&\left. \frac{k}{4} \sum_{m=1,3,5,\ldots}^{\infty} \frac{A_m}{\beta_m} \left(\tanh \beta_m + \frac{\beta_m}{\cosh^2 \beta_m} \right) (-1)^{\frac{m-1}{2}} \cos \pi m m y \right\} = 0,
\end{aligned} \tag{2.257}$$

$$\begin{aligned}
&-(1-\varepsilon) \sum_{m=1,3,5,\ldots} (-1)^{\frac{m-1}{2}} A_m \cos \frac{\pi m}{k} x + \\
&\varepsilon \left\{ \frac{q}{8k} \sum_{m=1,3,5,\ldots} \frac{1}{\alpha_m^4} \left(\frac{\alpha_m}{\cosh^2 \alpha_m} - \tanh \alpha_m \right) \times \cos \frac{\pi m}{k} x - \right. \\
&\frac{1}{4} \sum_{m=1,3,5,\ldots} (-1)^{\frac{m-1}{2}} \frac{A_m}{\alpha_m} \left(\tanh \alpha_m + \frac{\alpha_m}{\cosh^2 \alpha_m} \right) \cos \frac{\pi m}{k} x - \\
&\left. \frac{k}{4} \sum_{m=1,3,5,\ldots} \frac{B_m}{\cosh \beta_m} \underline{\underline{\left(\tanh \beta_m \cosh \frac{2}{k} \beta_m x - \frac{2}{k} x \sinh \frac{2}{k} \beta_m x \right)}} \right\} = 0.
\end{aligned} \tag{2.258}$$

Underlined terms in Equation (2.257) represent even functions of y, which are equal to zero for $y = \pm 0.5$, and which can be presented in the series form with respect to $\cos \pi i y$, $i = 1, 3, 5, \ldots$. The underlined term in Equation (2.258) is developed into a series with respect to $\cos \pi i x/k$, $i = 1, 3, 5, \ldots$. By substituting the mentioned series into Equations (2.257), (2.258) and collecting terms standing by $\cos \pi i y$ and $\cos \pi i x/k$, the following infinite system of LAE with respect to coefficients A_i and B_i is obtained:

$$B_i \left\{ 1 + \varepsilon \left[\frac{k^2}{4} \frac{1}{\beta_i} \left(\tanh \beta_i + \frac{\beta_i}{\cosh^2 \beta_i} \right) - 1 \right] \right\} + \varepsilon \frac{4}{\pi^2} i \sum_{m=1,3,5,\dots} A_m \frac{m}{\left(\frac{m^2}{k^2} - i^2 \right)^2} =$$

$$\varepsilon \frac{q}{8} \frac{1}{\beta_i^4} \left(\frac{\beta_i}{\cosh^2 \beta_i} - \tanh \beta_i \right), \quad i = 1, 3, 5, \dots, \tag{2.259}$$

$$A_i \left\{ 1 + \varepsilon \left[\frac{k^2}{4} \frac{1}{\alpha_i} \left(\tanh \alpha_i + \frac{\alpha_i}{\cosh^2 \alpha_i} \right) - 1 \right] \right\} + \varepsilon \frac{4}{\pi^2} \frac{i}{k} \sum_{m=1,3,5,\dots} B_m \frac{m}{\left(m^2 + \frac{i^2}{k^2} \right)^2} =$$

$$\varepsilon \frac{q}{8k} \frac{1}{\alpha_i^4} \left(\frac{\alpha_i}{\cosh^2 \alpha_i} - \tanh \alpha_i \right), \quad i = 1, 3, 5, \dots. \tag{2.260}$$

Let us present the coefficients A_i and B_i in the following form:

$$A_i = \sum_{j=0}^{\infty} A_{i(j)} \varepsilon^j, \qquad B_i = \sum_{j=0}^{\infty} B_{i(j)} \varepsilon^j. \tag{2.261}$$

After substitution of Ansatzes (2.261) into the system of LAE (2.259) and (2.260), the following formulas for determination of the j-th approximations of the unknown coefficients are derived:

$$A_{i(0)} = 0, \quad B_{i(0)} = 0,$$

$$A_{i(1)} = \frac{q}{8k} \frac{1}{\alpha_i^4} \left(\frac{\alpha_i}{\cosh^2 \alpha_i} - \tanh \alpha_i \right), \quad B_{i(1)} = \frac{q}{8} \frac{1}{\beta_i^4} \left(\frac{\beta_i}{\cosh^2 \beta_i} - \tanh \beta_i \right),$$

$$A_{i(j)} = A_{i(j-1)} \left[1 - \frac{1}{4} \frac{1}{\alpha_i} \left(\tanh \alpha_i + \frac{\alpha_i}{\cosh^2 \alpha_i} \right) \right] - \frac{4}{\pi} i \times$$

$$\sum_{m=1,3,5,\dots} B_m(j-1) \frac{m/k}{\left(\frac{i^2}{k^2} + m^2 \right)^2}, \quad i = 1, 3, 5, \dots, \tag{2.262}$$

$$B_{i(j)} = B_{i(j-1)} \left[1 - \frac{k^2}{4} \frac{1}{\beta_i} \left(\tanh \beta_i + \frac{\beta_i}{\cosh^2 \beta_i} \right) \right] -$$

$$\frac{4}{\pi^2} i \sum_{m=1,3,5,\dots} A_m(j-1) \frac{m}{\left(\frac{m^2}{k^2} + i^2 \right)^2}, \quad i = 1, 3, 5, \dots \quad . \tag{2.263}$$

Let us consider the SSS of the square plate. Distribution of bending moments is equal regarding its edges; therefore $A_i = B_i$ and systems (2.259) and (2.260) became identity. After determination of first four coefficients A_i with respect to formulas (2.262) and (2.263), we apply the PA:

$$A_{i[1/1]}(\varepsilon) = \varepsilon \frac{a_0 + a_1 \varepsilon}{b_0 + b_1 \varepsilon}, \tag{2.264}$$

where $a_0 = A_{i(1)}$, $b_0 = 1$, $a_1 = A_{i(2)} + b_1 A_{i(1)}$, $b_1 = -A_{i(3)}/A_{i(2)}$.

Estimated via PA in (2.264) for $\varepsilon = 1$ deflection in the plate center achieves $1.275 \cdot 10^{-3}q$ (known value [76] - $1.260 \cdot 10^{-3}q$, which gives the error of 1.2%). Displacement, obtained using PS yields $1.797 \cdot 10^{-3}q$, which corresponds to the error of 42.6%.

The bending moment in the plate center estimated through the PA in (2.264) for $\varepsilon = 1$, $\nu = 0.3$ has a value of $5.173 \cdot 10^{-2}q$ (known value [76] - $5.130 \cdot 10^{-2}q$; error - 0.83%). The same quantities regarding the PS estimation yield: $3.883 \cdot 10^{-2}q$ (error - 24.2%).

The bending moments M_x and M_y achieve their maximal values in the center of the clamped plate side. The maximal bending moment defined via the PA in (2.264) is of $2.400 \cdot 10^{-2}q$ (known value [76] - $2.310 \cdot 10^{-2}q$, error - 3.90%). The corresponding result yielded by the PS for $\varepsilon = 1$ is $2.941 \cdot 10^{-2}q$ (error - 27.3%).

For the coefficients A_i, obtained on the basis of the PS by (2.262)–(2.263) formulas for $\varepsilon = 1$ (curves 1) and those computed with the help of the PA (2.264) (curves 2). Dependencies $W|_{x=y=0}$, $M_x|_{x=y=0} = M_y|_{x=y=0}$, $M_x|_{x=\pm 0.5k}$ versus ε are shown in Figures 2.27, 2.28.

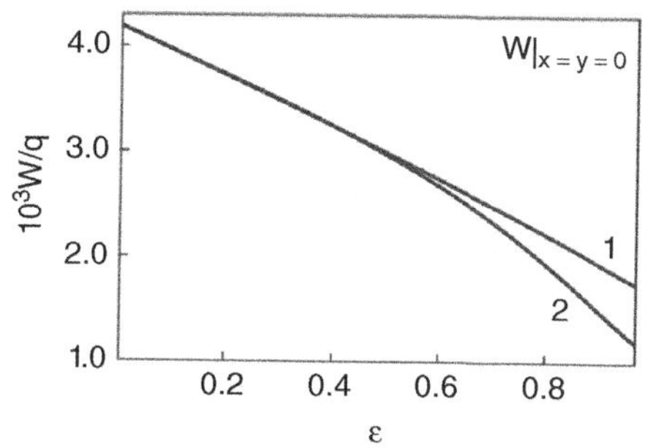

Figure 2.27 Estimation of the PS accuracy in determination of the plate normal displacements

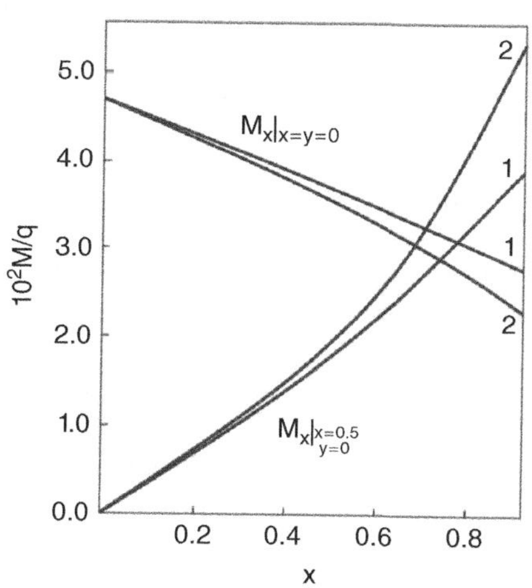

Figure 2.28 Estimation of the PS accuracy while bending moment determination

Reported graphs allow us to trace the dependencies of the SSS factors versus parameter ε. For all of the presented curves two zones are remarkable. First one begins at $\varepsilon = 0$ and ends at $\varepsilon = 0.6$. In this zone the difference between results obtained via the PS and PA is less than 5%. In the second zone (from $\varepsilon = 0.6$ up to $\varepsilon = 1.0$) one may observe the rather large difference of both types of curves (the largest difference is achieved for $\varepsilon = 1$). Results obtained via the PA can be practically treated as exact ones. Consequently, in the case studied we have reliable results corresponding to the practical application of the PS only for $\varepsilon = 0.6$.

Interestingly the following peculiarity of our proposed method is exhibited if one carries out computation only using the PS, then the largest error is achieved in the plate center (errors regarding bending moments estimation in both plate center and its edges are smaller). Application of the PA decreases the error in SSS factors estimation, making it almost the same for either deflection or bending moment estimations.

This can be explained in the following manner. Location of poles of the PA in the unit circle of the parameter ε is conserved during differentiation of the function found with respect to coordinates x and y. Therefore, series improvement using the PA either for deflections or bending moments is of the same order. This property belongs to one of the important PA benefits.

2.4.4 SSS of a Plate with Free Edges

Let us consider the SSS of the plate $(-0.5a \le \overline{x} \le 0.5a; -0.5b \le \overline{y} \le 0.5b)$ being under action of the self-balanced load $\overline{q}(x,y) = \overline{q}_0 \cos\frac{\pi m}{k} x \cos \pi n y,\ m,n = 2,4,6,\ldots$. The non-dimensonalized governing PDE can be cast to the following form:

$$\nabla^4 W = q_0 \cos \frac{\pi m}{k} x \cos \pi n y, \qquad (2.265)$$

where $q_0 = \overline{q}_0 b^4/D$.

BCs regarding the free contour in the transformed form have the following form:

$$\begin{aligned} &W_{xxx} + (2-\nu)W_{yyx} = 0, \\ &(1-\varepsilon)W_x \pm \varepsilon k(W_{xx} + \nu W_{yy}) = 0, \end{aligned} \qquad \text{for} \quad x = \pm 0.5k, \qquad (2.266)$$

$$\begin{aligned} &W_{yyy} + (2-\nu)W_{xxy} = 0, \\ &(1-\varepsilon)W_y \pm \varepsilon(W_{yy} + \nu W_{xx}) = 0, \end{aligned} \qquad \text{for} \quad y = \pm 0.5. \qquad (2.267)$$

After substitution of the perturbation series into Equation (2.265), and into BCs (2.266), (2.267), and after splitting with respect to powers of ε, the following recurrent system of the BVPs is obtained:

$$\nabla^4 W_0 = q_0 \cos \frac{\pi m}{k} x \cos \pi n y,$$

$$W_{0x} = 0, \quad W_{0xxx} = 0 \quad \text{for} \quad x = \pm 0.5k,$$

$$W_{0y} = 0, \quad W_{0yyy} = 0 \quad \text{for} \quad y = \pm 0.5,$$

$$\nabla^4 W_j = 0,$$

$$\left.\begin{array}{l} W_{jxxx} + (2-\nu)W_{jyyx} = 0, \\ W_{jx} = \mp k \sum_{i=0}^{j-1}(W_{ixx} + \nu W_{iyy}), \end{array}\right. \quad \text{for} \quad x = \pm 0.5k.$$

$$\left.\begin{array}{l} W_{jyyy} + (2-\nu)W_{jxxy} = 0, \\ W_{jy} = \mp \sum_{i=0}^{j-1}(W_{iyy} + \nu W_{ixx}), \end{array}\right. \quad \text{for} \quad y = \pm 0.5.$$

A few first coefficients of the PS

$$W = W_0 + W_1\varepsilon + W_2\varepsilon^2 + \ldots \tag{2.268}$$

follow

$$W_0 = \frac{q_0}{\pi^4\left(\frac{m^2}{k^2} + n^2\right)} \cos\frac{\pi m}{k}x \cos\pi n y,$$

$$W_1 = -\frac{q_0(1-\nu)}{2\pi^2\left(\frac{m^2}{k^2} + n^2\right)} \left\{ k\frac{\left(\frac{m^2}{k^2} + \nu n^2\right)}{\pi n \sinh \pi n k/2}(-1)^{\frac{m}{2}} \left[\left(\frac{\pi n k}{2} C \tanh\frac{\pi n k}{2} - \right.\right.\right.$$

$$\left.\left.\frac{1+\nu}{1-\nu}\right)\cosh\pi n x - \pi n x \sinh\pi n x\right] \cos\pi n y + \frac{n^2 + \nu\frac{m^2}{k^2}}{\frac{\pi m}{k}\sinh\frac{\pi m}{2k}}(-1)^{\frac{n}{2}} \left[\left(\frac{\pi m}{2k}\coth\frac{\pi m}{2k} - \right.\right.$$

$$\left.\left.\left.\frac{1+\nu}{1-\nu}\right)\coth\frac{\pi m}{k}y - \frac{\pi m}{k}y\sinh\frac{\pi m}{k}y\right]\cos\frac{\pi m}{k}x\right\},$$

$$W_2 = -\frac{q_0(1-\nu)^2}{4\pi^2\left(\frac{m^2}{k^2} + n^2\right)} \left\{ k\frac{\left(\frac{m^2}{k^2} + \nu n^2\right)}{\sinh \pi n k/2}(-1)^{\frac{m}{2}} \left[(1-\nu)\frac{\pi n k}{2}\frac{1}{\sinh^2 \pi n k/2} - \right.\right.$$

$$\left.(3+\nu)\coth\frac{\pi n k}{2}\right]\left[\left(\frac{\pi n k}{2}\coth\frac{\pi n k}{2} - \frac{1+\nu}{1-\nu}\right)\cosh\pi n x - \pi n x\sinh\pi n x\right]\cos\pi n y +$$

$$\frac{n^2 + \nu\frac{m^2}{k^2}}{\sinh\frac{\pi m}{2k}}(-1)^{\frac{n}{2}}\left[(1-\nu)\frac{\pi m}{2k}\frac{1}{\sinh^2\frac{\pi m}{2k}} - (3+\nu)\coth\frac{\pi m}{2k}\right]$$

$$\left[\left(\frac{\pi m}{2k}\coth\frac{\pi m}{2k} - \frac{1+\nu}{1-\nu}\right)\cosh\frac{\pi m}{k}y - \frac{\pi m}{k}y\sinh\frac{\pi m}{k}y\right]\cos\frac{\pi m}{k}x -$$

$$\frac{2q_0(1-\nu)^3(-1)^{\frac{m}{2}}(-1)^{\frac{n}{2}}}{\pi^5\left(\frac{m^2}{k^2} + n^2\right)^2}\left\{k\left(\frac{\pi m}{k}\right)^2\right.$$

$$\sum_{i=2,4,6,\ldots}\frac{i(-1)^{\frac{i}{2}}}{\sinh\frac{\pi i k}{2}\left(\frac{m^2}{k^2} + i^2\right)^2}\left[\left(\frac{\pi i k}{2}\coth\frac{\pi i k}{2} - \frac{1+\nu}{1-\nu}\right)\cosh\pi i x - \right.$$

$$\pi ix \sinh \pi ix] \cos \pi iy + (\pi n^2) \sum_{i=2,4,6,\ldots} \frac{(-1)^{\frac{i}{2}}(i/k)}{\sinh \frac{\pi i}{2k}\left(n^2+\frac{i^2}{k^2}\right)^2} \left[\left(\frac{\pi i}{2k}\coth\frac{\pi i}{2k} - \right.\right.$$

$$\left.\left.\frac{1+\nu}{1-\nu}\right)\cosh\frac{\pi i}{k}y - \frac{\pi i}{k}y\sinh\frac{\pi i}{k}y\right]\cos\frac{\pi i}{k}x\Bigg\}.$$

Further, we apply PA to the truncated PS (2.268). In what follows we give results of computations of some SSS factors for the plate with the following parameters: $m = n = 2$, $k = 1$, $\nu = 0.3$. Figures 2.29 and 2.30 show the plate displacement W as well as bending moments M_x, M_y for $y = 0$.

Here we apply an approach differing from that of the so far illustrated and discussed. Namely, although this problem could be reduced to that of an infinite system of LAEs, now the SSS is constructed through successive solutions to the recurrent BVPs with the successive application of the PA to a PS that is found. Even though two of the mentioned approaches yield the same results, the approach now applied is more suitable, since solutions corresponding to the first order approximation possess the same structure.

2.4.5 *SSS of the Plate with Mixed Boundary Conditions "Clamping–Simple Support"*

We consider a rectangular plate simply supported on its two edges, and with mixed BCs on the two remaining edges of the type of "clamping–simple support" (Figure 2.8). The plate is subjected to the uniformly distributed load of q intensity. In what follows we will study a symmetric problem (Figure 2.8a). Finding its solution is reduced to a solution of the infinite system of LAEs that is analogous to that already considered during studies of the SSS of

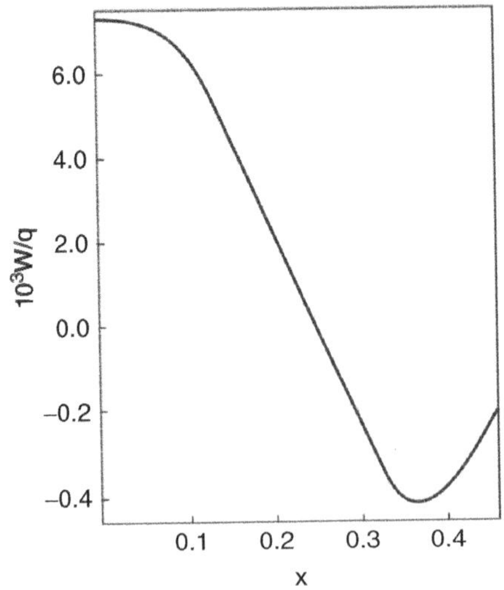

Figure 2.29 Normal displacement versus x for $y = 0$

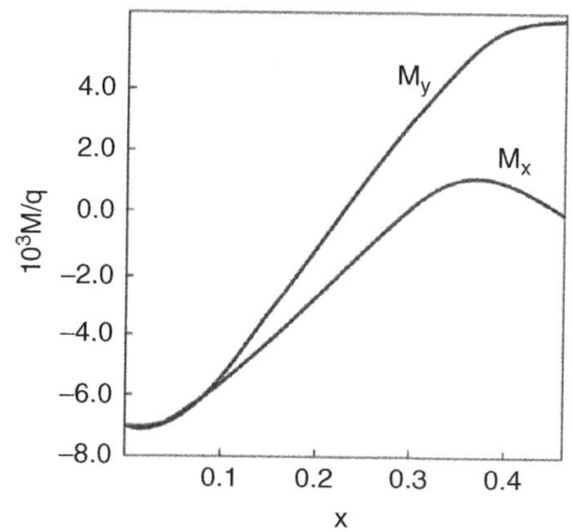

Figure 2.30 Bending moments versus x for $y = 0$

the rectangular clamped plate. After introduction of nondimensional quantities the basic PDE takes the form (2.252), and BCs correspond to (2.114), (2.115). Plate displacement has the following form

$$W = W_1 + W_2,$$

$$W_1 = \frac{q}{8k} \sum_{m=1,3,5,\ldots} \frac{(-1)^{\frac{m-1}{2}}}{\alpha_m} \left(1 - \frac{\alpha_m \tanh \alpha_m + 2}{2 \cosh \alpha_m} \cosh 2\alpha_m y + \right.$$

$$\left. \frac{\alpha_m}{\cosh \alpha_m} y \sinh 2\alpha_m y \right) \cos \frac{\pi m}{k} x, \tag{2.269}$$

$$W_2 = \frac{1}{8} \sum_{m=1,3,5,\ldots} \frac{(-1)^{\frac{m-1}{2}}}{\alpha_m^2} \frac{A_m}{\cosh \alpha_m} (\alpha_m \tanh \alpha_m \cosh 2\alpha_m y -$$

$$2\alpha_m y \sinh 2\alpha_m y) \cos \frac{\pi m}{k} x, \tag{2.270}$$

where $\alpha_m = \frac{\pi m}{2k}$.

Similarly to the case of the clamped plate, formula (2.269) describes the displacement of simply supported plate along its contour, whereas formula (2.270) yields the displacement of the simply supported plate which is subjected to the bending moments loading M_y acting on edges $y = \pm 0.5$:

$$M_y|_{y=\pm 0.5} = \sum_{m=1,3,5,\ldots} A_m (-1)^{\frac{m-1}{2}} \cos 2\alpha_m x.$$

Ansatzes (2.269), (2.270) satisfy the condition of equality to zero of the vertical displacements on the plate contour, as well as the BCs (2.114).

Satisfaction to BCs (2.115) yields the following equations:

$$\sum_{m=1,3,5,\ldots} A_m(-1)^{\frac{m-1}{2}} \cos 2\alpha_m x = \varepsilon \overline{H}(x) \left\{ \sum_{m=1,3,5,\ldots} (-1)^{\frac{m-1}{2}} A_m \cos 2\alpha_m x + \right.$$

$$\sum_{m=1,3,5,\ldots} \frac{q}{8k} \frac{(-1)^{\frac{m-1}{2}}}{\alpha_m^4} \left(\frac{\alpha_m}{\cosh^2 \alpha_m} - \tanh \alpha_m \right) \cos 2\alpha_m x -$$

$$\left. \sum_{m=1,3,5,\ldots} \frac{(-1)^{\frac{m-1}{2}}}{4\alpha_m} \left(\frac{\alpha_m}{\cosh^2 \alpha_m} + \tanh \alpha_m \right) \cos 2\alpha_m x \right\}, \quad i = 1, 3, 5, \ldots \tag{2.271}$$

In order to obtain a system of LAEs the following series is applied

$$\overline{H}(x) \cos 2\alpha_m x = \sum_{i=1,3,5,\ldots} \gamma_{im} \cos 2\alpha_i x, \tag{2.272}$$

$$\gamma_{im} = \begin{cases} 2\left[0.5 - \mu - \frac{1}{2\pi m} \sin 2\pi \mu m\right] & \text{for} \quad i = m, \\ \frac{4}{\pi} \frac{1}{(m^2 - i^2)} [i \sin \pi \mu i \cos \pi \mu m - m \sin \pi \mu m \cos \pi \mu i] & \text{for} \quad i \neq m. \end{cases}$$

Substituting Ansatz (2.272) into Equation (2.271) and collecting coefficients standing by $\cos 2\alpha_i x$, the following infinite system of LAEs is obtained:

$$A_i(-1)^{\frac{i-1}{2}} = \varepsilon \sum_{m=1,3,5,\ldots} \gamma_{im}(-1)^{\frac{m-1}{2}} A_m \left[1 - \frac{1}{4\alpha_m} \left(\frac{\alpha_m}{\cosh^2 \alpha_m} + \tanh \alpha_m \right) \right] +$$

$$\varepsilon \frac{q}{8k} \sum_{m=1,3,5,\ldots} \gamma_{im} \frac{(-1)^{\frac{m-1}{2}}}{\alpha_m^4} \left(\frac{\alpha_m}{\cosh^2 \alpha_m} - \tanh \alpha_m \right), \quad i = 1, 3, 5, \ldots. \tag{2.273}$$

Now perturbation analysis can be applied. We present coefficients A_i in a PS (2.261), and after their substitution to Equations (2.273) and splitting with respect to ε, the following recurrent system of equations is obtained

$$A_{i(0)} = 0, \tag{2.274}$$

$$A_{i(1)} = (-1)^{\frac{m-1}{2}} \frac{q}{8k} \sum_{m=1,3,5,\ldots} \gamma_{im} \frac{(-1)^{\frac{m-1}{2}}}{\alpha_m^4} \left(\frac{\alpha_m}{\cosh^2 \alpha_m} - \tanh \alpha_m \right), \tag{2.275}$$

$$A_{i(j)} = (-1)^{\frac{i-1}{2}} \sum_{m=1,3,5,\ldots} \gamma_{im}(-1)^{\frac{m-1}{2}} A_{m(j-1)} \left[1 - \frac{1}{4\alpha_m} \left(\frac{\alpha_m}{\cosh^2 \alpha_m} + \tanh \alpha_m \right) \right]. \tag{2.276}$$

Note that the PA applied for definition of coefficients A_i has the form (2.264).

Let us analyse the obtained solution (2.274)–(2.276) in limiting cases. The value $\mu = 0.5$ corresponds to a simple support plate on the edges $y = \pm 0.5$, and $\gamma_{im} \equiv 0$. Second limiting case ($\mu = 0$) corresponds to clamped edges $y = \pm 0.5$, and $\gamma_{im} = \delta_{im}$, where δ_{im} is the Kronecker delta

$$\delta_{im} = \begin{cases} 1, & \text{for} \quad i = m, \\ 0, & \text{for} \quad i \neq m. \end{cases}$$

Recurrent relations (2.274)–(2.276) take the following form:

$$A_{i(0)} = 0, \tag{2.277}$$

$$A_{i(1)} = \frac{q}{8k}\frac{1}{\alpha_i^4}\left(\frac{\alpha_i}{\cosh^2\alpha_i} - \tanh\alpha_i\right), \tag{2.278}$$

$$A_{i(2)} = \frac{q}{8k}\frac{1}{\alpha_i^4}\left(\frac{\alpha_i}{\cosh^2\alpha_i} - \tanh\alpha_i\right)\left[1 - \frac{1}{4\alpha_i}\left(\frac{\alpha_i}{\cosh^2\alpha_i} + \tanh\alpha_i\right)\right], \tag{2.279}$$

$$A_{i(3)} = \frac{q}{8k}\frac{1}{\alpha_i^4}\left(\frac{\alpha_i}{\cosh^2\alpha_i} - \tanh\alpha_i\right)\left[1 - \frac{1}{4\alpha_i}\left(\frac{\alpha_i}{\cosh^2\alpha_i} + \tanh\alpha_i\right)\right]^2. \tag{2.280}$$

For the truncated PS (2.277)–(2.280) and for $\varepsilon = 1$ PA follows:

$$A_{i[1/1]} = \frac{q}{2\alpha_i^3}\frac{\alpha_i - \tanh\alpha_i(\alpha_i\tanh\alpha_i + 1)}{\alpha_i - \tanh\alpha_i(\alpha_i\tanh\alpha_i - 1)}. \tag{2.281}$$

It should be emphasized that formula (2.281) serving for determination of A_i coincides completely with the S.P. Timoshenko solution [76].

Computation of the SSS components is carried out for the squared plate taking into account first ten coefficients A_i obtained via formula (2.264) for $\varepsilon = 1$. Displacement and bending moments of the plate in its center versus various values of the μ parameter are shown in Figures 2.31, 2.32 (solid curves–our solution; dashed curves–FEM solution).

Figure 2.33 reported the computational results of the bending moment M_y distributed on edges $y = \pm 0.5$ and for different values of the μ parameter.

Let us briefly discuss the following problem. It is obvious that some of the SSS components have singularities in the places of the BCs change (see Table 2.1). One may find the following asymptotics: $W = O(r^{\varphi+1})$, $M_n = O(r^{\varphi-1})$, $Q_n = O(r^{\varphi-2})$. Here $r = \sqrt{x^2 + y^2}$, M_n and Q_n are the bending moment and generalized transversal force in n plate direction, respectively. Note that for BCs 1–5 given in Table 2.1 we have $\varphi = 0.5$, whereas in the case of BCs 6 the parameter φ is defined via the following transcendental equation:

$$\cos 2\pi\varphi = -\frac{4 + (1 + \nu)^2}{4 - (1 - \nu)^2}.$$

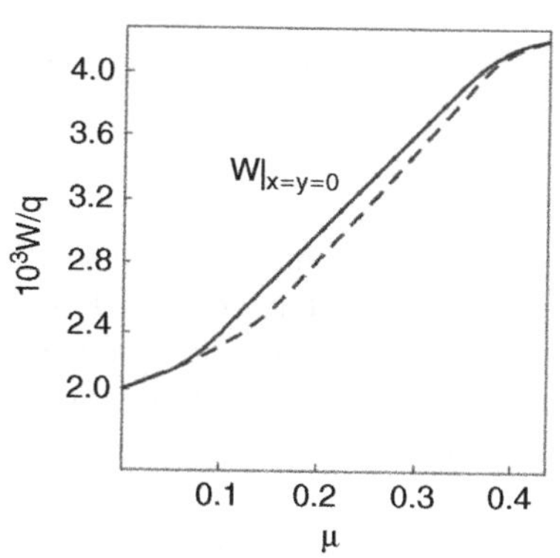

Figure 2.31 Comparison of FEM solutions with ours for plate displacement

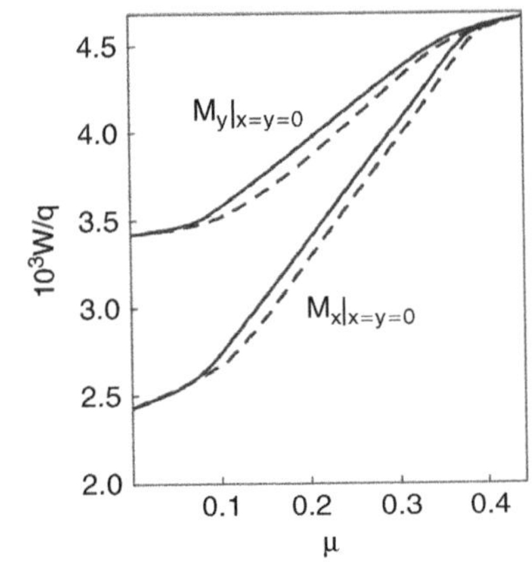

Figure 2.32 Comparison of FEM computational results with ours regarding bending moments

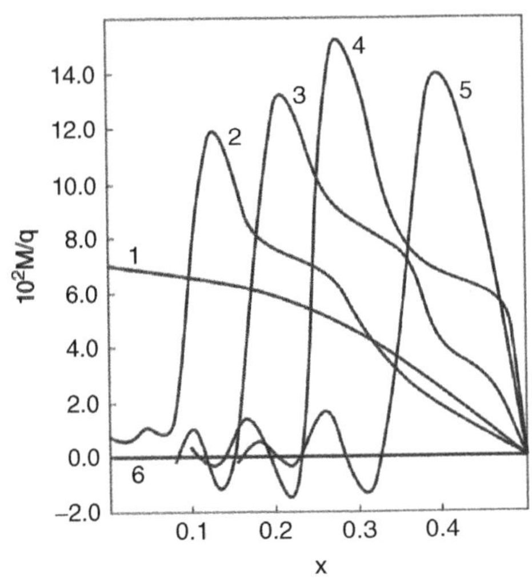

Figure 2.33 Bending moment M_y for $y = \pm 0.5$ for different values of μ : $1 - \mu = 0$, $2 - 0.1$, $3 - 0.2$; $4 - 0.3$; $5 - 0.4$, $6 - 0.5$

The constructed previously approximate solution does not suffer for the quoted singularities, and hence they should be added through the known procedure. Namely, let us assume that we are going to improve moments M_y in the neighborhood of a point of changes of BCs from the simple support to clamping. Let us introduce polar coordinates with a pole in the BCs change point (see Figure 2.34). Singular component of the bending moment M_y can be written in the following way:

$$M_y^{(c)} = r^{0.5}C(\theta) = r^{0.5}(C_0 + C_1\theta + \ldots), \tag{2.282}$$

where $C(\theta)$ is a function depending on angle θ.

Table 2.1 Type of mixed BCs

	Computational scheme	BCs
1	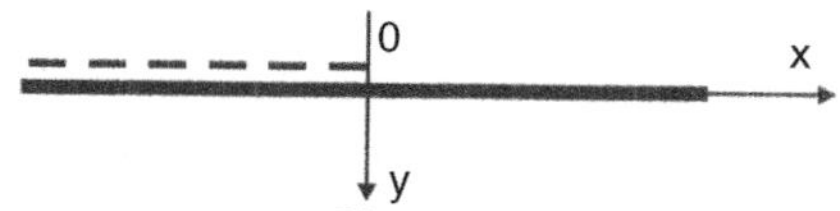	$x<0: \quad M_y=0, \quad W=0,$ $x>0: \quad M_y=0, \quad Q_y=0,$
2	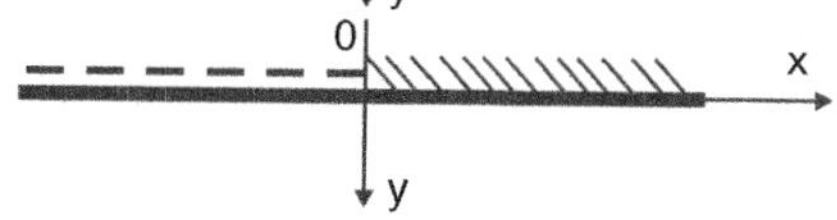	$x<0: \quad M_y=0, \quad W=0,$ $x>0: \quad W=0, \quad dW/dy=0,$
3	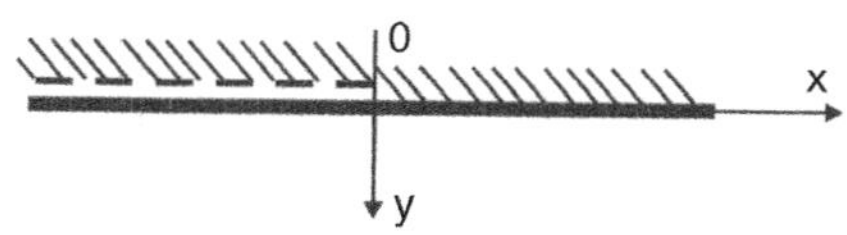	$x<0: \quad Q_y=0, \quad dW/dy=0,$ $x>0: \quad W=0, \quad dW/dy=0,$
4	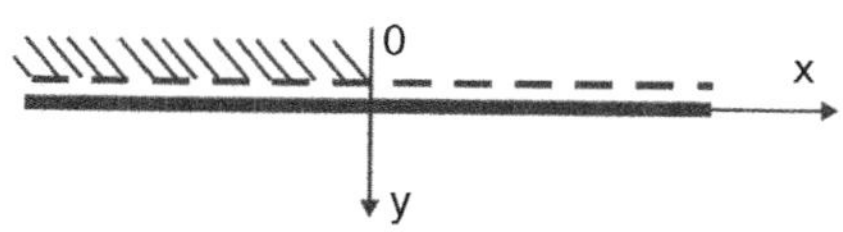	$x<0: \quad Q_y=0, \quad dW/dy=0,$ $x>0: \quad M_y=0, \quad W=0,$
5	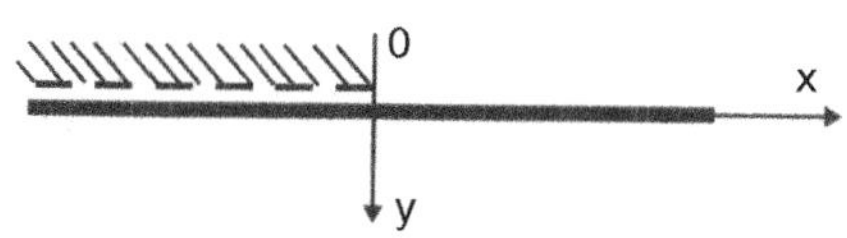	$x<0: \quad Q_y=0, \quad dW/dy=0,$ $x>0: \quad M_y=0, \quad Q_y=0,$
6	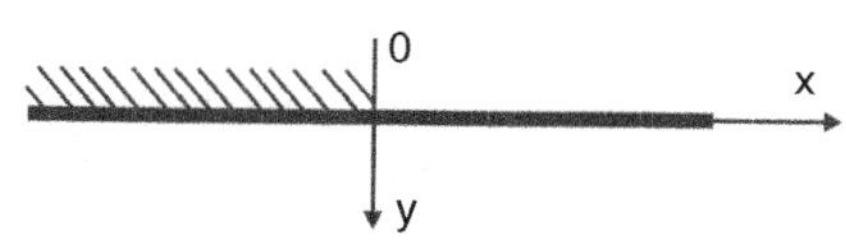	$x<0: \quad W=0, \quad dW/dy=0,$ $x>0: \quad Q_y=0, \quad M_y=0,$

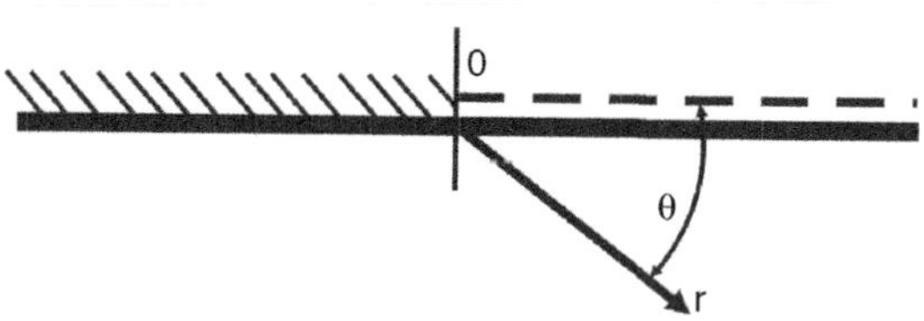

Figure 2.34 Local coordinates in a point of BCs change

In formula (2.282) while developing the function $C(\theta)$ into a series with respect to θ one may leave only C_0.

Let us construct the following equations:

$$M_{y|_{y=\pm 0.5}} = M^{(c)}_{y|_{y=\pm 0.5}} \quad \text{for} \quad \theta = 0, \tag{2.283}$$

$$Q_{y|_{y=\pm 0.5}} = Q^{(c)}_{y|_{y=\pm 0.5}} \quad \text{for} \quad \theta = 0, \tag{2.284}$$

where M_y, Q_y is the bending moment and shearing force, respectively, defined via perturbation series.

Formulas (2.283), (2.284) yield the constant C_0 and the point y_c of matching of singular and regular components of bending moments.

Let us now calculate the SSS of a plate having its clamping part nonsymmetrically located with respect to the plate center (Figure 2.8b). PDE (2.252) and BCs (2.114) keep their forms, whereas in BCs (2.115) the function $\overline{H}(x)$ should be defined in the following way: $\overline{H}(x) = H(x) - H(x - \mu k)$. In Equations (2.269), (2.270) $\cos \frac{\pi m}{k} x$ is substituted by $\sin \frac{\pi m}{k} x$, $(-1)^{\frac{m-1}{2}}$ is substituted by 1, and summation is carried out through all m. Ansatz (2.272) can be recast to the following form:

$$\overline{H}(x) \sin 2\alpha_m x = \sum_{i=1,2,3,\ldots} \gamma_{im} \sin 2\alpha_i x,$$

where γ_{im} are defined by Equation (2.134).

System of LAEs takes the following form:

$$A_i = \varepsilon \sum_{m=1,2,3,\ldots} A_m \gamma_{im} \left[1 - \frac{1}{4\alpha_m} \left(\frac{\alpha_m}{\cosh^2 \alpha_m} + \tanh \alpha_m \right) \right] +$$

$$\varepsilon \frac{q}{8k} \sum_{m=1,2,3,\ldots} \gamma_{im} \frac{1}{\alpha_m^4} \left(\frac{\alpha_m}{\cosh^2 \alpha_m} - \tanh \alpha_m \right), \qquad i = 1, 2, 3, \ldots .$$

In reccurrent relations (2.277)–(2.280) for A_i it is necessary to carry out the following changes: both $(-1)^{\frac{i-1}{2}}$ and $(-1)^{\frac{i-1}{2}}$ are substituted by 1, and summation is carried out through all i and m.

PA regarding coefficients A_i keep the form of (2.264).

Computation of SSS components of the squared plate is carried out taking into account the first ten coefficients A_i for various values of the μ parameter. Figures 2.35 and 2.36 demonstrate displacement and moments in the plate center in dependence on μ parameter. In Figure 2.37 the distribution of the bending moment M_y for different values of the μ parameter is presented. Solid curves correspond to our solutions, whereas dashed ones to FEM solution.

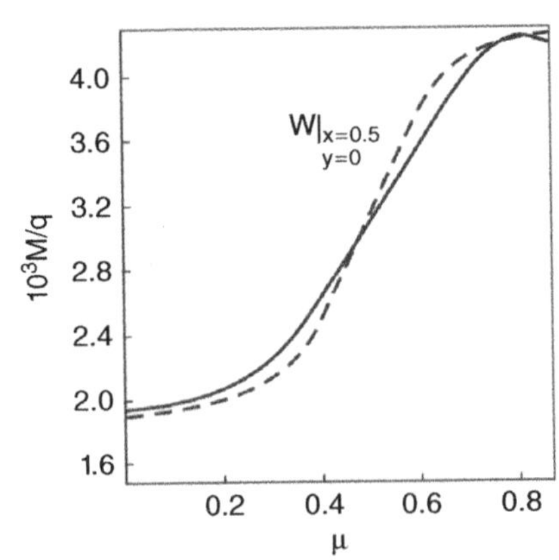

Figure 2.35 Comparison of the normal displacement solutions using our and FEM

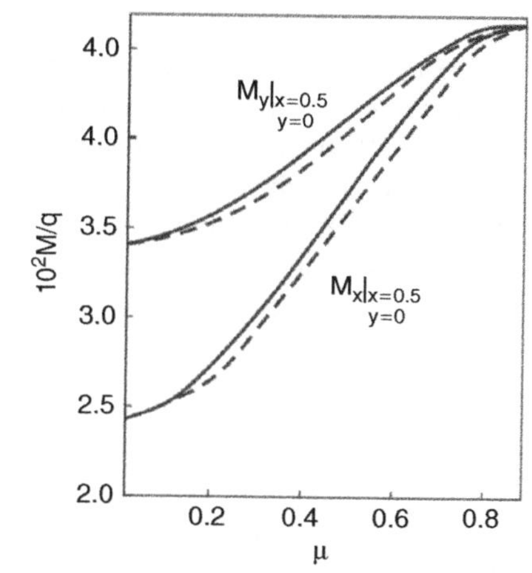

Figure 2.36 Comparison of the bending moment solutions using our and FEM

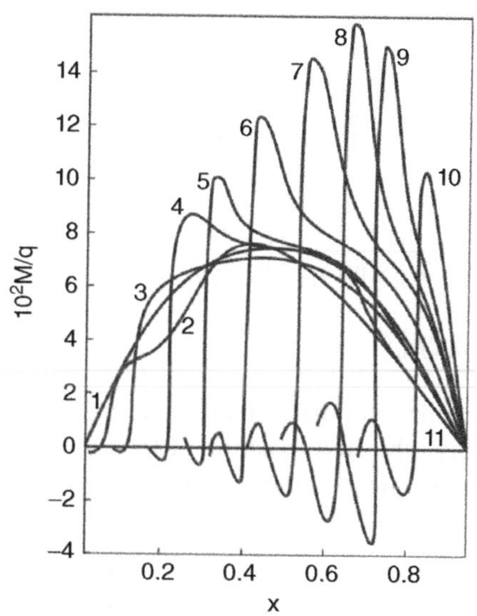

Figure 2.37 Bending moment M_y for $y = \pm 0.5$ for various values of the parameter μ: 1 - $\mu = 0$, 2–0.1; 3–0.2; 4–0.3; 5–0.4; 6–0.5; 7–0.6; 8–0.7; 9–0.8; 10–0.9; 11–1.0

Analysis of the obtained results shows that the application of HPM yields reliable determination of the SSS components of the plate with mixed BCs in places located relatively far from the singularities.

Analysis of the drawings of the SSS factors versus the geometric size of the mixed BCs allows us to distinguish three characteristic zones. The first zone concerns either the symmetric location of the clamped plate edges and is located in the interval from $\mu = 0$ to $\mu = 0.1$ or nonsymmetric location in the interval from $\mu = 0$ to $\mu = 0.4$. In this zone observed are rather negligible changes of the SSS factors. Deviation from the basic state is less than 5%. In the second zone with increase of the parameter μ the significant almost linear increase of the SSS factors is observed. The zone boundaries are located in intervals $\mu = 0.1$ to $\mu = 0.4$ for the symmetric and from $\mu = 0.4$ to $\mu = 0.9$ for the nonsymmetric problem. Third zone begins with $\mu = 0.4$ and ends at $\mu = 0.5$. The zone from $\mu = 0.9$ to $\mu = 1.0$ is characterized by a rather negligible increase of SSS factors with the increase of μ.

Therefore, rather negligible inclusions of nonhomogenous BCs (simple support and clamping) does not influence SSS of the plate significantly.

2.4.6 *SSS of a Plate with Mixed Boundary Conditions "Free Edge–Moving Clamping"*

Let us consider a computation of the rectangular plate simply supported on two edges, and with mixed BCs "free edge - moving clamping" on two other (see Figure 2.17). Plate is subjected to action of the uniformly distributed load action of intensity q. In what follows we study the symmetric case with respect to the origin coordinates (Figure 2.17a).

The input nondimensional BVP has the form (2.270), (2.159), (2.160). Plate displacement has the following form:

$$W = W_1 + W_2,$$

$$W_1 = \frac{q}{24}\left(x^4 - \frac{3}{2}k^2x^2 + \frac{5}{16}k^4\right), \tag{2.285}$$

$$W_2 = \frac{1}{4}\sum_{m=1,3,5,\ldots}^{\infty} \frac{A_m(\nu-1)}{\alpha_m \sinh\alpha_m}\left[\left(\frac{\nu+1}{\nu-1} + \alpha_m \coth\alpha_m\right)\cosh 2\alpha_m y - \right.$$

$$\left. 2\alpha_m y \sinh 2\beta_m y\right]\cos 2\alpha_m x. \tag{2.286}$$

Formula (2.285) describes the cylindrical plate bending, if the plate is simply supported on the edges $x = \pm 0.5k$ and is subjected to action of the uniformly distributed load. Formula (2.286) governs the plate displacement which is simply supported on edges $x = \pm 0.5k$, whereas on edges $y = \pm 0.5$ the following BCs hold:

$$W_{yyy} + (2-\nu)W_{xxy} = 0, \qquad W_y = \sum_{m=1,3,5,\ldots} A_m \cos 2\alpha_m x.$$

Satisfaction to BCs (2.159), (2.160) yields the following equation:

$$\sum_{m=1,3,5,\ldots} A_m \cos 2\alpha_m x =$$

$$\varepsilon \overline{H}(x)\left\{\sum_{m=1,3,5,\ldots} A_m\left[1 - \alpha_m(\nu-1)(\alpha_m(1-\nu)\frac{1}{\sinh^2\alpha_m} - \right.\right.$$

$$\left.\left.(3+\nu)C\tanh\alpha_m)\right]\cos 2\alpha_m x + \underline{\underline{\nu\frac{q}{8}(4x^2-k^2)}}\right\}, \qquad m = 1,3,5,\ldots.$$

Substituting the underlined terms and function $\overline{H}(x)\cos 2\alpha_m x$ in the form of the series with respect to $\cos 2\alpha_i x$ gives the following infinite system of LAEs regarding coefficients A_i:

$$A_i = \varepsilon \sum_{m=1,3,5,\ldots} \gamma_{im} A_m \left\{ 1 - \alpha_m(\nu - 1) \left[\alpha_m(1-\nu)\frac{1}{\sinh^2\alpha_m} \right] \right\} -$$

$$\varepsilon\nu q \left(4\mu \frac{k^2}{\pi^2 t^2 \cos \pi\mu i} + 2\frac{k^2}{\pi i}\left[\left(\mu^2 - \frac{1}{4}\right) - \frac{2}{\pi^2 t^2} \right] \right), \quad i = 1, 3, 5, \ldots,$$

$$\gamma_{im} = \begin{cases} 2\mu + \frac{1}{4\pi m}\sin 2\pi\mu m & i = m, \\ \frac{4}{\pi}\frac{1}{(m^2-i^2)}[m \sin \pi\mu m \cos \pi\mu i - i \sin \pi\mu i \cos \pi\mu m] & i \neq m. \end{cases}$$

Further we apply the perturbation approach. As a result the following recurrent conditions for coefficients A_i determination are obtained:

$$A_{i(0)} = 0,$$

$$A_{i(1)} = \nu q \left(4\mu \frac{k^2}{\pi^2 i^2}\cos \pi\mu i + \frac{2k^2}{\pi i}\left[\left(\mu^2 - \frac{1}{4}\right) - \frac{2}{\pi^2 i^2} \right] \right),$$

$$A_{i(j)} = \sum_{m=1,3,5,\ldots} A_{m(j-1)}\gamma_{im} \left\{ 1 - \alpha_m(\nu - 1) \left[\alpha_m(1-\nu)\frac{1}{\sinh^2\alpha_m} - \right.\right.$$

$$\left.\left. (3+\nu)\coth \alpha_m \right] \right\}.$$

In the next step coefficients are recast by PA (2.264).

Let us analyse the obtained solution in limiting cases. Values of $\mu = 0$ correspond to movable clamping on the plate edges $y = \pm 0.5$, where $\gamma_{im} \equiv 0$. The second limiting case corresponds to $\mu = 0.5$ and is associated with completely free edges $y = \pm 0.5$, where $\gamma_{im} = 1 - \delta_{im}$. In this case the PA in (2.264) for $\varepsilon = 1$ yields the following exact solution:

$$W = \frac{q}{24}\left(x^4 - \frac{3}{2}k^2x^2 + \frac{5}{16}k^4 \right) + \nu\frac{q}{8k} \sum_{m=1,3,5,\ldots} \frac{(-1)^{\frac{m-1}{2}}}{\alpha_m^5} \times$$

$$\frac{1}{\sinh \alpha_m(\alpha_m(1-\nu)\sinh^{-2}\alpha_m + (3+\nu)\coth \alpha_m)} \times$$

$$\left[\left(\frac{\nu+1}{\nu-1} + \alpha_m \coth \alpha_m \right)\cosh 2\alpha_m y - 2\alpha_m y \sinh 2\alpha_m y \right] \cos 2\alpha_m x.$$

Computation of the SSS components has been carried out for a square plate taking into account ten first coefficients A_i obtained via PA (2.264) for $\varepsilon = 1$. Both displacements and bending moments are defined in the plate center for certain values of the parameter μ. Results are presented in Figures 2.38, 2.39, where solid (dashed) curves are obtained by our (FEM) method.

In a similar way the problem regarding a computation of the SSS of a plate having a part of the movable clamping nonsymmetrically located with respect to the plate center is solved

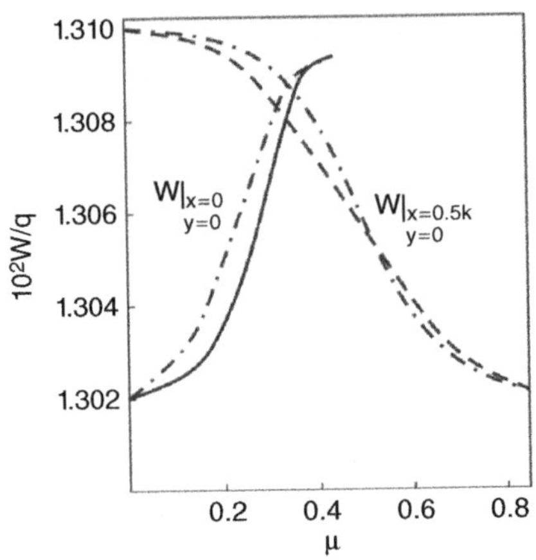

Figure 2.38 Comparison of the plate normal deflection using different methods

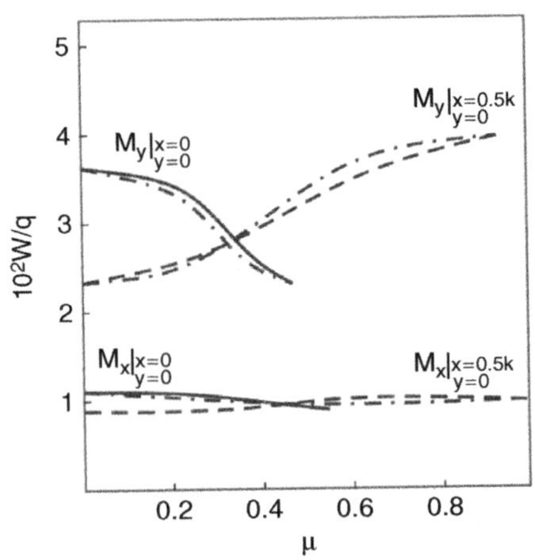

Figure 2.39 Comparison of the plate bending moments using different methods

(Figure 2.17b). In this case the basic PDE (2.252) and BCs (2.159), (2.160) remain valid. In BCs (2.159), (2.160) the function $\overline{H}(x)$ takes the following form:

$$\overline{H}(x) = H(x) - H(x - \mu k).$$

The formula describing the cylindrical plate bending has the following form:

$$W_1 = \frac{q}{24} x(x^3 2kx^2 + k^3).$$

In Ansatz (2.286) summation with respect to odd m is substituted by summation for all m.

Infinite system of LAEs regarding the coefficients A_i has the following form:

$$A_i = \varepsilon \sum_{m=1,3,5,\ldots} \left\{ \gamma_{im} A_m (\nu - 1) \left[\alpha_m (1-\nu) \frac{1}{\sinh^2 \alpha_m} - (3+\nu) \coth \alpha_m \right] \right\} +$$
$$\varepsilon \nu \frac{q}{2} \left\{ 2 \frac{k^2}{\pi^2 i^2} \sin \pi \mu i (2\mu - 1) - \frac{2k^2}{\pi i} \left(\mu^2 + \mu - \frac{2}{\pi^2 i^2} \right) \cos \pi i \mu - 4 \frac{k^2}{\pi^3 i^3} \right\},$$

where γ_{im} is defined by the formula (2.134).

Recurrent relations for the coefficients of the trigonometric series are as follows:

$$A_{i(0)} = 0,$$

$$A_{i(1)} = \nu \frac{q}{2} \left\{ 2 \frac{k^2}{\pi^2 i^2} \sin \pi \mu i (2\mu - 1) - \right.$$
$$\left. \frac{2k^2}{\pi i} \left(\mu^2 + \mu - \frac{2}{\pi^2 i^2} \right) \cos \pi i \mu - 4 \frac{k^2}{\pi^3 i^3} \right\},$$

$$A_{i(j)} = \sum_{m=1,3,5,\ldots} A_{m(j-1)} \gamma_{im} \left\{ 1 - \alpha_m (\nu - 1) \left[\alpha_m (1-\nu) \frac{1}{\sinh^2 \alpha_m} - \right. \right.$$
$$\left. \left. (3+\nu) \coth \alpha_m \right] \right\}.$$

PA coefficients A_i in (2.264) remain valid.

Let us consider the solution obtained in the limiting cases. Value $\mu = 0$ corresponds to the movable clamping on edges $y = \pm 0.5$, and hence $\gamma_{im} \equiv 0$. Second limiting case ($\mu = 1.0$) corresponds to free edges $y = \pm 0.5$ and hence γ_{im}. In this case the PA in (2.264) for $\varepsilon = 1$ yields the exact solution:

$$W = \frac{q}{24} x(x^3 - 2kx^2 + k^3) +$$
$$\nu \frac{q}{8k} \sum_{m=1} \frac{1}{\alpha_m^5} \frac{1}{\sinh \alpha_m (\alpha_m (1-\nu) \sinh^{-2} \alpha_m + (3+\nu) \coth \alpha_m)} \times$$
$$\left[\left(\frac{\nu+1}{\nu-1} + \alpha_m \coth \alpha_m y \right) \cosh 2\alpha_m y - 2\alpha_m y \sinh 2\alpha_m y \right] \sin 2\alpha_m x.$$

Results devoted to computation of the deflection and bending moments in the square plate center are presented in Figures 2.38, 2.39 by the dashed curve. Computation is carried out using the first ten coefficients A_i. Dashed curves correspond to results obtained via FEM.

Also in this case three important zones of the parameter μ are distinguished. The first one corresponds to the interval of $[0, 0.15]$ (symmetric) and $[0, 0.3]$ (nonsymmetric) problems. The second zone concerns intervals $[0.15, 0.45]$ and $[0.3, 0.9]$, whereas the third zone is associated with intervals $[0.45, 0.5]$ and $[0.9, 1.0]$. In the first and third zones changes of the parameter μ have negligible influence on the plate SSS factors. In the second zone small changes of the parameter μ lead to essential changes of all SSS plate factors. The plate SSS is not changed essentially by small parts of nonhomogeneous BCs.

2.5 Forced Vibrations of Beams and Plates

2.5.1 Forced Vibrations of a Clamped Beam

We consider the clamped beam $(-0.5l \le \overline{x} \le 0.5l)$ subjected to the periodic external load action of the form $P(\overline{x}, y) = \overline{P}(x)\sin(\omega t + \alpha)$. Basic PDE has the following form:

$$EIy_{\overline{xxxx}} + \rho y_{tt} = \overline{P}(x)\sin(\omega t + \alpha). \tag{2.287}$$

Solution to Equation (2.287) is sought in the following form:

$$y = W(x)\sin(\omega t + \alpha), \tag{2.288}$$

where $x = \overline{x}/l$.

Substituting Ansatz (2.288) into Equation (2.287) yields

$$W^{IV} - \lambda^4 W = P(x), \tag{2.289}$$

where $\lambda^4 = \frac{\rho\omega^2 l^2}{EI}$, $P(x) = \frac{\overline{P}(x)l^3}{EI}$.

BCs are taken in the form of (2.15).

Let us present the beam displacement as PS. Substituting it to Equation (2.289) and BCs (2.15), and after splitting with respect to ε the following recurrent sequence of the BVPs is obtained:

$$W_0^{IV} - \lambda^4 W_0 = P(x),$$
$$W_0 = 0, \quad W^{II} = 0 \quad \text{for} \quad x = \pm 0.5, \tag{2.290}$$

$$W_j^{IV} - \lambda^4 W_j = 0,$$
$$W_j = 0, \quad W_j^{II} = \mp \sum_{i=0}^{j-1} W_i^{I} \quad \text{for} \quad x = \pm 0.5. \tag{2.291}$$

Let us consider the case when the frequency of excitation does not coincide with any of the eigenfrequencies of the simply supported beam. PDE (2.290) in the zero order approximation has the following form:

$$W_0^{IV} - \lambda^4 W_0 = \sum_{n=1,3,5,\ldots} A_n \cos \pi n x + \sum_{n=2,4,6,\ldots} B_n \sin \pi n x,$$

where

$$A_n = 2\int_{-0.5}^{0.5} P(x)\cos \pi n y \quad \text{for} \quad n = 1, 3, 5, \ldots,$$
$$B_n = 2\int_{-0.5}^{0.5} P(x)\sin \pi n x \quad \text{for} \quad n = 2, 4, 6, \ldots.$$

In this case we obtain:

$$W = \sum_{n=1,3,5,\ldots} \frac{A_n}{\pi^4 n^4 - \lambda^4}\cos \pi n x + \sum_{n=2,4,6,\ldots} \frac{B_n}{\pi^4 n^4 - \lambda^4}\sin \pi n x +$$
$$\sum_{j=0} \varepsilon^{j+1}\left\{ \sum_{n=1,3,5,\ldots} \frac{\pi n A_n (-1)^{\frac{n-1}{2}}}{2\lambda^2(\pi^4 n^4 - \lambda^4)}\left[1 - \frac{1}{2\lambda^2}\left(\tanh\frac{\lambda}{2} + \tan\frac{\lambda}{2}\right)\right]^j \times \right. \tag{2.292}$$

$$\left(\frac{\cosh\lambda x}{\cosh\lambda/2}-\frac{\cos\lambda x}{\cos\lambda/2}\right)-\sum_{n=2,4,6,\ldots}\frac{\pi n B_n(-1)^{\frac{n}{2}}}{2\lambda^2(\pi^4n^4-\lambda^4)}\left[1-\frac{1}{2\lambda^2}\left(\coth\frac{\lambda}{2}\cot\frac{\lambda}{2}\right)\right]^j\times$$
$$\left.\left(\frac{\sinh\lambda x}{\sinh\lambda/2}-\frac{\sin\lambda x}{\sin\lambda/2}\right)\right\}.$$

Changing the summation order with respect to n and j in formula (2.292), the following expression is obtained:

$$W=\sum_{n=1,3,5,\ldots}\frac{A_n}{\pi^4n^4-\lambda^4}\cos\pi nx+\sum_{n=2,4,6,\ldots}\frac{B_n}{\pi^4n^4-\lambda^4}\sin\pi nx+$$
$$\sum_{n=1,3,5,\ldots}\frac{\pi n A_n(-1)^{\frac{n-1}{2}}}{2\lambda^2(\pi^4n^4-\lambda^4)}\frac{\varepsilon}{1-\left[1-\frac{1}{2\lambda^2}\left(\coth\frac{\lambda}{2}+\cot\frac{\lambda}{2}\right)\right]\varepsilon}\times$$
$$\left(\frac{\cosh\lambda x}{\cosh\lambda/2}-\frac{\cos\lambda x}{\cos\lambda/2}\right)-\sum_{n=2,4,6,\ldots}\frac{\pi n B_n(-1)^{\frac{n}{2}}}{2\lambda^2(\pi^4n^4-\lambda^4)}\frac{\varepsilon}{1-\left[1-\frac{1}{2\lambda^2}\left(\cot\frac{\lambda}{2}\cot\frac{\lambda}{2}\right)\right]\varepsilon}\times$$
$$\left.\left(\frac{\sinh\lambda x}{\sinh\lambda/2}-\frac{\sin\lambda x}{\sin\lambda/2}\right)\right\}.$$

The same result is obtained if in Ansatz (2.292) we keep only three first terms, and then recast them through the PA.

Let us estimate how the solution obtained via the PA differs from results found through PS. Let us study the particular case:

$$P(x)=q\cos\pi mx,\quad m=1,3,5,\ldots.$$

Then, in the zeroth order approximation we get

$$W_0=\frac{q}{\pi^4m^4-\lambda^4}\cos\pi mx. \tag{2.293}$$

In the next approximation we find

$$W_1=\frac{xmq(-1)^{\frac{m-1}{2}}}{2\lambda^2(\pi^4m^4-\lambda^4)}\left(\frac{\cosh\lambda x}{\cosh\lambda/2}-\frac{\cos\lambda x}{\cos\lambda/2}\right), \tag{2.294}$$

$$W_2=\frac{\pi mq(-1)^{\frac{m-1}{2}}}{2\lambda^2(\pi^4m^4-\lambda^4)}\left(1-\frac{1}{2\lambda}\left(\tanh\frac{\lambda}{2}+\tan\frac{\lambda}{2}\right)\right)\left(\frac{\cosh\lambda x}{\cosh\lambda/2}-\frac{\cos\lambda x}{\cos\lambda/2}\right). \tag{2.295}$$

Let us compute the beam displacement using the PS (Figure 2.40), where we have fixed the value $k_m=\lambda^4/(\pi m)^4=0.9$ during computations. It is clear that PS for $\varepsilon=1$ gives a rather high error of the beam displacement. In Figure 2.41 graphs of errors associated with the beam displacement estimation for various values of the parameter k_m are shown.

Analogous results are obtained if one computes the bending moment for each step of the approximations, and then the results obtained are compared with the exact solution (Figure 2.42, $k_m=0.9$).

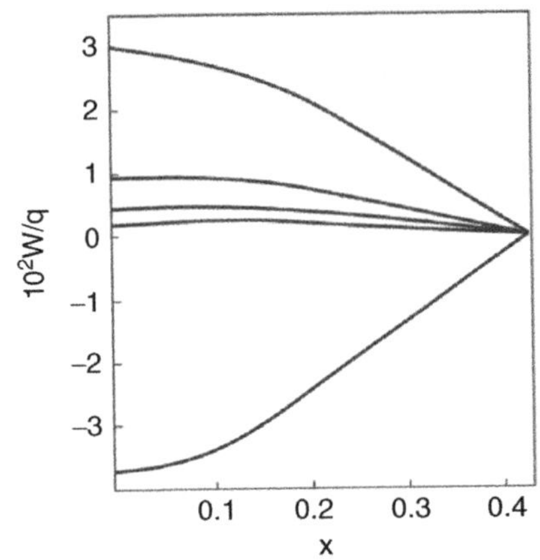

Figure 2.40 Computation of the beam displacement using PS and PA: 1 - W_0 (2.293); 2 - W_1 (2.294); 3 - W_2 (2.295); 4 - $W_0 + W_1 + W_2$; 5 - exact solution and PA

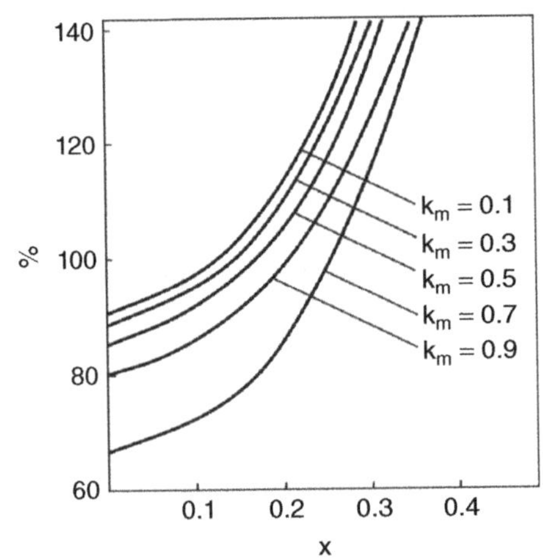

Figure 2.41 Error in estimation of the beam deflection versus parameter $k_m = \lambda^4/(\pi m)^4$

In Figure 2.43 it is shown how the error of the bending moment determination defined via PS changes for $\varepsilon = 1$ in different beam cross sections for some values of the parameter k_m.

The solution constructed so far possesses the following drawback: secular terms appear when the exciting frequency coincides with one of the simple supported beam frequencies. However, this effect can be removed by developing the external load into a series regarding modes of the clamped beam (2.29). One may verify that if in Equation (2.29) we take $c = \sqrt{2}$, then

$$\int_{-0.5}^{0.5} W_i W_j dx = \int_{-0.5}^{0.5} W_{0i} W_{0j} dx + \varepsilon \int_{-0.5}^{0.5} (W_{0i} W_{1j} + W_{1i} W_{0j}) dx + \varepsilon^2 \int_{-0.5}^{0.5} (W_{0i} W_{2j} + W_{1i} W_{1j} + W_{2i} W_{2j}) dx = \delta_{ij}. \tag{2.296}$$

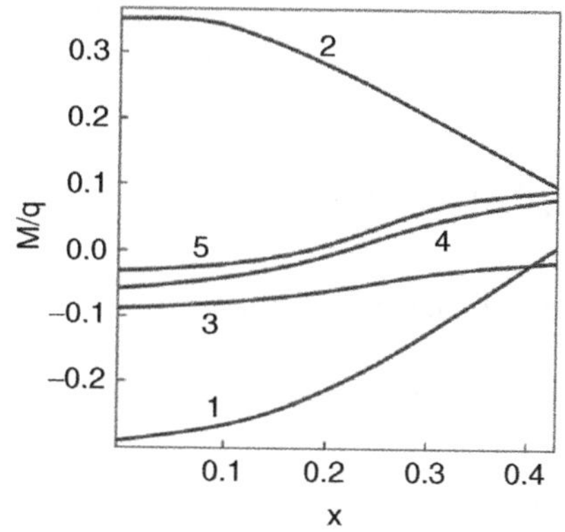

Figure 2.42 Errors in bending moment estimations by PS and PA: 1−M_0; 2−M_1; 3−M_2; 4−$M_0 + M_1 + M_2$; 5−exact solution and PA

We apply the method of splitting with respect to eigenmodes W_n and we look for the beam deflection in the following form:

$$W = \sum_{n=1}^{\infty} A_n W_n. \tag{2.297}$$

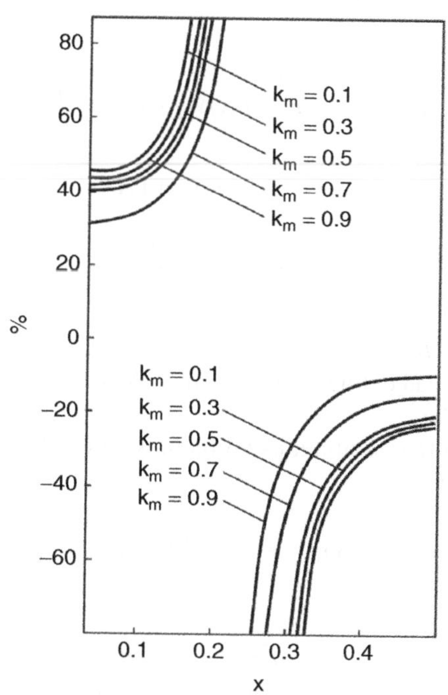

Figure 2.43 Error in bending moment estimation by PS

Substituting Ansatz (2.297) into Equation (2.289), multiplying it by the eigenform W_k, and carrying out the integration with respect to x from -0.5 to 0.5, we get

$$\int_{-0.5}^{0.5} \sum_{n=1}^{\infty} A_n(W_n^{IV} - \lambda^4 W_n)W_k dx = \int_{-0.5}^{0.5} P(x)W_k dx. \tag{2.298}$$

Since an arbitrary eigenform satisfies the equation

$$W_n^{IV} - \lambda_n^4 W_n = 0,$$

where λ_n^4 is the eigenvalue of the BVP (2.12)–(2.15), Equation (2.298) takes the form:

$$\int_{-0.5}^{0.5} \sum_{n=1}^{\infty} A_n(\lambda_n^4 - \lambda^4)W_n W_k dx = \int_{-0.5}^{0.5} P(x)W_k dx.$$

Taking into account Equation (2.296) we get

$$A_k = \frac{1}{\lambda_n^4 - \lambda^4} \int_{-0.5}^{0.5} f(x)(W_{0k} + W_{1k}\varepsilon + W_{2k}\varepsilon^2 + \ldots)dx.$$

Therefore, coefficient A_k is defined in the form of truncated PS:

$$A_k = A_0 + A_1\varepsilon + A_2\varepsilon^2 + \ldots =$$
$$\frac{1}{\lambda_n^c - \lambda^4}\left[\int_{-0.5}^{0.5} f(x)W_{0k}dx + \varepsilon \int_{-0.5}^{0.5} f(x)W_{1k}dx + \varepsilon^2 \int_{-0.5}^{0.5} f(x)W_{2k}dx + \ldots\right].$$

Further, coefficients A_k are presented in the form of PA

$$A_{k[1/1]}(\varepsilon) = \frac{a_0 + a_1\varepsilon}{1 + b_1\varepsilon},$$

where $a_0 = A_{0k}$, $a_1 = A_{1k} + b_1 A_{0k}$, $b_1 = A_{2k}/A_{1k}$.

Now the beam deflection can defined using formula Ansatz (2.297), and eigenforms W_n are constructed in the form of PA.

One can mention that the free oscillations problem is reduced to the forced oscillations one because the initial conditions

$$w = w^{(0)}(x, y), \quad w_t = w^{(1)}(x, y); \quad \text{at} \quad t = 0$$

are equivalent to the load

$$q = w^{(0)}\frac{d^2\delta(t)}{dt^2} + w^{(1)}\frac{d\delta(t)}{dt},$$

where $\delta(t)$ is the Dirac function.

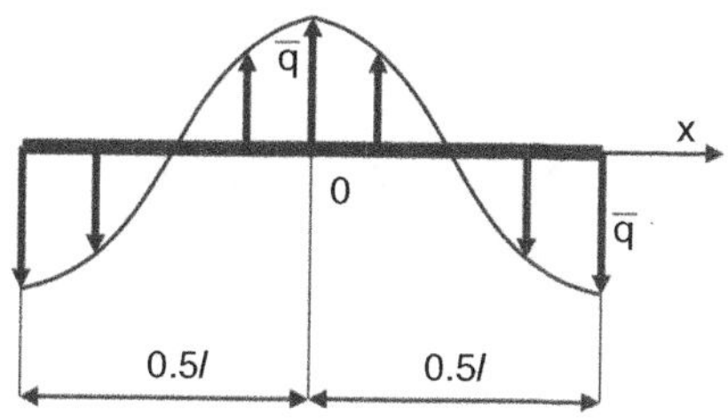

Figure 2.44 Beam subjected to a self-balanced load

2.5.2 *Forced Vibrations of Beam with Free Edges*

We consider computation of the dynamical SSS of a beam with free edges ($-0.5l \leq x \leq 0.5l$), subjected to action of the periodic self-balanced load $q(\overline{x}, t) = \overline{q} \cos \frac{\pi m}{l} \overline{x} \sin(\omega t + \alpha)$, $m = 2, 4, 6, \ldots$ (Figure 2.44). PDE governing the beam dynamics obtained via introduction of the nondimensional quantities (2.8), and then time and space variables separation (2.288) takes the following form:

$$W^{IV} - \lambda^4 W = q \cos \pi m x, \qquad m = 2, 4, 6, \ldots . \tag{2.299}$$

We apply BCs (2.41) to the ODE (2.299).

Note that exact solution to the BVP (2.299), (2.41) follows:

$$W = \frac{q}{\pi^4 n^4 - \lambda^4} \cos \pi m x + C_1 \cosh \lambda x + C_2 \cos \lambda x.$$

Satisfaction to the BCs (2.41) allows us to define arbitrary constants C_1 and C_2:

$$C_1 = \frac{q\pi^2 m^2 (-1)^{\frac{m}{2}}}{\lambda(\pi^4 n^4 - \lambda^4)} \cdot \frac{\varepsilon}{[2(1-\varepsilon) + \varepsilon\lambda(\coth \lambda/2 + \cot \lambda/2)]} \cdot \frac{1}{\sinh \lambda/2},$$

$$C_2 = \frac{q\pi^2 m^2 (-1)^{\frac{m}{2}}}{\lambda(\pi^4 n^4 - \lambda^4)} \cdot \frac{\varepsilon}{[2(1-\varepsilon) + \varepsilon\lambda(\coth \lambda/2 + \cot \lambda/2)]} \cdot \frac{1}{\sin \lambda/2}.$$

Finally, the exact solution has the following form:

$$W = \frac{q}{\pi^4 m^4 - \lambda^4} \left[\cos \pi m x + \frac{\pi^2 m^2 (-1)^{\frac{m}{2}} \varepsilon}{\lambda[2(1-\varepsilon) + \varepsilon\lambda(\coth \lambda/2 + \cot \lambda/2)]} \times \left(\frac{\cosh \lambda x}{\sinh \lambda/2} - \frac{\cos \lambda x}{\cos \lambda/2} \right) \right]. \tag{2.300}$$

Substituting the displacement W in the form of PS in (2.16), and then substituting it to the ODE (2.299) and BCs (2.41), and splitting with respect to ε, the following recurrent sequence of the BVPs is obtained:

$$W_0^{IV} - \lambda^4 W_0 = q \cos \pi m x, \qquad m = 2, 4, 6, \ldots,$$

$$W_0^{III} = 0, \quad W_0^{I} = 0 \quad \text{for} \quad x = \pm 0.5,$$

$$W_j^{IV} - \lambda^4 W_j = 0,$$

$$W_j^{III} = 0, \quad W_j^{I} = \mp \sum_{i=0}^{j-1} W_i^{II} \quad \text{for} \quad x = \pm 0.5, \quad j = 1, 2, 3, \ldots .$$

Taking into account the successive approximations, one may obtain the form of a general term of PS:

$$W_0 = \frac{q}{\pi^4 m^4 - \lambda^4} \cos \pi m x, \qquad m = 2, 4, 6, \ldots, \tag{2.301}$$

$$W_1 = \frac{q\pi^2 m^2 (-1)^{\frac{m}{2}}}{2\lambda(\pi^4 m^4 - \lambda^4)} \left(\frac{\cosh \lambda x}{\sinh \lambda/2} - \frac{\cos \lambda x}{\sin \lambda/2} \right), \tag{2.302}$$

$$W_2 = \frac{q\pi^2 m^2 (-1)^{\frac{m}{2}}}{2\lambda(\pi^4 m^4 - \lambda^4)} \left[1 - \frac{\lambda}{2} \left(\coth \frac{\lambda}{2} + \cot \frac{\lambda}{2} \right) \right] \left(\frac{\cosh \lambda x}{\sinh \lambda/2} - \frac{\cos \lambda x}{\sin \lambda/2} \right), \tag{2.303}$$

$$W_j = \frac{q\pi^2 m^2 (-1)^{\frac{m}{2}}}{2\lambda(\pi^4 m^4 - \lambda^4)} \left[1 - \frac{\lambda}{2} \left(\coth \frac{\lambda}{2} + \cot \frac{\lambda}{2} \right) \right]^{j-1} \left(\frac{\cosh \lambda x}{\sinh \lambda/2} - \frac{\cos \lambda x}{\sin \lambda/2} \right). \tag{2.304}$$

After summing up the obtained approximations the solution takes the form:

$$W = \frac{q}{\pi^4 m^4 - \lambda^4} \cos \pi m x + \frac{q\pi^2 m^2 (-1)^{\frac{m}{2}}}{2\lambda(\pi^4 m^4 - \lambda^4)} \times,$$
$$\left\{ \sum_{j=0}^{\infty} \varepsilon^{j+1} \left[1 - \frac{\lambda}{2} \left(\coth \frac{\lambda}{2} + \cot \frac{\lambda}{2} \right) \right]^{j} \right\} \left(\frac{\cosh \lambda x}{\sinh \lambda/2} - \frac{\cos \lambda x}{\sin \lambda/2} \right). \tag{2.305}$$

Since the series appearing in (2.305) is a geometric progression, its sum is

$$\sum_{j=0}^{\infty} \varepsilon^{j+1} \left[1 - \frac{\lambda}{2} \left(\coth \frac{\lambda}{2} + \cot \frac{\lambda}{2} \right) \right] = \frac{\varepsilon}{1 - \left[1 - \frac{\lambda}{2} \left(\coth \frac{\lambda}{2} + \cot \frac{\lambda}{2} \right) \right] \varepsilon}. \tag{2.306}$$

Observe that taking into account Equation (2.306) the solution (2.305) coincides with the exact one (2.300).

Exact solution can be obtained taking into account only three terms of the series regarding excitation and applying PA to the truncated PS.

Let us compare how the solution constructed using the truncated PS differs from the exact one. For this purpose we compute the deflection for $n = 2$, $r_m = \lambda^4/(\pi m)^4$ (Figure 2.45).

In Figure 2.46 bending moments computational results obtained via PS are reported. It is evident that the PS cannot be applied for the bending moment determination. However, application of PA essentially improves the results.

2.5.3 *Forced Vibrations of a Clamped Plate*

We study vibrations of the rectangular plate $(-0.5a \le \overline{x} \le 0.5a; -0.5b \le \overline{y} \le 0.5b)$ clamped along its contour and loaded by a normal periodic force of the following form

$$q(\overline{x}, \overline{y}, t) = \overline{q}_0 \cos \frac{\pi m}{a} \overline{x} \cos \frac{\pi n}{b} \overline{y} \sin(\omega t + \alpha), \quad m, n = 1, 3, 5, \ldots .$$

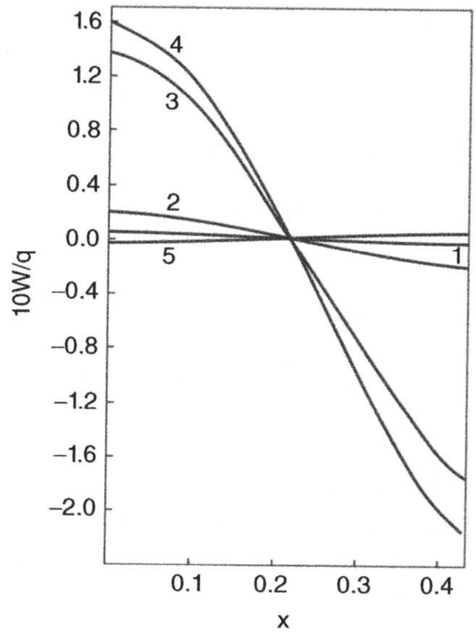

Figure 2.45 Comparison of efficiency of various approximations: 1 – W_0 (2.301); 2 – W_1 (2.302); 3 – W_2 (2.303); 4 – $W_0 + W_1 + W_2$; 5 – exact solution (2.300) and PA

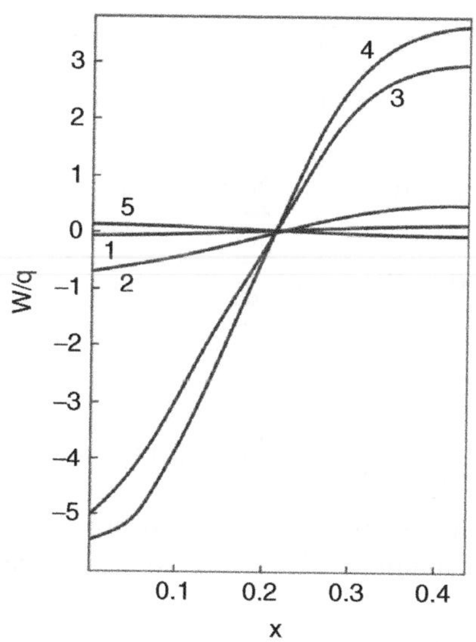

Figure 2.46 Comparison of efficiency of PS and PA: 1 – M_0; 2 – M_1; 3 – M_2; 4 – $M_0 + M_1 + M_2$; 5 – exact solution and PA

Basic PDE has the following form:

$$D\nabla^4\overline{W} + \rho\overline{W}_{tt} = \overline{q}_0 \cos\frac{\pi m}{a}\overline{x}\cos\frac{\pi n}{b}\overline{y}\sin(\omega t + \alpha). \tag{2.307}$$

Solution to Equation (2.299) is sought in the form:

$$\overline{W} = W(x, y)\sin(\omega t + \alpha), \tag{2.308}$$

where $x = \overline{x}/b$, $y = \overline{y}/b$.

Substituting Ansatz (2.308) into Equation (2.307) we obtain

$$\nabla^4 W - \lambda^4 W = q\cos\frac{\pi m}{k}x\cos\pi ny, \quad m, n = 1, 3, 5, \dots, \tag{2.309}$$

where $\lambda^4 = \rho\omega^2 b^2/D$; $q = \overline{q}_0 b^3/D$.

We add to Equation (2.308) BCs (2.66), (2.67).

We present the plate displacement as PS (2.16). Then we substitute the PS into Equation (2.309) and BCs (2.66), (2.67), and after a splitting procedure with respect to ε, the following recurrent set of the BVPs is obtained:

$$\nabla^4 W_0 - \lambda^4 W_0 = q\cos\frac{\pi m}{k}x\cos\pi ny, \quad m, n = 1, 3, 5, \dots,$$

$$W_0 = 0, \quad W_{0xx} = 0 \quad \text{for} \quad x = \pm 0.5k,$$

$$W_0 = 0, \quad W_{0yy} = 0 \quad \text{for} \quad y = \pm 0.5,$$

$$\nabla^4 W_j - \lambda^4 W_j = 0,$$

$$W_j = 0, \quad W_{jxx} = \mp k\sum_{i=0}^{j-1} W_{ix} \quad \text{for} \quad x = \pm 0.5k,$$

$$W_j = 0, \quad W_{jyy} = \mp k\sum_{i=0}^{j-1} W_{iy} \quad \text{for} \quad y = \pm 0.5.$$

In zeroth order approximation we get:

$$W_0 = \frac{q}{\pi^4\alpha^2 - \lambda^4}\cos\frac{\pi m}{k}x\cos\pi ny, \quad \alpha = n^2 + \frac{m^2}{k^2}. \tag{2.310}$$

In successive approximations we obtain problems governed by homogeneous equations and nonhomogeneous BCs. Their solutions with respect to the first- and second-order approximations follow:

$$W_1 = \frac{\pi q}{2\lambda^2(\pi^4\alpha^2 - \lambda^4)}\left\{ m(-1)^{\frac{m-1}{2}}\left[\frac{\cosh\beta_{1n}x}{\cosh\beta_{1n}/2} - \frac{\begin{Bmatrix}\cos\beta_{2n}x\\ \cosh\beta_{3n}x\end{Bmatrix}}{\begin{Bmatrix}\cos\beta_{2n}k/2\\ \cosh\beta_{3n}k/2\end{Bmatrix}}\right]\cos\pi ny + \right.$$

$$n(-1)^{\frac{n-1}{2}}\left[\frac{\cosh\gamma_{1m}y}{\cosh\gamma_{1m}/2}-\frac{\begin{Bmatrix}\cos\gamma_{2m}y\\ \cosh\gamma_{3m}y\end{Bmatrix}}{\begin{Bmatrix}\cos\gamma_{2m}/2\\ \cosh\gamma_{3m}/2\end{Bmatrix}}\right]\cos\frac{\pi m}{k}x;\quad \begin{Bmatrix}\lambda>\pi n, & \lambda>\frac{\pi m}{k}\\ \lambda<\pi n, & \lambda<\frac{\pi m}{k}\end{Bmatrix};$$

$$W_2=\frac{\pi q}{2\lambda^2(\pi^4\alpha^2-\lambda^4)}\left[m(-1)^{\frac{m-1}{2}}\left\{1-\frac{1}{2\lambda^2}\left[\beta_{1n}\tanh\beta_{1n}/2+\right.\right.\right.$$

$$\left.\left.\begin{Bmatrix}\beta_{2n}\tan\beta_{2n}k/2\\ -\beta_{2n}\tanh\beta_{3n}k/2\end{Bmatrix}\right]\right\}\times\left[\frac{\cosh\beta_{1n}x}{\cosh\beta_{1n}/2}-\frac{\begin{Bmatrix}\cos\beta_{2n}x\\ \cosh\beta_{3n}x\end{Bmatrix}}{\begin{Bmatrix}\cos\beta_{2n}k/2\\ \cosh\beta_{3n}k/2\end{Bmatrix}}\right]\cos\pi ny+$$

$$n(-1)^{\frac{n-1}{2}}\left\{1-\frac{1}{2\lambda^2}\left[\gamma_{1m}\tanh\gamma_{1m}k/2+\right.\right.$$

$$\left.\left.\left.\begin{Bmatrix}\gamma_{2m}\tan\gamma_{2m}/2\\ -\gamma_{3m}\tanh\gamma_{3m}/2\end{Bmatrix}\right]\right\}\left[\frac{\cosh\gamma_{1m}y}{\cosh\gamma_{1m}/2}-\frac{\begin{Bmatrix}\cos\gamma_{2m}y\\ \cosh\gamma_{3m}y\end{Bmatrix}}{\begin{Bmatrix}\cos\gamma_{2m}/2\\ \cosh\gamma_{3m}/2\end{Bmatrix}}\right]\cos\frac{\pi m}{k}x\right]-$$

$$\frac{2\pi^3 qmn}{\lambda^2(\pi^4\alpha^2-\lambda^4)}(-1)^{\frac{m-1}{2}}\times(-1)^{\frac{n-1}{2}}\sum_{i=1,3,5,\ldots}(-1)^{\frac{i-1}{2}}i\left[\frac{1}{\pi^4\left(\frac{m^2}{k^2}+i^2\right)^2}\times\right.$$

$$\left\{\frac{\cosh\beta_{1i}x}{\cosh\beta_{1i}k/2}-\frac{\begin{Bmatrix}\cos\beta_{2i}x\\ \cosh\beta_{3i}x\end{Bmatrix}}{\begin{Bmatrix}\cos\beta_{2i}k/2\\ \cosh\beta_{3i}k/2\end{Bmatrix}}\right\}\cos\pi iy-\frac{1}{\pi^4\left(\frac{i^2}{k^2}+n^2\right)^2-\lambda^4}\left\{\frac{\cosh\gamma_{1i}y}{\cosh\gamma_{1i}/2}-\right.$$

$$\left.\frac{\begin{Bmatrix}\cos\gamma_{2i}y\\ \cosh\gamma_3 y\end{Bmatrix}}{\begin{Bmatrix}\cos\gamma_{2i}/2\\ \cosh\gamma_{3i}/2\end{Bmatrix}}\right\}\cos\frac{\pi i}{k}x\Bigg];\qquad \begin{Bmatrix}\lambda>\pi i(n), & \lambda>\frac{\pi i(m)}{k}\\ \lambda<\pi i(n), & \lambda<\frac{\pi i(m)}{k}\end{Bmatrix}, \tag{2.311}$$

where $\beta_{1i}=\sqrt{\lambda^2+\pi^2 i^2}$, $\beta_{2i}=\sqrt{\lambda^2-\pi^2 i^2}$, $\beta_{3i}=\sqrt{\pi^2 i^2-\lambda^2}$, $\gamma_{1i}=\sqrt{\lambda^2+\pi^2 i^2/k^2}$, $\gamma_{2i}=\sqrt{\lambda^2-\pi^2 i^2/k^2}$, $\gamma_{3i}=\sqrt{\pi^2 i^2/k^2-\lambda^2}$.

We construct PA using the first three coefficients of the series W. In Figure 2.47 graphs of deflection changes for $y=0$, $m=n=1$, $k_m=\lambda^4/(\pi^4\lambda)$ are shown. In Figure 2.48 results concerning a computation of the bending moment are reported.

2.5.4 *Forced Vibrations of Plates with Free Edges*

We consider vibrations of the rectangular plate $(-a/2\le\bar{x}\le a/2; -b/2\le\bar{y}\le b/2)$, loaded by the self-balanced normal periodic load

$$\bar{q}(\bar{x},\bar{y},t)=\bar{q}\cos\frac{\pi m}{a}\bar{x}\cos\frac{\pi n}{b}\bar{y}\sin(\omega t+\alpha),\quad m,n=2,4,6,\ldots.$$

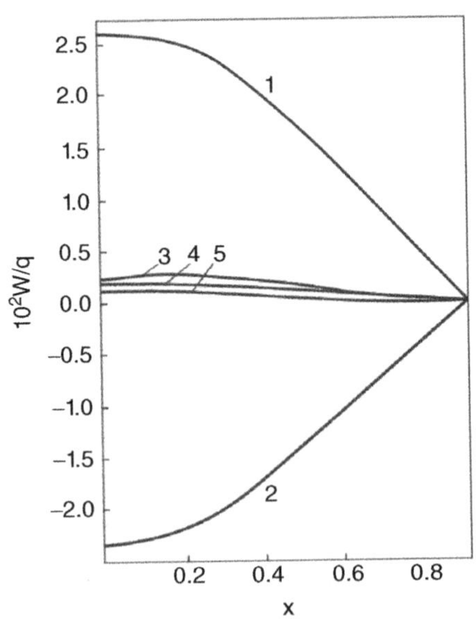

Figure 2.47 Computational results of deflections using PS and PA: 1–W_0 (2.310); 2–W_1 (2.311); 3–W_2 (2.311); 4–$W_0+W_1+W_2$; 5–PA

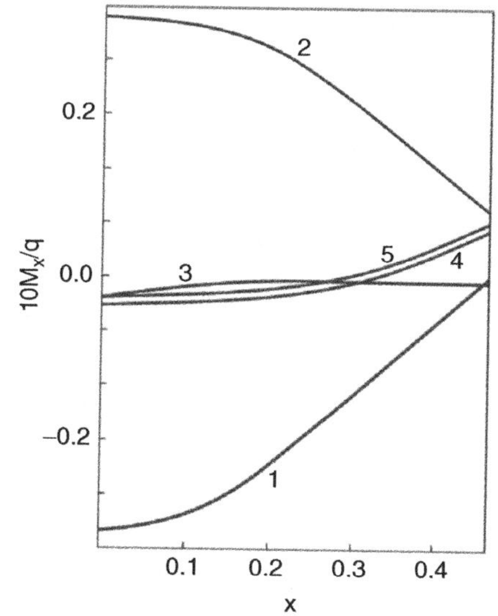

Figure 2.48 Computation of the bending moment using PS and PA: $1-M_0$; $2-M_1$; $3-M_2$; $4-M_0+M_1+M_2$; $5-$ PA

After introduction of the nondimensional quantities and after separation of spatial and time variables the input equations have the following form:

$$\nabla^4 W - \lambda^4 W = q\cos\frac{\pi m}{k}x\cos\pi ny, \quad m,n = 2,4,6,\ldots. \tag{2.312}$$

We attach BCs (2.284), (2.285) to the PDE (2.312). Furthermore, in order to solve the BVP (2.312), (2.284), (2.285) the PS () is applied. Plate deflection is presented in the form of (2.16). After its substitution into PDE (2.312) and BCs (2.284), (2.285), and after splitting with respect to ε, the following recurrent sequence of the BVPs is obtained

$$\nabla^4 W_0 - \lambda^4 W_0 = q\cos\frac{\pi m}{k}x\cos\pi ny, \quad m,n = 2,4,6,\ldots,$$

$$W_0 = 0, \quad W_{0xxx} = 0 \quad \text{for} \quad x = \pm 0.5k,$$

$$W_0 = 0, \quad W_{0yyy} = 0 \quad \text{for} \quad y = \pm 0.5,$$

$$\nabla^4 W_j - \lambda^4 W_j = 0,$$

$$W_{jxxx} + (2-\nu)W_{jyyx} = 0, \quad W_{jx} = \mp k\sum_{i=0}^{j-1}(W_{ixx} + \nu W_{iyy}) \quad \text{for} \quad x = \pm 0.5k,$$

$$W_{jyyy} + (2-\nu)W_{jxxy} = 0, \quad W_{jy} = \mp \sum_{i=0}^{j-1}(W_{iyy} + \nu W_{ixx}) \quad \text{for} \quad y = \pm 0.5.$$

Zeroth order approximation yields:

$$W_0 = \frac{q}{\pi^4\alpha^2 - \alpha^4}\cos\frac{\pi m}{k}x\cos\pi n y.$$

We find the PS coefficients successively solving the obtained sequence of approximations:

$$W_1 = \frac{q\pi^2}{2\lambda^2(\pi^4\alpha^2-\lambda^4)}\left\{\left(n^2+\nu\frac{m^2}{k^2}\right)(-1)^{\frac{n}{2}}\left[\left(\lambda^2+(1-\nu)\pi^2\frac{m^2}{k^2}\right)\frac{1}{\gamma_{1m}}\times\right.\right.$$

$$\left.\frac{\cosh\gamma_{1m}y}{\sinh\gamma_{1m}/2}+\left(\lambda^2-(1-\nu)\pi^2\frac{m^2}{k^2}\right)\begin{Bmatrix}-\frac{1}{\gamma_{2m}}\cdot\frac{\cos\gamma_{2m}y}{\sin\gamma_{2m}/2}\\[2ex] \frac{1}{\gamma_{3m}}\cdot\frac{\cosh\gamma_{3m}y}{\sinh\gamma_{3m}/2}\end{Bmatrix}\right]\cos\frac{\pi m}{k}x+$$

$$\left(\frac{m^2}{k^2}+\nu n^2\right)(-1)^{\frac{m}{2}}\left[(\lambda^2+(1-\nu)\pi^2n^2)\frac{1}{\beta_{1n}}\cdot\frac{\cosh\beta_{1n}x}{\sinh\beta_{1n}k/2}+\right.$$

$$\left.\left.(\lambda^2-(1-\nu)\pi^2n^2)\begin{Bmatrix}-\frac{1}{\beta_{2n}}\cdot\frac{\cos\beta_{2n}x}{\sin\beta_{2n}k/2}\\[2ex] \frac{1}{\beta_{3n}}\cdot\frac{\cosh\beta_{3n}x}{\sinh\beta_{3n}k/2}\end{Bmatrix}\right]\cos\pi n y\right\},$$

$$W_2 = W_1 - \frac{q\pi^2}{2\lambda^2(\pi^4\alpha^2-\lambda^4)}\left\{k(-1)^{\frac{n}{2}}\left(n^2+\nu\frac{m^2}{k^2}\right)\left[\frac{1}{\gamma_{1m}}(\lambda^2+\right.\right.$$

$$\left.(1-\nu)\pi^2\frac{m^2}{k^2}\right)^2\coth\gamma_{1m}/2+\left(\lambda^2-(1-\nu)\pi^2\frac{m^2}{k^2}\right)^2\left.\begin{Bmatrix}\frac{1}{\gamma_{2m}}\cot\gamma_{2m}/2\\[2ex] -\frac{1}{\gamma_{3m}}\coth\gamma_{3m}/2\end{Bmatrix}\right]\times$$

$$\left[\left(\lambda^2+(1-\nu)\pi^2\frac{m^2}{k^2}\right)\frac{1}{\gamma_{1m}}\cdot\frac{\cosh\gamma_{1m}y}{\sinh\gamma_{1m}/2}-\left(\lambda^2-(1-\nu)\pi^2\frac{m^2}{k^2}\right)\cdot\right.$$

$$\left.\begin{Bmatrix}-\frac{1}{\gamma_{2m}}\cdot\frac{\cos\gamma_{2m}y}{\sin\gamma_{2m}/2}\\[2ex] \frac{1}{\gamma_{3m}}\cdot\frac{\cosh\gamma_{3m}y}{\sinh\gamma_{3m}/2}\end{Bmatrix}\right]\cos\frac{\pi m}{k}x+(-1)^{\frac{m}{2}}\left(\frac{m^2}{k^2}+\gamma n^2\right)\times$$

$$\left[(\lambda^2+(1-\nu)\pi^2n^2)\frac{1}{\beta_{1n}}\coth\beta_{1n}k/2+(\lambda^2-(1-\nu)\pi^2n^2)\begin{Bmatrix}\frac{1}{\beta_{2n}}\cdot\frac{\cos\beta_{2n}x}{\sin\beta_{2n}k/2}\\[2ex] -\frac{1}{\beta_{3n}}\cdot\frac{\cosh\beta_{3n}x}{\sinh\beta_{3n}k/2}\end{Bmatrix}\right]\times$$

$$\left.\cos\pi n y\right\}+\frac{2\pi q(-1)^{\frac{m}{2}}(-1)^{\frac{n}{2}}}{\lambda^4(\pi^4\alpha^2-\lambda^4)}\sum_{i=2,4,6,\ldots}(-1)^{\frac{i}{2}}\left\{k\left(n^2+\nu\frac{m^2}{k^2}\right)\times\right.$$

$$\left[\frac{\left(\lambda^2+(1-\nu)n^2\frac{m^2}{k^2}\right)\left((1-\nu)\pi^2\frac{m^2}{k^2}+\nu\lambda^2\right)}{\lambda^2+\pi^2\left(\frac{m^2}{k^2}+i^2\right)}-\right.$$

$$\left.\frac{\left(\lambda^2-(1-\nu)\pi^2\frac{m^2}{k^2}\right)\left((1-\nu)\pi^2\frac{m^2}{k^2}+\nu\lambda^2\right)}{\lambda^2-\pi^2\left(i^2+\frac{m^2}{k^2}\right)}\right]\times$$

$$\left[\left(\lambda^2(1-\nu)\pi^2i^2\right)\frac{1}{\beta_{1i}}\cdot\frac{\coth\beta_{1i}x}{\sinh\beta_{1i}k/2}+(\lambda^2-(1-\nu)\pi^2i^2)\times\right.$$

$$\left.\left\{\begin{array}{c}-\frac{1}{\beta_{2i}}\cdot\frac{\cos\beta_{2i}x}{\sin\beta_{2i}k/2}\\ \frac{1}{\beta_{3i}}\cdot\frac{\cosh\beta_{3i}x}{\sinh\beta_{3i}k/2}\end{array}\right\}\right]\cos\pi iy+\frac{1}{k}\left(\frac{m^2}{k^2}+\nu n^2\right)\times$$

$$\left[\frac{\left(\lambda^2+(1-\nu)\pi^2n^2\right)((1-\nu)\pi^2n^2+\nu\lambda^2)}{\lambda^2+\pi^2\left(n^2+\frac{i^2}{k^2}\right)}-\right.$$

$$\left.\frac{\left(\lambda^2-(1-\nu)\pi^2n^2\right)((1-\nu)\pi^2n^2+\nu\lambda^2)}{\lambda^2-\pi^2\left(n^2+\frac{i^2}{k^2}\right)}\right].$$

$$\cdot\left[\left(\lambda^2+(1-\nu)\pi^2\frac{i^2}{k^2}\right)\frac{1}{\gamma_{1i}}\frac{\cosh\gamma_{1i}y}{\sinh\gamma_{1i}/2}+\left(\lambda^2+(1-\nu)\pi^2\frac{i^2}{k^2}\right)\times\right.$$

$$\left.\left.\left\{\begin{array}{c}-\frac{1}{\gamma_{2i}}\cdot\frac{\cos\gamma_{2i}y}{\sin\gamma_{2i}/2}\\ \frac{1}{\gamma_{3i}}\cdot\frac{\cosh\gamma_{3i}y}{\sinh\gamma_{3i}/2}\end{array}\right\}\right]\cos\frac{\pi i}{k}x\right\},$$

where

$$\beta_{1i}=\sqrt{\lambda^2+\pi^2i^2},\quad \beta_{2i}=\sqrt{\lambda^2-\pi^2i^2},\quad \beta_{3i}=\sqrt{\pi^2i^2-\lambda^2},$$

$$\gamma_{1i}=\sqrt{\lambda^2+\pi^2i^2/k^2},\quad \gamma_{2i}=\sqrt{\lambda^2-\pi^2i^2/k^2},\quad \gamma_{3i}=\sqrt{\pi^2i^2/2-\lambda^2}.$$

Knowing three terms of the PS we construct PA in (2.268). The displacement and bending moments of the squared plate computed via PA for $y=0$, $k=1$, $\nu=0.3$, $\eta_m=\lambda^4/\pi^4\alpha^2=0.9$ are shown in Figure 2.49 and 2.50.

2.5.5 *Forced Vibrations of Plate with Mixed Boundary Conditions "Clamping-Simple Support"*

Let us study vibrations of the squared plate $(-0.5a\le\overline{x}\le0.5a;-0.5b\le\overline{y}\le0.5b)$ simply supported on its edges $x=\pm0.5a$, having mixed BCs "clamping – simple support" on its sides

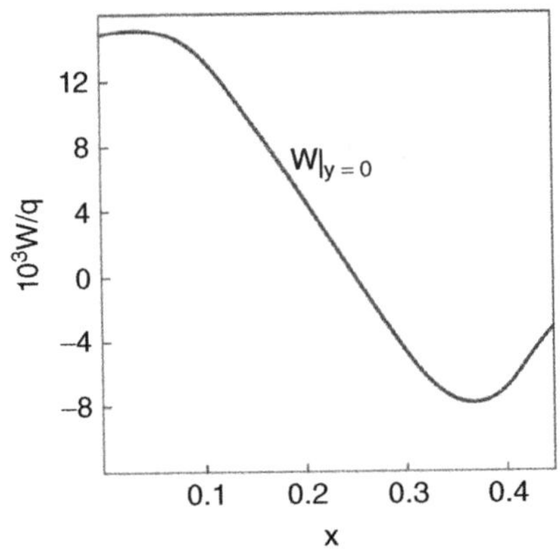

Figure 2.49 Displacement of the squared plate with free edges

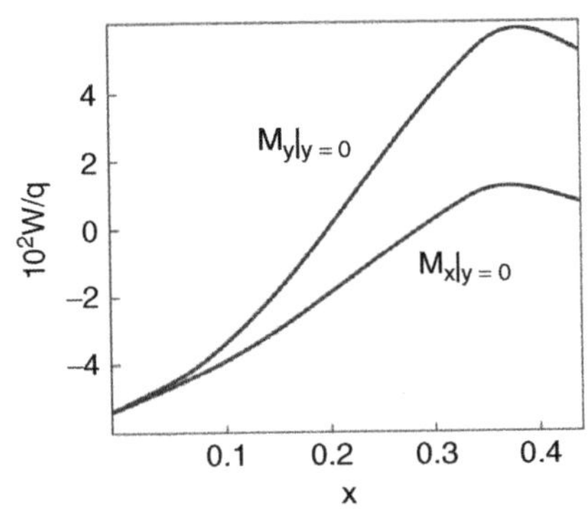

Figure 2.50 Bending moments of the squared plate with free edges

$y = \pm 0.5b$, and symmetrically located with respect to the axis y (Figure 2.8a). Plate is subjected to action of the normal periodic load of the form:

$$\overline{q}(\overline{x}, \overline{y}, t) = \overline{q}_0 \cos \frac{\pi m}{k} \overline{a} \cos \frac{\pi m}{b} \overline{y} \sin(\omega t + \alpha), \quad m, n = 1, 2, 3, \ldots .$$

BCs are taken in the form (2.114)–(2.115).

BVP (2.309), (2.114), (2.115) is reduced to an infinite system of LAE in a way analogous to the one in Section for a static case.

Boundary moment occurring on the clamping part can be approximated by the following series

$$M_{y|_{y=\pm 0.5}} = \sum_{i=1,3,5,\ldots} A_i \cos \frac{\pi i}{k} x,$$

where A_i are unknown coefficients.

Solution is sought in the following form

$$W = W_1 + W_2,$$

where $W_1 = \frac{q}{\pi^4\alpha^2-\lambda^4}\cos\frac{\pi m}{k}x\cos\pi n y$, and W_1 is a particular solution to Equation (2.309) for homogeneous BCs

$$W_1 = 0, \quad W_{1xx} = 0 \quad \text{for} \quad x = \pm 0.5k,$$

$$W_1 = 0, \quad W_{1yy} = 0 \quad \text{for} \quad y = \pm 0.5.$$

Function W_2 takes the following form:

$$W_2 = \sum_{1,3,5,\ldots} \frac{A_i}{2\lambda^2}\left[\frac{\cosh\beta_{1i}y}{\cosh\beta_{1i}/2} - \frac{\begin{Bmatrix}\cos\beta_{2i}y\\ \cosh\beta_{3i}y\end{Bmatrix}}{\begin{Bmatrix}\cos\beta_{2i}/2\\ \cos\beta_{3i}/2\end{Bmatrix}}\right]\cos\frac{\pi i}{k}x, \quad \begin{matrix}\lambda > \frac{\pi i}{k}\\ \lambda < \frac{\pi i}{k}\end{matrix},$$

where $\beta_{1i} = \sqrt{\lambda^2 + \pi^2 i^2/k^2}$, $\beta_{2i} = \sqrt{\lambda^2 - \pi^2 i^2/k^2}$, $\beta_{3i} = \sqrt{\pi^2 i^2/k^2 - \lambda^2}$. It describes a solution to the following BVP:

$$\nabla^4 W_2 - \lambda^4 W_2 = 0,$$

$$W_2 = 0, \quad W_{2xx} = 0 \quad \text{for} \quad x = \pm 0.5k,$$

$$W_2 = 0, \quad W_{2yy} = \sum_{i=1,3,5,\ldots}^{\infty} A_i\cos\frac{\pi i}{k}x \quad \text{for} \quad y = \pm 0.5.$$

Satisfaction to BCs (2.115) yields the following equations:

$$\sum_{i=1,3,5,\ldots} A_i\cos\frac{\pi i}{k}x = \varepsilon\overline{H}(x)\left[\sum_{i=1,3,5,\ldots} A_i\cos\frac{\pi i}{k}x + \frac{q\pi n(-1)^{\frac{n-1}{2}}}{\pi^4\alpha^2-\lambda^4}\cos\frac{\pi m}{k}x - \right.$$
$$\left.\sum_{i=1,3,5,\ldots}\frac{A_i}{2\lambda^2}\left(\beta_{1i}\tanh\beta_{1i}/2 + \begin{Bmatrix}\beta_{2i}\tan\beta_{2i}/2\\ -\beta_{3i}\tanh\beta_{3i}/2\end{Bmatrix}\right)\cos\frac{\pi i}{k}x\right], \quad i = 1,3,5,\ldots. \tag{2.313}$$

Splitting of the r.h.s. of Equation (2.313) into series regarding $\cos\frac{\pi j}{k}x$, $j = 1,3,5,\ldots$, yields the infinite system of LAE:

$$A_j = \varepsilon\sum_{i=1,3,5,\ldots}\gamma_{ji}A_i\left[1 - \frac{1}{2\lambda^2}\left(\beta_{1i}\tanh\beta_{1i}/2 + \begin{Bmatrix}\beta_{2i}\tan\beta_{2i}/2\\ -\beta_{3i}\tanh\beta_{3i}/2\end{Bmatrix}\right)\right] +$$
$$\varepsilon\gamma_{jm}\frac{q\pi n(-1)^{\frac{n-1}{2}}}{\pi^4\alpha^2-\lambda^4}, \quad j = 1,3,5,\ldots. \tag{2.314}$$

Here γ_{ji} coefficients are defined by Equation (2.272) ($j \to i, i \to m$).

Occurrence in the infinite system (2.314) of the parameter ε allows us to present coefficients A_j in the form of (2.261). After substitution of PS in (2.261) into system (2.314) and after comparison of coefficients standing by the same powers of ε, the following formulas for determination of unknown coefficients are obtained:

$$A_{j(0)} = 0, \tag{2.315}$$

$$A_{j(1)} = \gamma_{jm} \frac{q\pi n(-1)^{\frac{n-1}{2}}}{\pi^4\alpha^2 - \lambda^4}, \tag{2.316}$$

$$A_{j(p)} = \sum_{i=1,3,5,\ldots}^{\infty} \gamma_{ji} A_{i(p-1)} \left[1 - \frac{1}{2\lambda^2} \left(\beta_{1i} \tanh \beta_{1i}/2 + \begin{Bmatrix} \beta_{2i} \tan \beta_{2i}/2 \\ -\beta_{3i} \tanh \beta_{3i}/2 \end{Bmatrix} \right) \right]. \tag{2.317}$$

PA for coefficients A_j takes the form of (2.264).

We consider behavior of the obtained solution in the limiting cases. Limiting case $\mu = 0$ corresponds to the simple support of plate edges $y = \pm 0.5$. Second limiting case $\mu = 0.5$ coresponds to the following BVP

$$\nabla^4 W - \lambda^4 W = q \cos \frac{\pi m}{k} x \cos \pi n y, \quad n, m = 1, 3, 5, \ldots,$$

$$W = 0, \quad W_{xx} = 0 \quad \text{for} \quad x = \pm 0.5k,$$

$$W = 0, \quad W_{yy} = 0 \quad \text{for} \quad y = \pm 0.5,$$

having the following exact solution:

$$W = \frac{q}{\pi^4\alpha^2 - \lambda^4} \cos \frac{\pi m}{k} x \cos \pi n y + \frac{q\pi n(-1)^{\frac{n-1}{2}}}{\pi^4\alpha^2 - \lambda^4} \times \tag{2.318}$$

$$\frac{1}{\beta_{1m} \tanh \beta_{1m}/2 + \begin{Bmatrix} \beta_{2m} \tan \beta_{2m}/2 \\ -\beta_{3m} \tanh \beta_{3m}/2 \end{Bmatrix}} \cdot \left[\frac{\cosh \beta_{1m} y}{\cosh \beta_{1m}/2} - \frac{\begin{Bmatrix} \cos \beta_{2m} y \\ \cosh \beta_{3m} y \end{Bmatrix}}{\begin{Bmatrix} \cos \beta_{2m}/2 \\ \cos \beta_{3m}/2 \end{Bmatrix}} \right] \cos \frac{\pi m}{k} x.$$

Let us check what kind of benefits can be obtained applying the HPM. For $\mu = 0.5$ we have $\gamma_{ji} = \delta_{ij}$, and recurrent formulas for coefficients A_i (2.315)–(2.317) can be written in the following form

$$A_j = A_{j(0)} + A_{j(1)}\varepsilon + A_{j(2)}\varepsilon^2 + A_{j(3)}\varepsilon^3 + \ldots, \tag{2.319}$$

where

$$A_{j(0)} = 0; \qquad A_{j(1)} = \frac{q\pi n(-1)^{\frac{n-1}{2}}}{\pi^4\alpha^2 - \lambda^4},$$

$$A_{j(2)} = \frac{q\pi n(-1)^{\frac{n-1}{2}}}{\pi^4\alpha^2 - \lambda^4}\left[1 - \frac{1}{2\lambda^2}\left(\beta_{1i}\tanh\beta_{1i}/2 + \left\{\begin{matrix}\beta_{2i}\tan\beta_{2i}/2\\ -\beta_{3i}\tanh\beta_{3i}/2\end{matrix}\right\}\right)\right],$$

$$A_{j(3)} = \frac{q\pi n(-1)^{\frac{n-1}{2}}}{\pi^4\alpha^2 - \lambda^4}\left[1 - \frac{1}{2\lambda^2}\left(\beta_{1i}\tanh\beta_{1i}/2 + \left\{\begin{matrix}\beta_{2i}\tan\beta_{2i}/2\\ -\beta_{3i}\tanh\beta_{3i}/2\end{matrix}\right\}\right)\right]^2.$$

We recast the truncated PS (2.319) into PA in (2.264), and we get

$$A_{j[1/1]}(\varepsilon) = \frac{q\pi n(-1)^{\frac{n-1}{2}}}{\pi^4\alpha^2 - \lambda^4}\cdot\frac{\varepsilon}{1 - \left[1 - \frac{1}{2\lambda^2}\left(\beta_{1i}\tanh\beta_{1i}/2 + \left\{\begin{matrix}\beta_{2i}\tan\beta_{2i}/2\\ -\beta_{3i}\tanh\beta_{3i}/2\end{matrix}\right\}\right)\right]\varepsilon}. \tag{2.320}$$

If we compare formula (2.320) for $\varepsilon = 1$ with the multiplier standing before a squared bracket in formula (2.318), it is evident that the PA gives the exact value. Therefore, in the second limiting transition we get the exact solution of the BVP (2.309), (2.114), (2.115).

In the similar way we calculate also the dynamic SSS of the plate with mixed BCs and with nonsymmetric location of the plate clamping parts (Figure 2.8b). We take symmetrically located, with respect to the plate center, the periodic load of the following form:

$$q(\overline{x}, \overline{y}, t) = \overline{q}_0 \sin\frac{\pi m}{a}\overline{x}\cos\frac{\pi n}{b}\overline{y}\sin(\omega t + \alpha), \quad m = 1, 2, 3, \ldots, \quad n = 1, 3, 5, \ldots.$$

PDE (2.309) is still valid assuming that we substitute $\cos\frac{\pi m}{k}x$ by $\sin\frac{\pi m}{k}x$. In the remaining terms we also substitute $\cos\frac{\pi m}{k}x$ by $\sin\frac{\pi m}{k}x$, $\cos\frac{\pi i}{k}x$ by $\sin\frac{\pi i}{k}x$, whereas the summation is carried out with respect to $i, m = 1, 2, 3, \ldots$. Coefficient γ_{ji} takes the form (2.134) $(i \to j, m \to i)$. For the infinite system of LAE (2.314) the summation is carried out with respect to $i = 1, 2, 3, \ldots$.

Computation of the dynamic SSS for the squared plate is carried out by taking into account the first ten coefficients A_j, obtained with a help of the PA (2.264) for $\varepsilon = 1$. Displacement and bending moments in the plate center for different values of the parameter μ are computed. Results are shown in Figures 2.51, 2.52. Similarly as in previous cases, solid (dashed) curve corresponds to symmetric (nonsymmetric) location of the plate clamping (we fixed the value of $k_m = \pi^4\alpha^2/\lambda^4 = 0.9$).

While analysing of the reported curves one may distinguish three characteristic zones regarding the μ parameter: [0, 0.1], (0.1, 0.45], (0.45, 0.5] for the symmetric case and [0, 0.5], (0.5, 0.7], (0.7, 1] - for the nonsymmetric case. In first and third zone an increase of the plate displacement and bending moments accompanied by increase of the parameter μ is rather negligible. In the second zone even for small changes of the parameter μ essential changes of the plate displacement and bending moments are observed.

Therefore, if the size of the mixed BCs are within the first zone, one may consider it as the plate simply supported along its contour. If the size of the mixed BCs are within the third zone, then one may consider it as the plate simply supported along its two edges and clamped along the remaining two. In other cases we must take into account the mixed BCs.

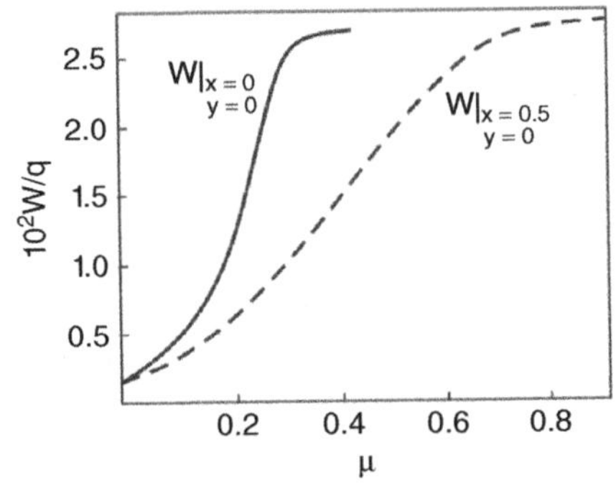

Figure 2.51 Dependence of the normal plate displacement versus length of simple support

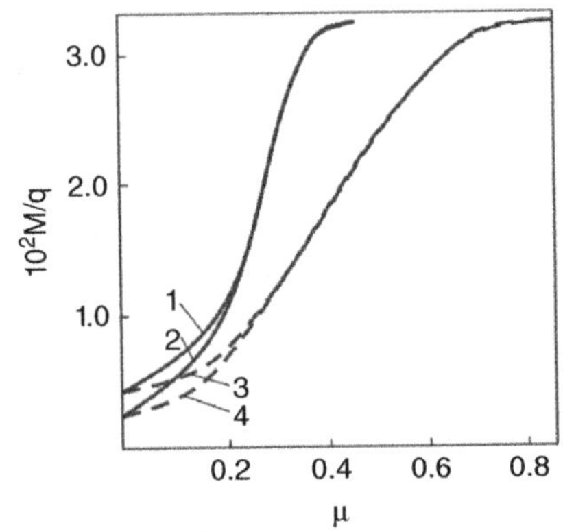

Figure 2.52 Dependence of the bending moment versus simple support length: $1 - M_{x|_{x=0,y=0}}$, $2 - M_{y|_{x=0,y=0}}$, $3 - M_{x|_{x=0.5,y=0}}$, $4 - M_{y|_{x=0.5,y=0}}$

2.5.6 Forced Vibrations of Plate with Mixed Boundary Conditions "Free Edge – Moving Clamping"

We consider forced vibrations of the rectangular plate ($-0.5a \le \overline{x} \le 0.5a; -0.5b \le \overline{y} \le 0.5b$) that is simply supported on its edges $\overline{x} = \pm 0.5a$ and has mixed BCs "free edge – moving clamping" on its edges $\overline{y} = \pm 0.5b$ symmetrically located with respect to the axis y (Figure 2.17a). Plate is loaded as follows

$$q(\overline{x}, \overline{y}, t) = \overline{q}_0 \cos \frac{\pi m}{a} x \sin \frac{\pi n}{b} y \sin(\omega t + \alpha), \quad m = 1, 3, 5, \ldots, \quad n = 2, 4, 6, \ldots.$$

Introducing the nondimensional parameters (2.59) we obtain basic PDE has the following form

$$\nabla^4 W - \lambda^4 W = q \cos \frac{\pi m}{k} x \cos \pi n y, \quad m = 1, 3, 5, \ldots, \quad n = 2, 4, 6, \ldots. \tag{2.321}$$

We attach to PDE (2.321) the BCs (2.159), (2.160).

Plate displacement follows:

$$W = W_1 + W_2,$$

where W_1 is the particular solution to nonhomogeneous PDE (2.321),

$$W_1 = \frac{q}{\pi^4\alpha^2 - \lambda^4} \cos\frac{\pi m}{k} x \cos \pi n y. \tag{2.322}$$

Function W_2 takes the form:

$$W_2 = \sum_{1,3,5,\ldots} \frac{A_i}{2\lambda^2}\left[\left(\lambda^2 + (1-\nu)\pi^2\frac{i^2}{k^2}\right)\frac{1}{\beta_{1i}} \cdot \frac{\cosh\beta_{1i}y}{\sinh\beta_{1i/2}} + \right. \tag{2.323}$$

$$\left.\left(\lambda^2 - (1-\nu)\pi^2\frac{i^2}{k^2}\right) \times \begin{Bmatrix} -\dfrac{1}{\beta_{2i}}\cdot\dfrac{\cos\beta_{2i}y}{\sin\beta_{2i}/2} \\ \dfrac{1}{\beta_{3i}}\cdot\dfrac{\cosh\beta_{3i}y}{\sinh\beta_{3i}/2}\end{Bmatrix}\right]\cos\frac{\pi i}{k}x, \quad \begin{Bmatrix}\lambda > \frac{\pi i}{k} \\ \lambda < \frac{\pi i}{k}\end{Bmatrix}.$$

It has been obtained by solving the following BVP:

$$\nabla^4 W_2 - \lambda^4 W = 0,$$

$$W = 0, \quad W_{xx} = 0 \quad \text{for} \quad x = \pm 0.5k,$$

$$W_{yyy} + (2-\nu)W_{xxy} = 0, \quad W_y = \sum_{i=1,3,5,\ldots} A_i \cos\frac{\pi i}{k}x \quad \text{for} \quad y = \pm 0.5.$$

Unknown coefficients A_i are defined from the BCs (2.159), (2.160):

$$\sum_{i=1,3,5,\ldots} A_i\cos\frac{\pi i}{k}x = \varepsilon\overline{H}(x)\left\{\sum_{i=1,3,5,\ldots} A_i \cos\frac{\pi i}{k}x + \frac{q\pi^2\left(n^2 + \nu\frac{m^2}{k^2}\right)}{\pi^4\alpha^2 - \lambda^4}\times\right.$$

$$(-1)^{\frac{n}{2}}\cos\frac{\pi m}{k}x - \sum_{i=1,3,5,\ldots}\frac{A_i}{2\lambda^2}\left[\frac{1}{\beta_{1i}}\left(\lambda^2 + (1-\nu)\pi^2\frac{i^2}{k^2}\right)\coth\beta_{1i}/2 + \right.$$

$$\left.\left.\left(\lambda^2 - (1-\nu)\pi^2\frac{i^2}{k^2}\right)^2\begin{Bmatrix}\frac{1}{\beta_{2i}}\cot\beta_{2i}/2 \\ -\dfrac{1}{\beta_{3i}\coth\beta_{3i}/2}\end{Bmatrix}\right]\right\}. \tag{2.324}$$

Splitting the r.h.s. of Equation (2.324) into series with respect to $\cos\frac{\pi j}{k}x, j = 1,3,5,\ldots$, the following infinite system of LAE is obtained:

$$A_j = \varepsilon\sum_{j=1,3,5,\ldots} A_j\gamma_{ji}\left[1 - \frac{1}{2\lambda^2}\left[\frac{1}{\beta_{1i}}\left(\lambda^2 + (1-\nu)\pi^2\frac{i^2}{k^2}\right)\coth\beta_{1i}/2 + \right.\right.$$

$$\left.\left.\left(\lambda^2 - (1-\nu)\pi^2\frac{i^2}{k^2}\right)\begin{Bmatrix}\frac{1}{\beta_{2i}}\cot\beta_{2i}/2 \\ -\frac{1}{\beta_{3i}}\coth\beta_{3i}/2\end{Bmatrix}\right]\right] + \varepsilon\gamma_{im}\frac{q\pi^2\left(n^2+\nu\frac{m^2}{k^2}\right)}{\pi^4\alpha^2 - \lambda^4}(-1)^{\frac{n}{2}}, \tag{2.325}$$

where γ_{ji} takes the form (2.135) ($m \to j$).

In what follows we recast coefficients A_j as PS (2.261), substitute them into the system (2.325) and we compare the coefficients standing by same powers of ε:

$$A_{j(0)} = 0, \tag{2.326}$$

$$A_{j(1)} = \gamma_{jm} \frac{q\pi^2 \left(n^2 + \nu \frac{m^2}{k^2}\right)(-1)^{\frac{n}{2}}}{\pi^4\alpha^2 - \lambda^4}, \tag{2.327}$$

$$A_{j(p)} = \sum_{j=1,3,5,\ldots} A_{i(p-1)}\gamma_{ji} \left[1 - \frac{1}{2\lambda^2}\left[\frac{1}{\beta_{1i}}\left(\lambda^2 + (1-\nu)\pi^2 \frac{i^2}{k^2}\right)\coth \beta_{1i}/2 + \left(\lambda^2 - (1-\nu)\pi^2\frac{i^2}{k^2}\right) \times \begin{Bmatrix} \frac{1}{\beta_{2i}}\cot \beta_{2i}/2 \\ -\frac{1}{\beta_{3i}}\coth \beta_{3i}/2 \end{Bmatrix}\right]\right], \quad j = 1,3,5,\ldots. \tag{2.328}$$

PA for coefficients A_j has the form (2.264).

We consider the behavior of the solution obtained so far in limiting cases. The first case corresponds to the movable clamping of plate sides $y = \pm 0.5$, the second case for $\mu = 0.5$ corresponds in full to free plate sides $y = \pm 0.5$. Observe that in the second case we get the exact solution:

$$W = \frac{q}{\pi^4\alpha^2 - \lambda^4}\cos\frac{\pi m}{k}x\cos \pi n y + \frac{q\left(n^2 + \nu\frac{m^2}{k^2}\right)(-1)^{\frac{n}{2}}}{\left[\frac{1}{\beta_{1m}}K_1^2 \coth\frac{\beta_{1m}}{2} + K_2^2\begin{Bmatrix}\frac{1}{\beta_{2m}}\cot \beta_{2m}/2 \\ -\frac{1}{\beta_{3m}}\coth \beta_{3m}/2\end{Bmatrix}\right]} \times \left[K_1\frac{1}{\beta_{1m}}\cdot\frac{\cosh \beta_{1m}y}{\sinh \beta_{1m}/2} + K_2\begin{Bmatrix}\frac{1}{\beta_{2m}}\cdot\frac{\cos\beta_{2m}y}{\sin\beta_{2m}/2} \\ -\frac{1}{\beta_{3m}}\cdot\frac{\cosh \beta_{3m}y}{\sinh\beta_{3m}/2}\end{Bmatrix}\right]. \tag{2.329}$$

where $K_1 = \left(\lambda^2 + (1-\nu)\pi^2\frac{m^2}{k^2}\right)$ and $K_2 = \left(\lambda^2 - (1-\nu)\pi^2\frac{m^2}{k^2}\right)$.

Our method for $\mu = 0.5$ yields $\gamma_{ji} = \delta_{ij}$, and the recurrent formulas for coefficients A_j (2.326)–(2.328) can be presented in the following form:

$$A_j = A_{j(1)}\varepsilon + A_{j(2)}\varepsilon^2 + A_{j(3)}\varepsilon^3 + \ldots, \tag{2.330}$$

where

$$A_{j(0)} = 0,$$

$$A_{j(1)} = \frac{q\pi^2\left(n^2 + \nu\frac{m^2}{k^2}\right)}{\pi^4\alpha^2 - \lambda^4}(-1)^{\frac{n}{2}},$$

$$A_{j(2)} = \frac{q\pi^2\left(n^2 + \nu\frac{m^2}{k^2}\right)}{\pi^4\alpha^2 - \lambda^4}(-1)^{\frac{n}{2}}\left[1 - \frac{1}{2\lambda^2}\left(\frac{1}{\beta_{1m}}K_1^2\coth\beta_{1m}/2+ K_2^2\left\{\begin{matrix}\frac{1}{\beta_{2m}}\cot\beta_{2m}/2\\ -\frac{1}{\beta_{3m}}\coth\beta_{3m}/2\end{matrix}\right\}\right)\right],$$

$$A_{j(3)} = \frac{q\pi^2\left(n^2 + \nu\frac{m^2}{k^2}\right)}{\pi^4\alpha^2 - \lambda^4}(-1)^{\frac{n}{2}}\left[1 - \frac{1}{2\lambda^2}\left(\frac{1}{\beta_{1m}}K_1^2\coth\beta_{1m}/2+ K_2^2\left\{\begin{matrix}\frac{1}{\beta_{2m}}\cot\beta_{2m}/2\\ -\frac{1}{\beta_{3m}}\coth\beta_{3m}/2\end{matrix}\right\}\right)\right]^2,$$

PA in (2.264) has the following form for PS (2.330):

$$A_{j[1/1]}(\varepsilon) = \frac{q\pi^2\left(n^2 + \nu\frac{m^2}{k^2}\right)}{\pi^4\alpha^2 - \lambda^4}(-1)^{\frac{n}{2}}\times$$

$$\frac{\varepsilon}{1 - \left[1 - \frac{1}{2\lambda^2}\left(\frac{1}{\beta_{1m}}K_1^2\coth\beta_{1m} + K_2^2\left\{\begin{matrix}\frac{1}{\beta_{2m}}\cot\beta_{2m}/2\\ -\frac{1}{\beta_{3m}}\coth\beta_{3m}/2\end{matrix}\right\}\right)\right]\varepsilon}.$$

If one compares the PA in (2.264) with the multiplier standing by the square bracket in formula (2.330), then it is clear that for $\varepsilon = 1$ the PA coefficients A_j in expression (2.264) yield the exact solution. Therefore, in the second limiting case we obtain the exact solution to the BVP (2.321), (2.159), (2.160).

In the analogous way we solve the BVP regarding the plate with nonsymmetric location of the clamping plate parts (Figure 2.17b). In this case we take

$$q(\overline{x},\overline{y},t) = \overline{q}_0\sin\frac{\pi m}{a}x\cos\frac{\pi n}{b}y\sin(\omega t+\alpha),\quad m = 1,2,3,\ldots,\quad n = 2,4,6,\ldots.$$

Consequently, assuming that we substitute $\cos\frac{\pi m}{k}x$ by $\sin\frac{\pi m}{k}x$ Equation (2.321) remains valid. In Equations (2.322)–(2.324) we substitute respectively $\cos\frac{\pi m}{k}x$ by $\sin\frac{\pi m}{k}x$, $\cos\frac{\pi i}{k}x$ by $\sin\frac{\pi i}{k}x$, and the summation is carried out on $m, i = 1,2,3,\ldots$. Coefficient γ_{ji} in this case takes the form of (2.134) ($m \to j$). In infinite system of LAE (2.325) the summation is carried out regarding $i = 1,2,3,\ldots$.

For the square plate computation of components of the plate dynamic SSS is carried out taking into account the first ten coefficients A_j, obtained with the help of PA in (2.264) for $\varepsilon = 1$. Both plate deflection and bending moments in the plate center are computed for different values of the μ parameter.

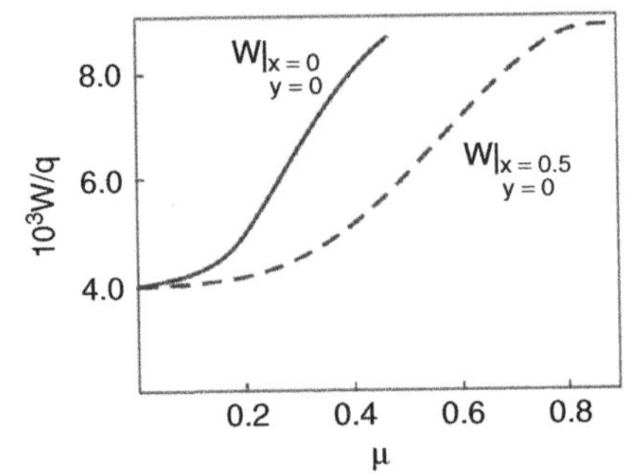

Figure 2.53 Influence of simple supported plate parts on its normal displacement

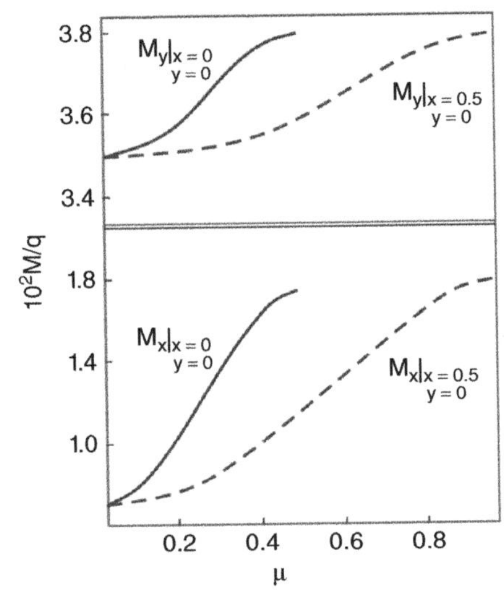

Figure 2.54 Dependence of bending moments versus length of a plate simple support

Results of computations are shown in Figures 2.53, 2.54. Solid (dashed) curves correspond to results with symmetric (nonsymmetric) location of the plate clamping parts.

Analyzing the results one may distinguish three characteristic zones regarding values of the parameter μ: in the case of symmetry – [0,0.05], [0.05,0.47], [0.47,0.5], and in the case of nonsymmetry – [0,0.47], [0.47,0.9]; [0.9,1]. In the case of first and third zones a small change of the plate SSS accompanies increase of the parameter μ. In the second case small change of the parameter μ yields significant changes of all SSS plate factors.

Therefore, if the size of the plate support parts belong to the first zone, then with a small error we may treat this plate as simply supported on edges $x = \pm 0.5k$ and edges $y = \pm 0.5$ are supported by moving clamping. If the size of the mixed BCs are related to the third zone, then that plate can be considered as simply supported on edges $x = \pm 0.5k$, and with free edges $y = \pm 0.5$. In transitional cases (second zone) mixed BCs should be taken into account.

2.6 Stability of Beams and Plates

2.6.1 Stability of a Clamped Beam

We begin with a study of a clamped beam ($-0.5l \leq \bar{x} \leq 0.5l$). The basic ODE governing its stability has the following form

$$EIW_{\overline{xxxx}} + NW_{\overline{xx}} = 0, \tag{2.331}$$

where N is the compressing force.

We introduce first the nondimensional quantities $x = \bar{x}/l$, then Equation (2.331) takes the following form:

$$W^{IV} + \alpha^2 W^{II} = 0, \tag{2.332}$$

where

$$\alpha^2 = Nl^2/EI. \tag{2.333}$$

In order to close the BVP we introduce BCs (2.15).

The solution to Equation (2.332) has the following form:

$$W = C_1 \cos \alpha x + C_2.$$

Satisfaction to BCs (2.15) reduces the problem to that of the following transcendental equation with respect to α:

$$(1-\varepsilon)\alpha \cos \frac{\alpha}{2} + \varepsilon \sin \frac{\alpha}{2} = 0. \tag{2.334}$$

The solution of this equation gives the buckling force.

Let us present parameter α as PS:

$$\alpha = 2\sum_{i=0}^{\infty} \alpha_i \varepsilon^i. \tag{2.335}$$

After substitution of Ansatz (2.335) into Equation (2.334) and after splitting with respect to powers of ε the following recurrent system is obtained:

$$\alpha_0 \cos \alpha_0 = 0,$$

$$-2\alpha_0\alpha_1 \sin \alpha_0 + (\alpha_1 - \alpha_0)\cos \alpha_0 + \sin \alpha_0 = 0,$$

$$-\alpha_0(2\alpha_2 \sin \alpha_0 + \alpha_1^2 \cos \alpha_0) - 2\alpha_1(\alpha_1 - \alpha_0)\sin \alpha_0 + 2(\alpha_2 - \alpha_1)\times$$

$$\cos \alpha_0 + \alpha_1 \cos \alpha_0 = 0,$$

$$\frac{\alpha_0}{3}(-3\alpha_3 \sin \alpha_0 - 6\alpha_1\alpha_2 \cos \alpha_0 + 4\alpha_1^3 \sin \alpha_0) - (2\alpha_2 \sin \alpha_0 - \alpha_1^2 \cos \alpha_0)\times$$

$$(\alpha_1 - \alpha_0) - 2\alpha_1 \sin \alpha_0(\alpha_2 - \alpha_1) + 2(\alpha_3 - \alpha_2)\cos \alpha_0 +$$

$$\left(\alpha_2 \cos \frac{\alpha_0}{2} - \frac{\alpha_1^2}{2} \sin \frac{\alpha_0}{2}\right) = 0,$$

$$\frac{\alpha_0}{12}(-6\alpha_4 \sin\alpha_0 - 4\alpha_1\alpha_3\cos\alpha_0 - 12\alpha_2^2\cos\alpha_0 + 12\alpha_1^2\alpha_2\sin\alpha_0 + \alpha_1^4\cos\alpha_0)+$$
$$(-2\alpha_3\sin\alpha_0 - 2\alpha_1\alpha_2\cos\alpha_0 + \alpha_1^3\sin\alpha_0)(\alpha_1 - \alpha_0) - (2\alpha_2\sin\alpha_0 + \alpha_1^2\cos\alpha_0)\times$$
$$(\alpha_1 - \alpha_0) - (2\alpha_2\sin\alpha_0 - \alpha_1^2\cos\alpha_0)(\alpha_2 - \alpha_1) - 2\alpha_1\sin\alpha_0(\alpha_3 - \alpha_2)+$$
$$2(\alpha_4 - \alpha_3)\cos\alpha_0 + \frac{1}{6}(6\alpha_3\cos\alpha_0 - 6\alpha_1\alpha_2\sin\alpha_0 - \alpha_3\cos\alpha_0) = 0.$$

Solving these problems successively we obtain eigenvalue:

$$\alpha = 2\left[\varphi + \frac{2}{\varphi}\varepsilon + \frac{2}{\varphi}\left(1 - \frac{2}{\varphi^2}\right)\varepsilon^2 + \frac{2}{\varphi}\left(1 - \frac{13}{\varphi^2} - \frac{8}{\varphi^4}\right)\varepsilon^3 + \frac{2}{\varphi}\left(1 - \frac{12}{\varphi^2} + \frac{110}{3\varphi^4} - \frac{40}{\varphi^6}\right)\varepsilon^4\right], \tag{2.336}$$

where $\varphi = \pi n$, $n = 1, 3, 5, \ldots$.

Application of the PA to the first 3 terms of truncated PS (2.336) yields:

$$\alpha_{[1/1]}(\varepsilon) = \frac{\varphi + \left(\frac{4}{\alpha_0} - \varphi\right)}{1 - \left(1 - \frac{2}{\varphi^2}\right)\varepsilon}. \tag{2.337}$$

For $n = 1$, $\varepsilon = 1$, we get $\alpha_{[1/1]} = 2\pi$, which coincides with the exact solution. Summation of the first 3 terms of truncated PS (2.336) gives the result of 32% below the exact value ($\alpha = 1.3642\pi$).

We consider higher order PA of the following form:

$$\alpha_{[2/2]}(\varepsilon) = \frac{\varphi + a_1\varepsilon + a_2\varepsilon^2}{1 + b_1\varepsilon + b_2\varepsilon^2}, \tag{2.338}$$

where

$$a_1 = 2(\alpha_1 + b_1\varphi), \qquad a_2 = 2(\alpha_2 + b_1\alpha_1 + b_2\varphi),$$

$$b_1 = \frac{\alpha_1\alpha_4 - \alpha_2\alpha_3}{\alpha_2^2 - \alpha_1\alpha_3}, \qquad b_2 = \frac{\alpha_3^2 - \alpha_2\alpha_4}{\alpha_2^2 - \alpha_1\alpha_3}.$$

For $n = 1$, $\varepsilon = 1$ the PA (2.338) gives exact solution, whereas truncated PS (2.336) gives the value of $\alpha = 1.8369\pi$ (error – 8.15%).

In Figure 2.55 dependencies of eigenvalue α versus the parameter ε for truncated PS having three and five terms (curves 1 and 2), and PA (2.337), (2.338) (curve 3) are reported. PA practically offer the same results for all values of $0 \le \varepsilon \le 1$. Numerical solution of the transcendental Equation (2.334) almost completely coincides the results obtained with the help of PA. Therefore, the following conclusion is formulated: there is no need to achieve high order PA, since they will not introduce any essentially important solution improvements. Furthermore, increase of the number of terms of truncated PS does not improve essentially the obtained results.

Finally, observe that eigenvalues of BVPs (2.332), (2.15) with the help of the proposed earlier method, can be obtained directly from the governing BVP without using a transcendental equation.

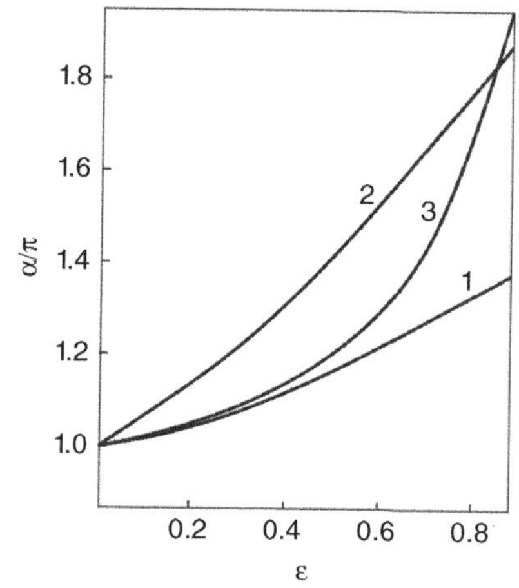

Figure 2.55 Comparison of efficiency of truncated PS and PA

2.6.2 *Stability of a Clamped Rectangular Plate*

We consider a clamped along its contour plate ($-a/2 \le \overline{x} \le a/2, -b/2 \le \overline{y} \le b/2$) being compressed along axis x by a continuously distributed load $\overline{N}$. The governing equation has the form

$$D\nabla^4 W + \overline{N} W_{\overline{xx}} = 0, \tag{2.339}$$

and its nondimensional form can be written as follows:

$$\nabla^4 W + N W_{xx} = 0, \tag{2.340}$$

where $N = \overline{N} b^2 / D$.

BCs are taken in the form (2.66), (2.67).

Buckling load N and plate deflection W are presented as PS:

$$N = \sum_{i=0}^{\infty} N_i \varepsilon^i, \qquad W = \sum_{i=0}^{\infty} W_i \varepsilon^i. \tag{2.341}$$

Substituting Ansatzes (2.341) into Equation (2.340) and BCs (2.66), (2.67) and splitting with respect to ε yields the following sequence of the BVPs:

$$\nabla^4 W_0 + N_0 W_{0xx} = 0, \tag{2.342}$$

$$W_0 = 0, \quad W_{0xx} = 0 \quad \text{for} \quad x = \pm 0.5k, \tag{2.343}$$

$$W_0 = 0, \quad W_{0yy} = 0 \quad \text{for} \quad y = \pm 0.5, \tag{2.344}$$

$$W_j + N_0 W_{jxx} = -\sum_{i=0}^{j-1} N_{j-1} W_{ixx},$$

$$W_j = 0, \quad W_{jxx} = \mp k \sum_{i=0}^{j-1} W_{ix} \quad \text{for} \quad x = \pm 0.5k,$$

$$W_j = 0, \quad W_{jyy} = \mp \sum_{i=0}^{j-1} W_{iy} \quad \text{for} \quad y = \pm 0.5.$$

BVP regarding the zero order approximation (2.342)–(2.344) describes stability of the simply supported plate, and its solution has the following form:

$$N_0 = \frac{\pi^2 k^2}{m^2}\left(\frac{m^2}{k^2} + n^2\right)^2, \tag{2.345}$$

$$W_0 = C \cos\frac{\pi m}{k} x \cos \pi n y, \quad n, m = 1, 3, 5, \ldots. \tag{2.346}$$

Here N_0 is the buckling load of simply supported plate; W_0 is the form of buckling. The first correction term to the buckling load follows:

$$N_1 = 4\frac{k^2}{m^2}\left(\frac{m^2}{k^2} + n^2\right), \tag{2.347}$$

whereas the first correction term to the mode of buckling has the following form:

$$W = \frac{n}{n^2\alpha}\left[\frac{(-1)^{\frac{n-1}{2}}}{2\cosh \pi\beta_1/2}\cosh \pi\beta_1 y - y \sin \pi n y\right]\cos\frac{\pi m}{k}x + \tag{2.348}$$

$$\left\{\begin{matrix} -\frac{2}{\pi}\cdot\frac{m^3/k^3}{\left(n^4 - \frac{m^4}{k^4}\right)}\left[\frac{(-1)^{\frac{n-1}{2}}}{2\cos\frac{\pi}{2}n^2\frac{k^2}{m}}\cos \pi n^2 \frac{k}{m}x - x\sin\frac{\pi m}{k}x\right], & n \neq \frac{m}{k} \\ -\frac{1}{2}x^2\cos\frac{\pi m}{k}x, & n = \frac{m}{k}\end{matrix}\right\}\cos \pi n y.$$

The expression regarding the second correcting term of the buckling load follows:

$$N_2 = N_{2x} + N_{2y}, \tag{2.349}$$

where

$$N_{2x} = \frac{4}{k}\cdot\frac{\pi^2 m^2}{k^2}\left\{1 + \left(\frac{n^2 - \frac{m^2}{k^2}}{2\pi^2\alpha^2} - \left\{\begin{matrix}\frac{2}{\pi}\cdot\frac{m^2/k^2}{\left(n^4 - \frac{m^4}{k^4}\right)}\left(\frac{\pi}{2}n^2\frac{k}{m}\tan\frac{\pi}{2}n^2\frac{k^2}{m} + 1\right) \\ \frac{k^2}{8}\end{matrix}\right\}\right)\right\} +$$

$$\frac{8}{k\pi^2}\cdot\frac{k^4}{m^4}\alpha\left\{\begin{matrix} -\frac{2}{\pi}\cdot\frac{m^3/k^3}{n^4 - \frac{m^4}{k^4}}\left[\frac{m^3/k^3}{n^4 - \frac{m^4}{k^4}}n^4\frac{k^2}{m^2}\tan\frac{\pi}{2}\frac{n^2}{m}k^2 + \frac{3}{4}m\right] \\ \frac{k}{8}\frac{\pi^2 m^2}{k^2}\left(\frac{k^2}{6} - 1\right)\end{matrix}\right\},$$

$$N_{2y} = 4n^2\frac{k^2}{m^2}\left[1 - \frac{1}{\pi^2\alpha}\left(\frac{\pi\beta_1}{2}\tanh\frac{\pi\beta_1}{2} - 1\right) + \frac{2}{\pi\alpha} - 4n(-1)^{\frac{n-1}{2}}\frac{k}{\pi m}\right] \times$$

$$\left\{\begin{matrix} -\frac{2}{\pi}\cdot\frac{m^3/k^3}{n^4-\frac{m^4}{k^4}}\left[\frac{2}{k}\cdot\frac{1}{\pi}\cdot\frac{m^3/k^3}{n^4-\frac{m^4}{k^4}}\tan\frac{\pi}{2}n^2\frac{k^2}{m} - \frac{k}{2\pi m}\right]\right] \\ \frac{1}{4}\left(\frac{k^2}{6} - 1\right) \end{matrix}\right\}, \quad \left\{\begin{matrix} n \neq \frac{m}{k} \\ n = \frac{m}{k} \end{matrix}\right\}.$$

Knowing three terms of the PS for N

$$N = N_0 + \varepsilon N_1 + \varepsilon^2 N_2, \tag{2.350}$$

we construct PA.

Let us compare obtained results with the numerical solution for the squared plate. The obtained results are as follows: for known solution–$N = 8.5540\pi^2$, for (2.350)–$N = 5.4512\pi^2$ (error–56.9%) and for PA–$N = 7.8662\pi^2$ (error–8.04%). In Figure 2.56 shown are curves of N versus ε. Curve 1 denotes results obtained with the (2.350); curve 2–results obtained via PA; point 3–numerical solution [87]. Drawings 1 and 2 for $\varepsilon < 0.5$ differ from each other on less than 5%.

2.6.3 Stability of Rectangular Plate with Mixed Boundary Conditions "Clamping-Simple Support"

We apply our approach to determine buckling loads for plates with mixed BCs of the form "clamping - simple support". We study two principally different computational shemes shown in Figure 2.57a,b and Figure 2.57c,d. BCs follow:

$$T_x = N, \quad T_{xy} = 0 \quad \text{for} \quad x = \pm 0.5k,$$
$$T_y = 0, \quad T_{xy} = 0 \quad \text{for} \quad y = \pm 0.5,$$

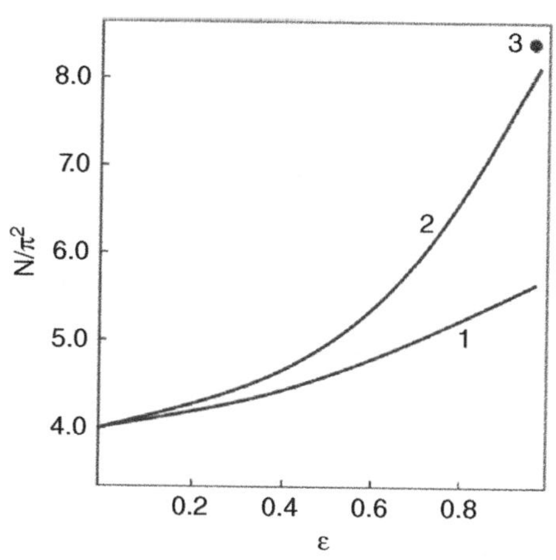

Figure 2.56 Comparison of efficiency of PS and PA

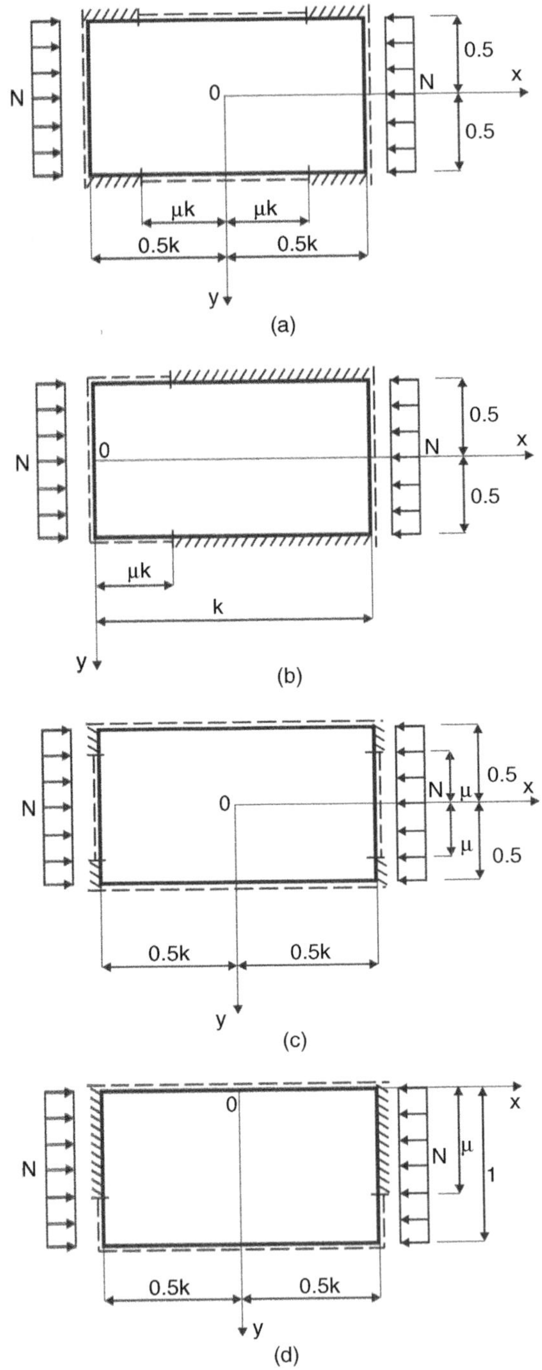

Figure 2.57 Plates with mixed BCs

where $T_x(T_y)$ is the force in $x(y)$ direction; T_{xy} is the shear force.

The basic equation has the form of (2.340).

Let us determine a buckling load for the plate shown in Figure 2.57a. BCs take the form (2.114), (2.115).

We substitute Ansatz (2.341) into Equation (2.340) and BCs (2.114)–(2.115), and splitting the problem regarding ε. As a result, the following sequence of BVPs is obtained:

$$\nabla^4 W_0 + N_0 W_{0xx} = 0,$$

$$W_0 = 0, \quad W_{0xx} = 0 \quad \text{for} \quad x = \pm 0.5k,$$

$$W_0 = 0, \quad W_{0yy} = 0 \quad \text{for} \quad y = \pm 0.5,$$

$$\nabla^4 W_j + N_0 W_{jxx} = -\sum_{i=0}^{j-1} N_{j-i} W_{ixx},$$

$$W_j = 0, \quad W_{jxx} = 0 \quad \text{for} \quad x = \pm 0.5k,$$

$$W_j = 0, \quad W_{jyy} = \mp \overline{H}(x) \sum_{i=0}^{j-1} W_{iy} \quad \text{for} \quad y = \pm 0.5.$$

In zero order approximation we get the problem for a plate simply supported on its contour and being compressed by the load N_0 in direction of the x axis.

BVP of the first order approximation has the form:

$$\nabla^4 W_1 + \pi^2 \frac{k^2}{m^2}\left(n^2 + \frac{m^2}{k^2}\right)^2 W_{xx} = N_1 \pi^2 \frac{m^2}{k^2} \cos\frac{\pi m}{k} x,$$

$$W_1 = 0, \quad W_{1xx} = 0 \quad \text{for} \quad x = \pm 0.5k,$$

$$W_1 = 0, \quad W_{1yy} = \pm \pi n (-1)^{\frac{n-1}{2}} \overline{H}(x) \cos\frac{\pi m}{k} x \quad \text{for} \quad y = \pm 0.5,$$

and its solution is

$$W_1 = \sum_{i=1,3,5,\ldots} y_{1i} \cos\frac{\pi i}{k} x.$$

The following two BVPs are obtained by developing what appeared in BCs functions into series with respect to $\cos(\pi i x/k)$, $i = 1, 3, 5, \ldots$:

for $i = m$:

$$Y_{1m}^{IV} - 2\frac{\pi^2 m^2}{k^2} Y_{1m}^{II} - \pi^4 n^2 \left(2\frac{m^2}{k^2} + n^2\right) Y_{1m} = N_1 \frac{\pi^2 m^2}{k^2} \cos \pi n y, \tag{2.351}$$

$$Y_{1m} = 0, \quad Y_{1m}^{II} = \pi n (-1)^{\frac{n-1}{2}} \gamma_{mm} \quad \text{for} \quad y = \pm 0.5, \tag{2.352}$$

for $i \neq m$:

$$Y_{1i}^{IV} - 2\frac{\pi^2 i^2}{k^2} Y_{1i}^{II} - \pi^4 \left[\frac{i^2}{m^2}\left(n^2 + \frac{m^2}{k^2}\right)^2 - \frac{i^4}{k^4}\right] Y_{1i} = 0, \tag{2.353}$$

$$Y_{1i} = 0, \quad Y_{1i} = \pm\pi n(-1)^{\frac{n-1}{2}}\gamma_{im} \quad \text{for} \quad y = \pm 0.5. \tag{2.354}$$

Here γ_{im} is defined by Equation (2.272), and N_1 follows:

$$N_1 = 4k^2\frac{n^2}{m}\gamma_{mm}. \tag{2.355}$$

Function Y_{im} takes the following form:

$$Y_{1m} = \frac{n}{\pi\alpha}\gamma_{mm}\left[\frac{(-1)^{\frac{n-1}{2}}}{2\text{ch}\pi\beta_1/2}\cosh\pi\beta_1 y - y\sin\pi n y\right],$$

where $\beta_1 = \sqrt{2m^2/k^2 + n^2}$.

The solution to BVP (2.353), (2.354) does not give a correction term to the buckling load, but improves the buckling form:

$$Y_{1i} = \pi n(-1)^{\frac{n-1}{2}}\sum_{i=1,3,5,\ldots}\left[\frac{\cosh\alpha_{2i}y}{\cosh\alpha_{2i}/2} - \frac{\begin{Bmatrix}\cosh\beta_{2i}y\\ \cos\varphi_{2i}y\end{Bmatrix}}{\begin{Bmatrix}\cosh\beta_{2i}/2\\ \cos\varphi_{2i}/2\end{Bmatrix}}\right],$$

$$\begin{Bmatrix}m(i-m) > n^2k^2\\ m(i-m) < n^2k^2\end{Bmatrix},$$

where $\alpha_{2i} = \pi\sqrt{i\left(\frac{i+m}{k^2} + \frac{n^2}{m}\right)}$, $\beta_{2i} = \pi\sqrt{i\left(\frac{i+m}{k^2} - \frac{n^2}{m}\right)}$, $\varphi_{2i} = \pi\sqrt{i\left(\frac{n^2}{m} - \frac{i-m}{k^2}\right)}$.

The first correcting term to a buckling form is as follows:

$$W_{1m} = \frac{n}{\pi\alpha}\gamma_{mn}\left[\frac{(-1)^{\frac{n-1}{2}}}{2\cosh\pi\beta_1/2}\cosh\pi\beta_1 y - y\sin\pi n y\right]\cos\frac{\pi m}{k}x +$$

$$\pi n(-1)^{\frac{n-1}{2}}\sum_{i=1,3,5,\ldots}\gamma_{im}\left[\frac{\cosh\alpha_{2i}y}{\cosh\alpha_{2i}/2} - \frac{\begin{Bmatrix}\cosh\beta_{2i}y\\ \cos\varphi_{2i}y\end{Bmatrix}}{\begin{Bmatrix}\cosh\beta_{2i}/2\\ \cos\varphi_{2i}/2\end{Bmatrix}}\right]\cos\frac{\pi i}{k}x.$$

We find the second correcting term to the buckling load by taking into account the second approximation:

$$N_2 = \frac{k^2}{\pi^2 m^2}\left\{4\pi^2 n^2 \gamma_{mm} - \frac{2n^2}{\alpha}\gamma_{mm}^2\left(\frac{\pi}{2}\beta_1 \tanh\frac{\pi}{2}\beta_1 - 1\right) - \right. \tag{2.356}$$

$$\left.\frac{n^2 m}{\alpha^2}\left(n^2 - \frac{m^2}{k^2}\right)\gamma_{mm}^2 - 4\pi^2 n^2 \sum_{i=1,3,5,\ldots}\gamma_{im}^2\left[\alpha_{2i}\tanh\frac{\alpha_{2i}}{2} - \left\{\begin{matrix}\beta_{2i}\tanh\frac{\beta_{2i}}{2}\\ -\varphi_{2i}\tan\frac{\varphi_{2i}}{2}\end{matrix}\right\}\right]\right\}.$$

For the case shown in Figure 2.57b, formula (2.310) remains valid when the summation with respect to odd values of i is extended into all values of i. The parameter γ_{im} in this case has the form of (2.134). Knowing three coefficients of truncated PS (2.350), we construct PA.

In the limiting case corresponding to completely clamping on sides $y = \pm 0.5$, the exact solution for the squared plate has been obtained numerically from the transcendental equation for $m = 1$ and it is equal to $N = 8.6044\pi^2$. PA yields $N = 8.7136\pi^2$ (error–1.27%), whereas (2.350) gives $N = 4.7757\pi^2$ (error - 44.5%). Numerical solution obtained from the transcendental equation for $m = 2$ gives $N = 7.6913\pi^2$, whereas PA - $N = 7.7156\pi^2$ (error–0.32%), (2.350)–$N = 6.4456\pi^2$ (error–16.2%).

The dependence of the critical force on the geometric dimensions of the mixed parts of BCs for the quadratic plate is given in Figure 2.58. Solid (dashed) curve corresponds to symmetric (nonsymmetric) location of clamping parts. Dashed–dotted curves correspond to results obtained via the R-function method [10]. Dots correspond to results obtained numerically for limiting cases $\mu = 0, 0.5, 1$ [84].

One may distinguish two zones of the parameter μ. First one for symmetric case begins at $\mu = 0$ and ends at $\mu = 0.15$ and second for nonsymmetric case from $\mu = 0$ up to $\mu = 0.55$. In this zone a buckling occurs with the appearance of two half-waves in direction x. In the second zone, from $\mu = 0.15$ to $\mu = 0.5$ for the symmetric case and from $\mu = 0.55$ to $\mu = 1$ for

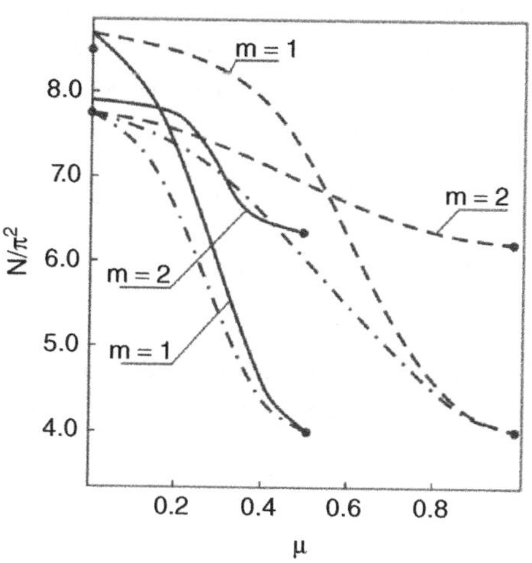

Figure 2.58 Buckling load versus length of simply support parts

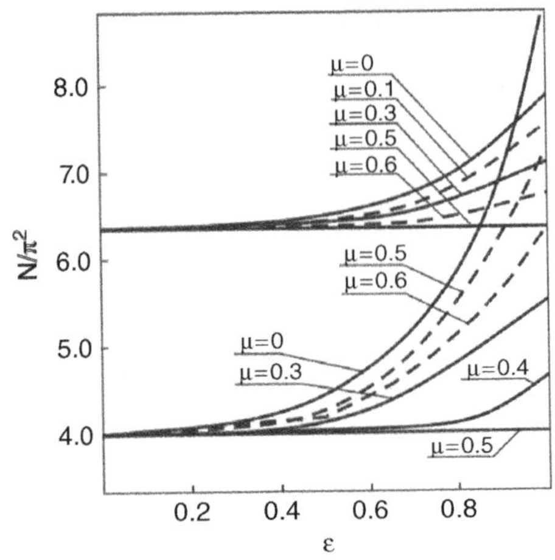

Figure 2.59 Buckling load versus stiffness of elastic clamping

the nonsymmetric case the plate buckles due to the occurrence of one half-wave in direction x. Therefore, for $\mu = 0.15$ (or $\mu = 0.55$) one may expect either the first or second buckling mode. Solutions obtained using the R-function method give practically the same results.

Our approach allows the investigate influence of the clamping stiffness on the buckling load N. In Figure 2.59 the dependence of the buckling load N versus the parameter ε for some values of μ is reported. Solid (dashed) curve correspond to symmetric (nonsymmetric) case. It should be emphasized that for the elastic clamping of the plate edges $y = \pm 0.5$ with a lack of mixed BCs related equilibrium forms occur for $\varepsilon = 0.96$, and the buckling associated with the occurrence of one half-wave in direction x is possible for $\varepsilon < 0.96$, and with two half-waves for $\varepsilon > 0.96$. The simultaneous occurrence of both buckling forms is possible only for mixed BCs, and for $\mu \to 0$ related buckling form appears for $\varepsilon \to 1$. The limiting value of μ, for which the plate buckling exhibiting one half-wave in direction x for $\varepsilon = 1$ corresponds to $\mu = 0.25$ ($\mu = 0.58$) for the case of symmetric (nonsymmetric) position of the mixed BCs.

Let us proceed now to the second case of the mixed BCs (Figure 2.57c,d). Basic Equation has the form of (2.340). We attach the following BCs:

$$W = 0, \quad W_{yy} = 0 \quad \text{for} \quad y = \pm 0.5, \tag{2.357}$$

$$W = 0, \quad W_{xx} = \overline{H}(y)\varepsilon(W_{xx} \mp kW_x) \quad \text{for} \quad x = \pm 0.5k, \tag{2.358}$$

where $\overline{H}(y) = H(y - \mu) - H(-y - \mu)$.

Let us present the displacement W and buckling load N as PS (2.341). Substituting them into Equation (2.340) and BCs (2.357), (2.358) we get the following recurrent sequence of the BVPs:

$$\nabla^4 W_0 + N_0 W_0 = 0,$$

$$W_0 = 0, \quad W_{0yy} = 0 \quad \text{for} \quad y = \pm 0.5,$$

$$W_0 = 0, \quad W_{0xx} = 0 \quad \text{for} \quad x = \pm 0.5k,$$

$$\nabla^4 W_j + N_0 W_{jxx} = -\sum_{i=0}^{j-1} N_{j-i} W_{ixx},$$
$$W_j = 0, \quad W_{jyy} = 0 \quad \text{for} \quad y = \pm 0.5,$$
$$W_j = 0, \quad W_{jxx} = \mp \overline{H}(y) k \sum_{i=0}^{j-1} W_x \quad \text{for} \quad x = \pm 0.5k.$$

First order BVP is

$$\nabla^4 W_1 + \pi^2 \frac{k^2}{m^2}\left(n^2 + \frac{m^2}{k^2}\right)^2 W_{1xx} = N_1 \pi^2 \frac{m^2}{k^2} \cos\frac{\pi m}{k} x \cos \pi n y, \tag{2.359}$$

$$W_1 = 0, \quad W_{1yy} = 0 \quad \text{for} \quad y = \pm 0.5, \tag{2.360}$$

$$W_1 = 0, \quad W_{1xx} = \pm k\frac{\pi m}{k}(-1)^{\frac{m-1}{2}} \overline{H}(y) \cos \pi n y \quad \text{for} \quad x = \pm 0.5k. \tag{2.361}$$

The assumed solution follows:

$$W_1 = \sum_{p=1,3,5,\ldots} X_{1p} \cos \pi p y. \tag{2.362}$$

Substituting series (2.362) into Equation (2.359) and BCs (2.360), (2.361) we obtain the following two one-dimensional problems:

for $p \neq n$:

$$X_{1n}^{IV} + \pi^2 \frac{k^2}{m^2}\left(\frac{m^2}{k^2} + n^2\right) X_{1n}^{II} + \pi^4 n^4 X_{1n} = N_1 \frac{\pi^2 m^2}{k^2} \cos\frac{\pi m}{k} x, \tag{2.363}$$

$$X_{1n} = 0, \quad X_{1n}^{II} = \pm k\frac{\pi m}{k}(-1)^{\frac{m-1}{2}} \gamma_{nn} \quad \text{for} \quad x = \pm 0.5k, \tag{2.364}$$

for $p = n$:

$$X_{1p}^{IV} - \pi^2 \left[2p^2 - \frac{k^2}{m^2}\left(\frac{m^4}{k^4} + n^4\right)^2\right] X_{1p}^{II} + \pi^4 p^4 X_{1p} = 0, \tag{2.365}$$

$$X_{1p} = 0, \quad X_{1p}^{II} = \pm k\frac{\pi m}{k}(-1)^{\frac{m-1}{2}} \gamma_{pm} \quad \text{for} \quad x = \pm 0.5k. \tag{2.366}$$

Here γ_{pn} is defined by formula (2.272) ($i \to p, m \to n$).
Solvability condition of the BVP (2.363), (2.364) yields

$$N_1 = 4\gamma_{nn}. \tag{2.367}$$

BVP (2.363), (2.364) yields corrections to the force and to the form of the buckling. Its solution is composed of two parts:

for $n \neq m/k$:

$$X_{1n} = -\frac{2}{\pi} \cdot \frac{m^3/k^3}{\left(n^4 - \frac{m^4}{k^4}\right)} \gamma_{nn} \left[\frac{(-1)^{\frac{m-1}{2}} k}{2\cos\frac{\pi}{2} n^2 \frac{k^2}{m}} \cos \pi n^2 \frac{k}{m} x - x \sin\frac{\pi m}{k} x\right],$$

for $n = m/k$:

$$X_{1n} = -\frac{\gamma_{nn}}{2}x^2 \cos\frac{\pi m}{k}x.$$

BVP (2.365), (2.366) gives only a correction to the buckling form

$$X_{1p} = \pi m(-1)^{\frac{m-1}{2}} \frac{\gamma_{pn}}{\alpha_p^2 - \beta_p^2}\left[\frac{\cosh\alpha_p x}{\cosh\alpha_p k/2} - \frac{\cosh\beta_p x}{\cosh\beta_p k/2}\right],$$

where

$$\left\{\begin{matrix}\alpha_p\\ \\ \beta_p\end{matrix}\right\} =$$

$$\sqrt{\frac{\pi^2}{2}\left(2p^2 + \frac{k^2}{m^2}\left(\frac{m^2}{k^2}+n^2\right)^2\right) \pm \frac{\pi^2 k}{2m}\left(\frac{m^2}{k^2}+n^2\right)\sqrt{\frac{k^2}{m^2}\left(\frac{m^2}{k^2}+n^2\right)^2 - 4p^2}}.$$

In the case where α_p and β_p are complex, $\cosh\alpha_p x$ and $\cosh\beta_p x$ should be substituted by $\cos\alpha_p x$ and $\cos\beta_p x$, respectively.

First, the correcting term to the buckling form follows:

$$W_1 = \left\{\begin{matrix} -\frac{2}{\pi}\cdot\frac{m^3/k^3}{\left(n^4 - \frac{m^4}{k^4}\right)}\gamma_{nn}\left[\frac{(-1)^{\frac{n-1}{2}}k}{2\cos\frac{\pi}{2}n^2\frac{k^2}{m}}\cos\pi n^2\frac{k}{m}x - x\sin\frac{\pi m}{k}x\right] \\ \\ -\frac{\gamma_{nn}}{2}x^2\cos\frac{\pi m}{k}x \end{matrix}\right\}\cos\pi n y +$$

$$\pi m(-1)^{\frac{m-1}{2}}\sum_{p=1,3,5,\ldots}\frac{\gamma_{pn}}{\alpha_p^2-\beta_p^2}\left[\frac{\cosh\alpha_p x}{\cosh\alpha_p^2 k/2} - \frac{\cosh\beta_p x}{\cosh\beta_p^2 k/2}\right]\cos\pi p y \quad \left\{\begin{matrix} n \neq \frac{m}{k} \\ \\ n = \frac{m}{k}\end{matrix}\right\}. \tag{2.368}$$

The second approximation gives

$$N_2 = \left\{\begin{matrix} 4\gamma_{nn}\frac{\pi^2 m^2}{k^2}\left\{1 - 2\gamma_{nn}\frac{m^2/k^2}{\pi^2\left(n^4-\frac{m^4}{k^4}\right)}\left[\frac{k}{2}n^2\tan\frac{\pi}{2}n^2\frac{k^2}{m^2}\left(1 - \frac{4n^2}{k\left(n^4-\frac{m^4}{k^4}\right)}\right) - \frac{3}{2}\right]\right\} \\ \\ \gamma_{nn}\left(4 - \frac{k^2}{3}\gamma_{nn} - \gamma_{nn}\right)\end{matrix}\right\} -$$

$$\frac{4k^2}{\pi m}\sum_{p=1,3,5,\ldots}\gamma_{pn}^2\frac{1}{\alpha_p^2-\beta_p^2}\left[\alpha_p\tanh\frac{\alpha_p k}{2} - \beta_p\tanh\frac{\beta_p k}{2}\right]. \tag{2.369}$$

In the case of nonsymmetric location of the clamping part (Figure 2.57g) formulas (2.367), (2.368) remain valid assuming that the previous summation with respect to the odd values of p should be substituted by the even values of p. Now the parameter γ_{pn} is defined by (2.134).

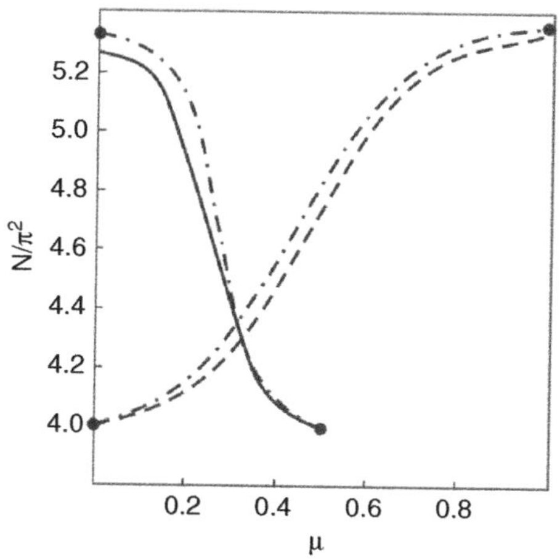

Figure 2.60 Buckling force versus length of simply supported parts

In the limiting case corresponding to the complete clamping of sides $x = \pm 0.5k$, the buckling force obtained numerically from the transcendental equation is equal to $N = 5.4390\pi^2$. (2.350) yields $N = 4.7654\pi^2$ (error - 14.01%), PA - $N = 5.2159\pi^2$ (error - 4.1%).

The dependence of buckling force N versus length μ of the squared plate clamping part is shown in Figure 2.60. A solid (dashed) curve corresponds to symmetric (nonsymmetric) location of the mixed BCs, whereas a dashed-dotted curve presents results obtained using the R-function method [10]. Points are associated with results obtained numerically in limiting cases for $\mu = 0, 0.5, 1$.

One may isolate three zones of different values of the parameter μ: case of symmetry – [0, 0.1], (0.1, 0.45], (0.45, 0.5]; case of nonsymmetry – (0.8, 1], (0.1, 0.8], [0, 0.1]. If size of the mixed BCs belong to the first zone, then this plate can be treated as simply supported on its two edges $y = \pm 0.5$, and clamped on two remaining ones. The buckling force is overestimated in comparison to the exact solution by no more than 5%. If the size of the mixed BCs are within the third zone, the studied plate can be treated as that being simply supported along its contour, and the error is less than 5%. If the size of the mixed BCs corresponds to the second zone, then the analyzed plate can be treated as that with the mixed BCs.

The plates buckling force versus the parameter ε is shown in Figure 2.61. In this case, contrary to the previously studied one, we cannot distinguish an intensive force increase part. There may be observed a remarkable influence of the clamped parts for arbitrary values of ε.

2.6.4 Comparison of Theoretical and Experimental Results

Let us study the stability of the plate shown in Figure 2.62. For the buckling load one has a formula (2.350), where N_0, N_1 and N_2 can be found from Equations (2.345), (2.355) and (2.356) respectively. Coefficients γ_{im} in this case are defined by the formula (2.135).

Next we recast (2.350) into PA. For the limiting case, corresponding to complete clamping of the plate edges $y = \pm 0.5$, the results obtained coincide with those presented and discussed in the previous section.

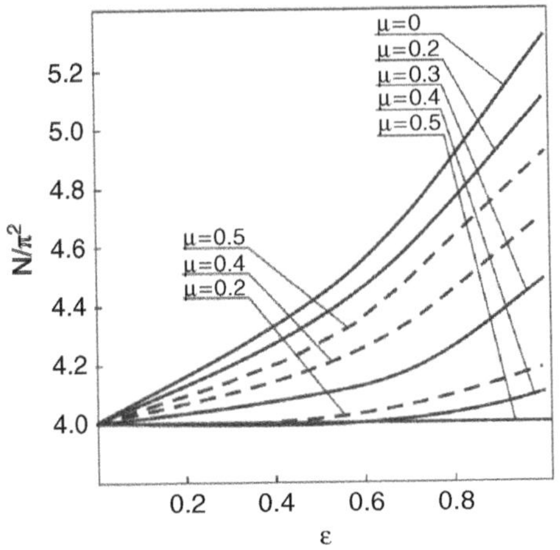

Figure 2.61 Influence of stiffness of an elastic support on plate buckling force

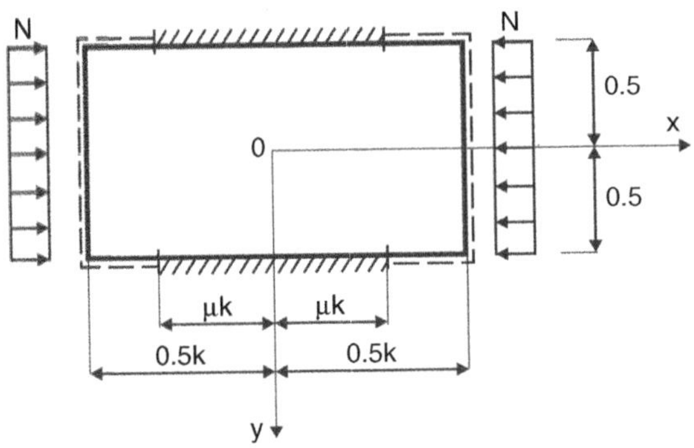

Figure 2.62 Computational plate scheme

In Figure 2.63 solid curves correspond to the buckling force dependent on the length of the squared plate simple support (μ). The dashed curve is associated with results obtained in reference [41]; dashed-dotted curve = results reported in [49]; points-experimental data [41].

Two zones, i.e. $[0, 0.08]$ and $[0.08, 1]$ can be distinguished. In the first (second) zone the plate buckling takes place through the occurrence of one (two) half-wave(s) in direction of the axis x. For $\mu = 0.08$ either occurrence of the first or second buckling form is expected.

Results obtained via our method have high coincidence with the results obtained by others, as well as with experimental results.

Observe that occurrence of related buckling forms is possible for various values of the parameter ε. For example, in the case of complete clamping of the plate sides $y = \pm 0.5$ related forms appear for $\varepsilon = 0.96$. If $\mu < 0.075$, the plate will buckle through the exhibition of only one half-wave in direction of x for an arbitrary value of ε.

The buckling force N versus elasticity of clamping of plate faces ε is shown in Figure 2.64.

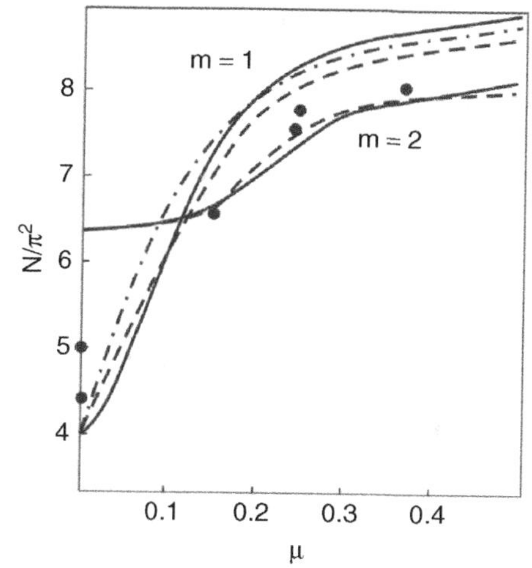

Figure 2.63 Buckling force versus length of plate simple support parts

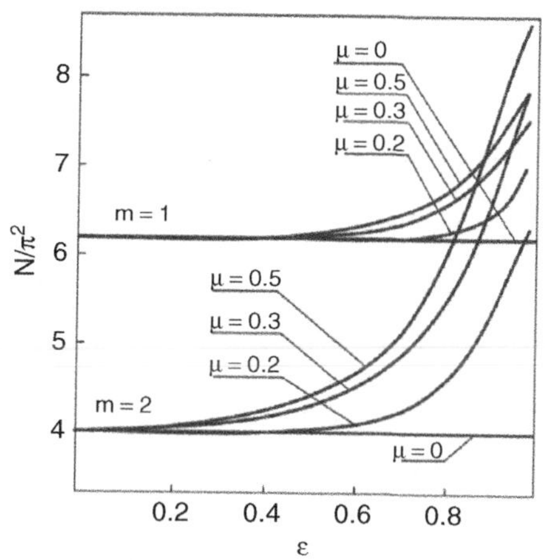

Figure 2.64 Buckling plate force versus plate stiffness support

2.7 Some Related Problems

2.7.1 Dynamics of Nonhomogeneous Structures

It is obvious that an analytical or numerical solution of dynamical problems for nonhomogeneous structures is associated with a lot of problems. Analytical methods, relying on splitting of an initial problem into subsystems with their successive connections, yield a system of higher order algebraic equations. On the other hand, numerical methods in many cases cannot be directly applied due to the change in the structures characteristics in threshold places, where different object characteristics meet, and therefore stiff problems are produced. Here we apply HPM.

We consider as an example longitudinal vibrations of a nonhomogeneous rod with the density ρ, the area of its cross section F with elasticity modulus E_1 in the interval $-1 \leq x < 0$ and in the interval $0 < x \leq 1$ - modulus E_2. Basic equations have the form

$$E_{1(2)}F\frac{\partial^2 u^{1(2)}}{\partial x^2} - \rho F\frac{\partial^2 u^{1(2)}}{\partial t^2} = 0, \tag{2.370}$$

where $u^{1(2)}$ is the longitudinal displacement.

Solutions to Equation (2.370) are sought in the following form:

$$u^{1(2)}(x) = u^{1(2)}(x)(A\cos\omega t + B\sin\omega t). \tag{2.371}$$

Substituting Ansatz (2.371) into Equation (2.370), we get

$$\frac{d^2u^{1(2)}}{dx^2} + \frac{\omega^2}{a^2_{1(2)}}u^{1(2)} = 0,$$

where $a^2_{1(2)} = E_{1(2)}/\rho$.

Solutions to these equations follow:

$$u^{1(2)} = C_1^{1(2)}\cos\frac{\omega}{a_{1(2)}}x + C_2^{1(2)}\sin\frac{\omega}{a_{1(2)}}x.$$

Arbitrary constants are defined by BCs $C_i^{1(2)}$ due to conditions

$$u^1(-1) = u^2(1) = 0$$

and conditions of equality in zero point of both displacements and longitudinal forces. These conditions yield a transcendental equation.

$$(1 - \sqrt{1+\varepsilon})\cos\omega^*\sin(\omega^*\sqrt{1+\varepsilon}) + \sin\omega^*\cos\omega^*(\sqrt{1+\varepsilon}) = 0. \tag{2.372}$$

Here $\varepsilon = (E_2 - E_1)/E_1$, $\omega^* = \omega/a_1$.

We develop the function $\omega^*(\varepsilon)$ into a Maclauring series up to the third term:

$$\omega^* \approx \omega_0 + \varepsilon\omega_1 + \varepsilon^2\omega_2. \tag{2.373}$$

Substituting the Ansatz (2.373) into Equation (2.372), carrying out elementary transformations and comparing coefficients standing by the same powers of ε we get

$$\omega_1 = \pi k\left(1 - \frac{\varepsilon}{4} + \frac{\varepsilon^4}{4}\right), \quad \omega_2 = \left(\frac{\pi}{2} + \pi k\right)\left(1 - \frac{\varepsilon}{4} + \frac{\varepsilon^2}{8}\right), \quad k = 1, 2, 3, \ldots.$$

For $k = 1$ the corresponding PA regarding ω_1 and ω_2 follow:

$$\omega_{1[1/1]} = \frac{\pi}{4}\left(\frac{4+3E}{1+E}\right),$$

$$\omega_{2[1/1]} = \frac{\pi}{4}\left(\frac{4+3E}{2+E}\right).$$

Table 2.2 Solutions to transcendental Equation (2.372)

ε	ω_n	$\omega_{1[1/1]}$	ω_n	$\omega_{2[0/1]}$
0.01	1.566888	1.566888	3.133800	3.133
0.1	1.532510	1.533390	3.068490	3.070
0.3	1.451702	1.468353	2.948120	2.960
0.5	1.398080	1.413716	2.851390	2.870
1.0	1.265672	1.308996	2.666309	2.740
2.0	1.079770	1.178097	2.397360	2.610

Also solutions ω_n of transcendental Equation (2.372) are found numerically with the accuracy of 10^{-6}. The results obtained are reported in Table 2.2. It can be observed that application of the PA improves asymptotic results essentially and allows for their application even for the large values of ε.

Let us now consider vibrations of a beam with density ρ and with static moment of the beam cross section I, having the elasticity modulus E_1 on the interval $-1 \leq x < 0$, whereas the modulus E_2 on the interval $0 < x \leq 1$. After the introduction of perturbation parameter $\varepsilon = (E_2 - E_1)/E_1$, the elasticity modulus of the whole interval [-1;1] can be rewritten as follows: $E = E_1 + \varepsilon H(x - 0.5)E_1$.

The basic equations can be presented in the following form:

$$EIw^{IV} - \rho F\omega^2 w = 0, \tag{2.374}$$

with the BCs

$$w = w_{xx} = 0 \quad \text{for} \quad x = -1, x = 1. \tag{2.375}$$

We assume the following solution form:

$$\omega^2 \approx \omega_1^2 + \varepsilon\omega_1^2 + \varepsilon^2\omega_2^2, \qquad \omega \approx \omega_0 + \omega_1\varepsilon + \omega_2\varepsilon^2, \tag{2.376}$$

and we substitute Ansatz (2.376) into Equations (2.374), (2.375). Comparison of coefficients standing by the same powers of ε gives:

$$w_0^{IV} - \frac{\rho F}{IE}\omega_0^2 w_0 = 0, \tag{2.377}$$

$$w_0 = w_{0xx} = 0 \quad \text{for} \quad x = -1, x = 1, \tag{2.378}$$

$$w_0^{IV} - \frac{\rho F}{IE_1}\omega_0^2 w_1 = \frac{\rho F}{IE_1}\omega_1^2 w_0 - H(x - 0.5)w_0^{IV},$$

$$w_0 = w_{1xx} = 0 \quad \text{for} \quad x = -1, x = 1.$$

A solution to BVPs (2.377), (2.378) takes the following form:

$$\omega_0 = (\pi k)^2\sqrt{\frac{E_1 I}{\rho F}}, \qquad w_0 = D\sin(\pi kx), \quad k = 1, 2, 3, \ldots.$$

Table 2.3 Computational results

λ/k	1	2	3	4	5
λ_I	2.64	5.28	7.92	10.56	13.2
λ	2.69	5.39	8.099	10.799	13.499
$\lambda_{[1/1]}$	2.69	5.38	8.07	10.77	13.46

For the next approximation one obtains

$$\omega_1^2 = 0.5\omega_0^2, \qquad \omega_2^2 = \omega_0^2/32,$$

which means that

$$\omega^2 \approx (\pi k)^4 \frac{E_1 I}{\rho F}\left(1 + \frac{\varepsilon}{2} + \frac{\varepsilon^2}{32}\right), \quad k = 1, 2, 3, \ldots . \tag{2.379}$$

PA for the given case is

$$\omega_{[1/1]}^2 = (\pi k)^2 \frac{E_1 I(16 + 7\varepsilon)}{\rho F(16 - \varepsilon)}, \quad k = 1, 2, 3, \ldots . \tag{2.380}$$

Computational results regarding $\lambda = \omega/\omega_0$ with the help of Equation (2.379), PA (2.380) and numerical solution with accuracy of 10^{-6} for various values of k for $\varepsilon = 1$ are reported in Table 2.3. One may conclude that an application of the HPM for $\varepsilon = 1$ is generally efficient. However, an additional investigation implies that for $\varepsilon \geq 2$ also the PA possesses a lot of benefits regarding the accuracy of computational results.

2.7.2 *Method of Ishlinskii-Leibenzon*

Ishlinskii [47] and Leibenzon [55] have proposed the method of stability investigation devoted to structures with free edges (or bodies with free surfaces). Namely, in the governing stability equations parametric terms are omitted and they are kept only for BCs. This method has been widely applied; however, there were many doubts regarding its accuracy. In what follows we show how one can treat the mentioned approach as a zeroth approximation of a certain asymptotic approach [17].

Firstly let us consider the model problem regarding the stability of a cantilever beam. Both stability equations and BCs are presented in the following form:

$$EIw_{xxx} + Tw_{xx} = 0, \tag{2.381}$$

$$w(0) = w_x(0) = 0, \tag{2.382}$$

$$w_{xx}(L) = 0, \tag{2.383}$$

$$EIw_{xxx}(L) + Tw_x(L) = 0, \tag{2.384}$$

where T denotes the compressing force.

The exact solution to the eigenvalue problem (2.381)–(2.384) has the form

$$T = \frac{\pi^2}{4}\frac{EI}{L^2}, \tag{2.385}$$

$$w = 1 - \cos\frac{\pi x}{2L}. \tag{2.386}$$

Direct comparison of parametric terms occurred in stability equations (2.381) and BCs (2.384) constitutes a rather difficult task, therefore we apply the variational-asymptotic method. The potential energy of the cantilever beam is

$$\Pi = \frac{1}{2}\int_0^1 (EIw_{xx}^2 + Tw_x^2)dx = \tag{2.387}$$

$$\frac{1}{2}EI\int_0^1 w_{xx}^2 dx + \frac{1}{2}T\int_0^1 ww_{xx}dx + \frac{1}{2}Tww_x|_{x=1}.$$

It appears that a ratio of the second and third terms in Equation (2.387), taking for w Ansatz (2.386), is a relatively small one: $1 - \pi/4 \approx 0.215$. Consequently, we can omit the parametric term in the stability equation, and let us try to construct an iteration procedure.

Namely, we introduce the homotopic parameter ε into Equation (2.381) in the following way:

$$EIw_{xxx} + \varepsilon Tw_{xx} = 0. \tag{2.388}$$

Both eigenfunction w and eigenvalue T are presented in the PS:

$$w = w_0 + w_1\varepsilon + w_2\varepsilon^2 + \ldots, \tag{2.389}$$

$$T = T_0 + T_1\varepsilon + T_2\varepsilon^2 + \ldots. \tag{2.390}$$

Substituting Ansatz (2.389) into BVP (2.388), (2.381)–(2.383) and carrying out the splitting regarding ε, we get

$$EIw_{0xxxx} = 0,$$

$$w_0(0) = w_{0x}(0) = 0, \quad w_{0xx}(L) = 0, \quad EIw_{0xxx}(L) + T_0w_{0x}(L) = 0,$$

$$EIw_{kxxxx} = -\sum_{i=0}^{k-1} T_i w_{k-1xx}, \quad k = 1, 2, \ldots,$$

$$w_k(0) = w_{kx}(0) = 0, \quad w_{kxx}(L) = 0, \quad EIw_{kxxx}(0) + \sum_{i=1}^{k} T_i w_{k-ix}(L) = 0.$$

In zeroth order case we have

$$T_0 = 2EI/L^2, \qquad w_0 = Ax^2(x - 3L).$$

In the case of first order approximation we get

$$T_1 = \frac{EI}{3L^2}, \quad w_1 = -5ALx^2 + Ax^3 + \frac{A}{2L}x^4 - \frac{A}{10L^2}x^5.$$

Second order approximation yields $T_2 = 4EI/(45L^2)$.

The obtained solution is improved via the PA, and we obtain

$$T = 2\frac{(15-1.5\varepsilon)EI}{(15-4\varepsilon)L^2}.$$

Note that for $\varepsilon = 1$ we have $T = 2.4545EI/L^2$, which differs from the exact solution $T = 0.25\pi^2 EI/L^2 \approx 2.4650EI/L^2$ by 0.42%.

As the second example we take a beam lying on an elastic foundation and having free ends. After the introduction of the homotopic parameter ε we get

$$EIw_{xxxx} + \varepsilon T w_{xx} + cw = 0,$$

where c is the elasticity coefficient of foundation support.

Let us investigate a symmetric form of buckling (in an analogous way one may investigate the nonsymmetric case). BCs follow:

$$w_x(0) = w_{xxx}(0) = 0,$$

$$w_{xx}(l) = 0, \qquad EIw_{xx}(l) + Tw_x(l) = 0.$$

Using Ansatz (2.389), (2.390) and after splitting regarding ε, the following BVPs are obtained:

$$EIw_{0xxxx} + cw_0 = 0,$$

$$w_{0xx}(0) = w_{0xxx}(0) = 0,$$

$$w_{0xx}(l) = 0, \quad EIw_{0xxx}(l) + T_0 w_{0x}(l) = 0,$$

$$EIw_{kxxxx} + cw_k = -k\sum_{i=0} T_i w_{k-(i+1)xx},$$

$$w_k(0) = w_{kixx}(0) = 0,$$

$$w_{kxx}(l) = 0, \quad EIw_{kxxx}(l) + \sum_{i=0}^{k} T_i w_{k-ix}(l) = 0, \quad k = 1, 2, 3, \ldots.$$

Zeroth order approximation gives

$$w_0 = C_1(a_1 \sinh kx \sin kx + \cosh kx \cos kx), \quad T_0 = (EI/l^2)\overline{T}_0, \tag{2.391}$$

where: $\overline{T}_0 = 2\omega^2(a_2 - a_1 a_3)/(a_3 + a_1 a_2)$, $\omega = lk$, $a_1 = \tanh\omega \tan\omega$, $a_2 = \coth\omega + \tan\omega$, $a_3 = \cot\omega - \tan\omega$.

First order approximation yields

$$T_1 = (EI/l^2)\overline{T}_1, \quad \overline{T}_1 = \omega^2(t_1 + t_2 - t_3 t_4)w_{0x}^{-1}(l),$$

$$t_1 = \frac{\overline{T}_0}{2\omega^2}(b_1(\coth\omega + k^2\omega a_4) + b_2(\tanh\omega + k^2\omega a_1)),$$

$$t_2 = b_2(3\cot\omega - \omega a_4) - b_1(\tan\omega + \omega a_5), \quad t_3 = \frac{1}{2a_1}\left(ka_2 + \frac{\overline{T}_0}{\omega l}a_3\right),$$

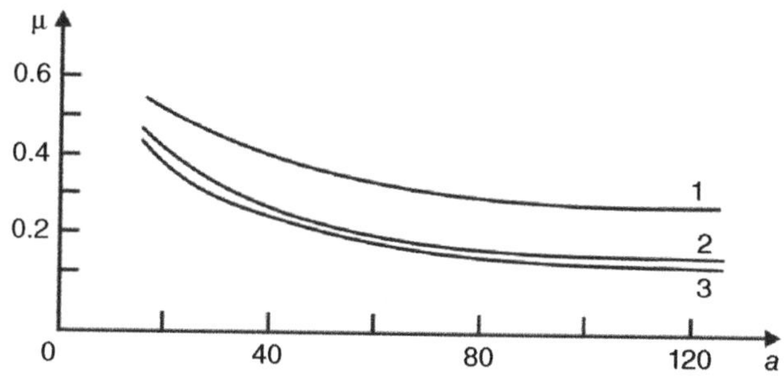

Figure 2.65 Comparison of various approaches for investigation of beam stability

$$t_4 = (b_1(a_4 - \omega \tan \omega) + b_2(a_5 + \omega \coth \omega))k^{-1},$$

$$b_1 = \frac{\overline{T}_0 a_4}{4\omega l}, \quad b_2 = \frac{\overline{T}_0 a_5}{4\omega l}, \quad a_4 = -a_1, \quad a_5 = 1 + a_1.$$

PA for $\varepsilon = 1$ gives:

$$T \approx T_0^2/(T_0 - T_1). \tag{2.392}$$

The results of calculations by using the formula (2.392) compared with the exact solution are shown in Figure 2.65, where $a = cl^4EI/16$ and $\mu = \pi/\overline{T}^{-1/2}$. In this figure, curve 1 was obtained in accordance with Equation (2.391), curve 2 was calculated by using formula (2.392), and curve 3 represents the exact solution [20].

Now let us investigate a stability of a thin isotropic rectangular plate ($0 \le x \le a; 0 \le y \le b$) being simply supported on three sides and compressed on its fourth side by the load P being parallel to two simply supported sectors. After introduction of the homotopic parameter ε the plate stability equation is

$$D\nabla^4 w + \varepsilon P w_{yy} = 0.$$

BCs follow:

$$w = 0, \quad w_{xx} = 0 \quad \text{for} \quad x = 0, a,$$

$$w = 0, \quad w_{yy} = 0 \quad \text{for} \quad y = 0,$$

$$w_{yy} + \nu w_{xx} = 0, \quad D[w_{yyy} + (2 - \nu)w_{xxy}] + Pw_y = 0 \quad \text{for} \quad y = b.$$

Let us present the function w in a PS (2.389), and the parameter P in the form of

$$P = P_0 + \varepsilon P_1 + \varepsilon^2 P_2 + \dots .$$

After splitting with respect to ε the following BVPs are obtained:

$$D\nabla^4 w_i + \sum_{i=0}^{i-1} P_{i-1} w_{k-ixx} = 0, \tag{2.393}$$

$$w_i = 0, \quad w_{ixx} = 0 \quad \text{for} \quad x = 0, a, \tag{2.394}$$

$$w_i = 0, \quad w_{iyy} = 0 \quad \text{for} \quad y = 0, \tag{2.395}$$

$$w_{iyy} + \nu w_{ixx} = 0, \quad D[w_{iyyy} + (2-\nu)w_{ixxy}] + \sum_{i=0}^{k} P_i w_{k-iy} = 0 \tag{2.396}$$
$$\text{for} \quad y = b, \quad i = 0, 1, 2, \ldots,$$

where $P_i = 0$.

Equations (2.393)–(2.396) for $i = 0$ are the Ishlinskii-Leibenzon equations. Their solutions is

$$w_0 = C(\sinh\lambda y - \lambda y C_1 \cosh\lambda y)\sin\lambda y, \quad \sigma_0 = H_0\varphi,$$

where

$$H_0 = [(1-\nu) + C_1(1+\nu) - (1-\nu)A\tanh A]\vartheta, \quad \vartheta = 1/[1 - C_1(1 + A\tanh A)],$$

$$\varphi = \frac{\pi^2 E}{12(1-\nu^2)}\left(\frac{h}{a}\right)^2, \quad \sigma_0 = \frac{P_0}{h}, \quad A = \frac{b\pi}{a}, \quad \tilde{N}_1 = (1-\nu)\psi,$$

$$\psi = 1/(2 + A(1-\nu)\coth A).$$

Solution to the first order approximation gives

$$w_1 = C\left\{\sinh\lambda y - \lambda(C_1C_2)y\cosh\lambda y + \frac{H_0\lambda^2}{8}\left[(C_2-1)y^2\sinh\lambda y + \frac{\lambda C_1}{3}y^3\cosh\lambda y\right]\right\}\sin\lambda x, \quad \sigma_1 = H_1\varphi,$$

where

$$C_2 = \frac{H_0}{8}\left\{[2 + 4A\coth A + (1-\nu)A^2] + \frac{C_1A}{3}[6\coth A + 6A + (1-\nu)A^2\coth A]\right\}\psi,$$

$$H_1 = \left\{-1 + (C_1 + C_2)(3 + A\tanh A) - \frac{H_0}{8}\left[(C_2-1)(6 + 6A\tanh A + A^2) + \frac{C_1}{3}(6 + 18A\tanh A + 9A^2 + A^3\tanh A)\right] - (H_0 - 2 + \nu)\left\{1 - (C_1 + C_2)(1 + A\tanh A) + \frac{H_0}{8}\left[(C_1-1)(2A\tanh A + A^2) + \frac{C_1}{3}(3A^2 + A^3\tanh A)\right]\right\}\right\}\vartheta.$$

We solve a second order approximation in a similar way and we obtain

$$\sigma_2 = H_2\varphi,$$

where

$$H_2 = \{[24\beta - 6(a - 2\beta A^2 + 3\gamma - 10\theta A^2) - A^2(\gamma + \theta A^2) + C_1 + C_3]\tanh A +$$
$$\{\beta A^4 - 1 - aA^2 - 6(a - 6\beta A^2 + \gamma - 10\theta A^2) + 3(C_1 + C_3 - A^2(3\gamma - 5\theta A^2))\} -$$
$$(H_0 - 2 + \nu)\{[2A(a - 2\beta A^2) - A(C_1 + C_3) + A^3(\gamma - \theta A^2)]\tanh A +$$
$$[1 - aA^2 - \beta A^4 - C_1 - C_3 + A^2(3\gamma - 5\theta A^2)] - H_1\{1 - (C_1 + C_3)(1 + A\tanh A +$$

$$\frac{H_0}{8}\left[(C_1-1)(2A\tanh A+A^2)+\frac{C_1}{3}(3A^2+A^3\tanh A)\right]\bigg\}\bigg\}\vartheta,$$

$$C_3=\{\{2a+A^2[(1-\nu)a-12\beta+6\gamma]-A^4[(1-\nu)\beta+10\theta]\}+\{2A(a+3\gamma)+$$

$$\{2A(a+3\gamma)+A^3[(1-\nu)\gamma-8\beta-20\theta]-A^5(1-\nu)\theta\}\coth A\}\varphi,$$

$$\alpha=\frac{1}{8}\left[H_0(C_1+C_3)+\frac{H_0^2}{16}(C_1-1)+H_1C_1-H_0-H_1\right],\quad \beta=\frac{H_0^2}{384}(2C_1-1),$$

$$\theta=\frac{H_0^2C_1}{1920},\quad \gamma=\frac{1}{24}\left[H_0(C_1+C_3)+\frac{H_0^2}{16}(4-3C_1)+H_1C_1\right].$$

Finally we get

$$\sigma\approx\varphi(H_0+\varepsilon H_1+\varepsilon^2H_2). \qquad (2.397)$$

Taking into account two first terms of truncated PS (2.397) and for $\varepsilon=1$ the PA gives

$$\gamma=\varphi H_0/(1-\beta_1), \qquad (2.398)$$

where $\beta_1=H_1/H_{01}$.

PA for $\varepsilon=1$ and taking into account three terms of truncated PS (2.397) yields

$$H=\varphi H_0\frac{1+(\beta_1+\beta_2)}{1+\beta_2}, \qquad (2.399)$$

where $\beta_2=-H_2/H_1$.

In Table 2.4 are shown results regarding parameter H computation and results are reported in [84] for plates with different ratio a/b for $\nu=0.3$. Also values of relative errors δ as a percentage are given.

Table 2.4 Computational results regarding plates with different sides ratio

a/b	1	1.5	2	3
Solution []	2.36	2.30	2.19	1.72
φH_0	1.76	1.7	1.64	1.54
	$\delta=25.3\%$	$\delta=25.7\%$	$\delta=25.1\%$	$\delta=10.5\%$
$\varphi(H_0+H_1)$	2.30	2.11	1.96	1.71
	$\delta=2.5\%$	$\delta=8.7\%$	$\delta=10.5\%$	$\delta=0.6\%$
Formula (2.397)	2.18	2.17	1.91	1.68
for $\varepsilon=1$	$\delta=2.41\%$	$\delta=10.1\%$	$\delta=12.9\%$	$\delta=2.26\%$
Formula (2.398)	2.30	2.19	1.98	1.07
	$\delta=2.41\%$	$\delta=4.87\%$	$\delta=9.37\%$	$\delta=0.43\%$
Formula (2.399)	2.36	2.25	2.02	1.71
	$\delta=0.16\%$	$\delta=2.17\%$	$\delta=7.94\%$	$\delta=0.43\%$

Therefore, two approximations matched with PA give good results. However, even either zero or first approximations can be used directly.

The Ishlinskii-Leibenzon simplification can be treated as zero approximation to a certain asymptotic process. If the error of zero order approximation is high, then the solution can be improved by applying the PA to the series part obtained so far. It appears that modified, in the way described so far, the Ishlinskii-Leibenzon method can be directly applied to stability investigation of elastic systems without the previously mentioned drawbacks.

2.7.3 *Vibrations of a String Attached to a Spring-Mass-Dashpot System*

In this section we study the dynamical system shown in Figure 2.66. The given problem can describe vibrations of suspended bridges, cables of power transmission lines, or shroud systems [28], [31], [60].

Basic BVP has the following form:

$$\rho \overline{u}_{\tau\tau} - T\overline{u}_{xx} = 0, \tag{2.400}$$

$$\overline{u}(0,\tau) = 0, \tag{2.401}$$

$$m_1\overline{u}_{\tau\tau}(1,\tau) + \gamma_2\overline{u}(1,\tau) + \overline{u}_x(1,\tau) = -\alpha\overline{u}_\tau(1,\tau), \tag{2.402}$$

$$\overline{u}(x,0) = \varphi(x), \qquad \overline{u}_\tau(x,0) = \psi_1(x), \tag{2.403}$$

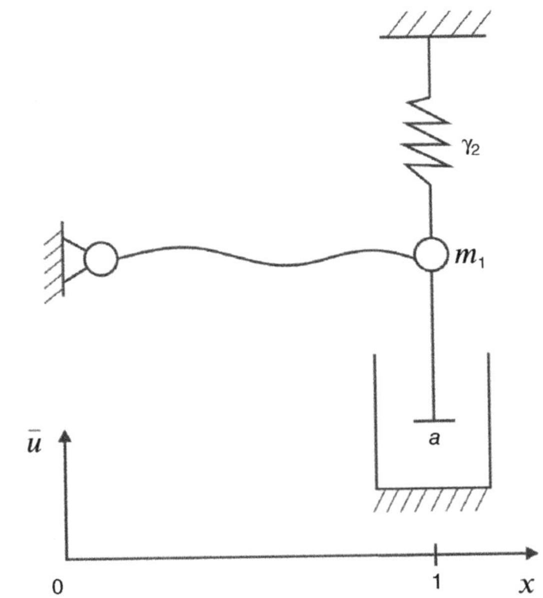

Figure 2.66 Model of a string fixed in the point $x = 0$ and having in the point $x = 1$ the attached mass and viscous damper

where ρ is the density of string material per its length, T is the string stretching force, τ is the time, and the remaining notation is illustrated in Figure 2.66.

Let us rewrite system (2.400)–(2.403) in new variables

$$\overline{u}_{tt} - \overline{u}_{xx} = 0, \tag{2.404}$$

$$\overline{u}(0, t) = 0, \tag{2.405}$$

$$m\overline{u}_{tt}(1, t) + \gamma_1 \overline{u}(1, t) + \overline{u}_x(1, t) = -\varepsilon_1 \overline{u}_t(1, t), \tag{2.406}$$

$$\overline{u}(x, 0) = \varphi(x), \qquad \overline{u}_t(x, 0) = \psi(x). \tag{2.407}$$

where $t = \tau\sqrt{\rho/T}$, $m = m_1/\rho$, $\gamma_1 = \gamma_2/T$, $\varepsilon_1 = \alpha/\sqrt{\rho T}$, $\psi = \psi_1\sqrt{\rho/T}$.

Let us present the function $\overline{u}(x, t)$ in the following form:

$$\overline{u} = \exp\left(-\frac{\varepsilon_1 t}{2m}\right) u(x, t). \tag{2.408}$$

Substituting Ansatz (2.408) into Equations (2.404) – (2.407) we get

$$u_{tt} - u_{xx} = \varepsilon u_t - 0.25\varepsilon^2 u, \tag{2.409}$$

$$u(0, t) = 0, \tag{2.410}$$

$$mu_{tt}(1, t) + \gamma u(1, t) + u_x(1, t) = 0, \tag{2.411}$$

$$u(x, 0) = \varphi(x), \qquad u_t(x, 0) = \psi(x) + 0.5\varepsilon\varphi(x), \tag{2.412}$$

where $\varepsilon = \varepsilon_1/m$, $\gamma = \gamma_1 - 0.25\varepsilon_1^2/m$.

Further we assume that $\varepsilon << 1$. A solution to BVP (2.409) – (2.412) can be presented as PS

$$u = u_0 + \varepsilon u_1 + \varepsilon^2 u_2 + \ldots. \tag{2.413}$$

Substituting Ansatz (2.413) into Equations (2.409)–(2.412) after splitting with respect to ε gives

$$u_{0tt} - u_{0xx} = 0, \tag{2.414}$$

$$u_0(0, t) = 0, \tag{2.415}$$

$$mu_{0tt}(1, t) + \gamma u_0(1, t) + u_{0x}(1, t) = 0, \tag{2.416}$$

$$u_0(x, 0) = \varphi(x), \qquad u_{0t}(x, 0) = \psi(x), \tag{2.417}$$

$$u_{1tt} - u_{1xx} = u_{0t},$$

$$u_1(0, t) = 0, \tag{2.418}$$

$$mu_{1tt}(1, t) + \gamma u_1(1, t) + u_{1x}(1, t) = 0, \tag{2.419}$$

$$u_1(x,0)=0, \qquad u_{1t}(x,0)=0.5\psi(x), \tag{2.420}$$

$$u_{2tt}-u_{2xx}=u_{1t}-0.25u_0,$$

$$u_2(0,t)=0,$$

$$mu_{2tt}(1,t)+\gamma u_2(1,t)+u_{2x}(1,t)=0,$$

$$u_2(x,0)=0, \qquad u_{2t}(x,0)=0,$$

....

BVP (2.414)–(2.417) allows for the variables separation. Eigenvalues to the problem are yielded by solutions to the transcendental equation

$$\cot\lambda = m\lambda - \gamma/\lambda. \tag{2.421}$$

One may check that $(n-1)\pi < \lambda_n < n\pi$, $n=1,2,3,\ldots$, and for $n\to\infty$ we have $\lambda_n \to n\pi$. Roots to Equation (2.421) are found numerically. Eigenfunctions $\sin\lambda_n x$ are orthogonal in the following sense:

$$\int_0^1 [1+m\delta(x-1)]\sin(\lambda_n x)\sin(\lambda_m x)dx = 0.$$

Finally, a solution to the BVP (2.414) – (2.417) has the following form:

$$u_0(x,t)=\sum_{n=1}^{\infty}[A_n\sin(\lambda_n t)+B_n\cos(\lambda_n t)]\sin(\lambda_n x), \tag{2.422}$$

where

$$A_n = A((\psi(x),\sin(\lambda_n x))/\overline{A}, \quad B_n = A((\varphi(x),\sin(\lambda_n x))/(\lambda_n\overline{A}),$$

$$\overline{A}=A(\sin(\lambda_n x),\sin(\lambda_n x)), \quad A(a(x),b(x))=\int_0^1 [1+m\delta(x-1)]abdx.$$

Equation of the first order approximation is

$$u_{1tt}-u_{1xx}=\sum_{n=1}^{\infty}\lambda_n[A_n\cos(\lambda_n t)-B_n\sin(\lambda_n t)]\sin(\lambda_n x),$$

and a particular solution to this equation follows:

$$u_1^{(0)} = -0.5t\sum_{n=1}^{\infty}[A_n\cos(\lambda_n t)+B_n\sin(\lambda_n t)]\sin(\lambda_n x), \tag{2.423}$$

and it is a secular one.

Let us apply the Pritulo method to remove singularities. We briefly revisit this approach presented in [62], [63], [69]. Let us suppose that as a result of the perturbation method application the following solution is obtained

$$x(t)=\cos\omega t+\beta t\varepsilon\sin\omega t.$$

One may apply the AEF in order to remove secular term

$$x(t) = \cos(\omega + \varepsilon\beta)t.$$

In fact, this defines the Pritulo method.

Pritulo approach applied to formulas (2.422), (2.423) allows to remove the secular terms

$$u_1^{(1)} = u_0 + \varepsilon u_1 \approx \sum_{n=1}^{\infty} [A_n \sin((\lambda_n + 0.5\varepsilon)t) + B_n \cos((\lambda_n + 0.5\varepsilon)t)] \sin(\lambda_n x).$$

Hence, a solution to equation

$$u_{1tt}^{(1)} - u_{1xx}^{(1)} = 0$$

with conditions (2.418) – (2.420) takes the following form:

$$u_1^{(1)} = \sum_{n=1}^{\infty} A_n^{(1)} \sin(\lambda_n t) \sin(\lambda_n x),$$

where $A_n^{(1)} = 0.5A(\varphi(x), \sin(\lambda_n x))/\overline{A}$.

Finally, the following approximating solution is obtained

$$u \approx u_0^{(1)} + \varepsilon u^{(1)}.$$

2.7.4 *Vibrations of a String with Nonlinear BCs*

In this section we study the linear wave equation

$$u_{tt} = u_{xx} \tag{2.424}$$

with the nonlinear BCs

$$u(0, t) = 0, \tag{2.425}$$

$$u_x(1, t) + u(1, t) + \alpha u^3(1, t) = 0. \tag{2.426}$$

In the case $\alpha << 1$ problem (2.424) – (2.426) is often solved via the Bubnov-Galerkin method or using the multiple-scales method. On the other hand during application of the variation approaches a nontrivial problem appears of reduction of an infinite system of ODEs [28], [29]–[31], [60], [74], [89].

Successive sequences of asymptotic solutions are constructed in references [29]–[31]. However, for $\alpha \sim 1$ the problem becomes difficult. If we consider solutions in the form

$$u(x, t) = \sum_{j=1,3,5,\ldots}^{\infty} A_j \sin\frac{\pi j x}{2} \sin\frac{\pi j t}{2},$$

then the problem is reduced to an infinite system of coupled LAEs regarding coefficients A_j. Even though it can be solved through a truncation, the solution is not satisfactory even for a relatively small number of equations.

Therefore, it seems to be promising an idea of seeking for another asymptotic parameter instead of α. Here we use the method of small delta (see Chapter 1.2.3). Following the mentioned approach the BC (2.426) can be presented in the following form:

$$u_x + u + \alpha u^{1+2\delta} = 0 \quad \text{for} \quad x = 1, \tag{2.427}$$

and a solution to the BVP (2.424), (2.425), (2.427) can be sought in the form of PS.

Let us consider the BVP (2.424), (2.425), (2.427) by using in the last formula $\alpha = 1$, and let us find its solution as PS

$$u = \sum_{k=0}^{\infty} \delta^k u_k. \tag{2.428}$$

Variational approaches are also widely used [28], [60], [74], [89].

We change the variable applying

$$t = \frac{\tau}{\omega}, \qquad \omega^2 = 1 + \alpha_1\delta + \alpha_2\delta^2 + \ \dots. \tag{2.429}$$

The nonlinear term in the BC (2.427) has the form

$$u^3 \equiv u^{1+2\delta} = u\left[1 + \delta \ln\ u^2 + \frac{\delta^2}{2}(\ln\ u^2)^2 + \dots\right]. \tag{2.430}$$

We take small δ in the solving process, whereas in the final result we put $\delta = 1$.

Substituting Ansatzes (2.429), (2.428), (2.430) into the BVP (2.424), (2.425), (2.427) and splitting with respect to δ the following recurrent sequence of BVPs is obtained:

$$u_{0tt} = u_{0xx}, \tag{2.431}$$

$$u_0 = 0 \quad \text{for} \quad x = 0, \tag{2.432}$$

$$u_{0x} + 2u_0 = 0 \quad \text{for} \quad x = 1, \tag{2.433}$$

$$u_{0tt} = u_{0xx} - \sum_{p=0}^{1} \alpha_{i-p} u_{ptt}, \tag{2.434}$$

$$u_1 = 0 \quad \text{for} \quad x = 0, \tag{2.435}$$

$$u_{1x} + 2u_1 = -u_0 \ln\ u_0^2 \quad \text{for} \quad x = 1, \tag{2.436}$$

$$u_{0tt} = u_{0xx} - \sum_{p=0}^{2} \alpha_{i-p} u_{ptt}, \tag{2.437}$$

$$u_2 = 0 \quad \text{for} \quad x = 0, \tag{2.438}$$

$$u_{2x} + 2u_2 = -u_1 \ln\ u_0^2 - 2u_1 - 0.5u_0(\ln\ u_0^2)^2 \quad \text{for} \quad x = 1, \tag{2.439}$$

....

where $\alpha_0 = 0$.

The solution to zeroth order BVP (2.431)–(2.433) has the following form:

$$u_0 = A \sin(\omega_0 x) \sin(\omega_0 t),$$

Table 2.5 Nonzero values of ω

$\omega_0^{(1)}$	$\omega_0^{(2)}$	$\omega_0^{(3)}$	$\omega_0^{(4)}$	$\omega_0^{(5)}$	$\omega_0^{(6)}$	$\omega_0^{(7)}$	$\omega_0^{(8)}$	$\omega_0^{(9)}$	$\omega_0^{(10)}$
2.289	5.087	8.096	11.17	14.28	17.39	20.52	23.65	26.78	29.91

where the frequency ω_0 is governed by the following transcendental equation:

$$\omega_0 = -2\tan\omega_0.$$

A few first nonzero values of ω are given in Table 2.7.4. For $k \to \infty$ we have the following asymptotics: $\omega_0^{(k)} \to 0.5\pi(2k+1)$.

Relations regarding the first order approximation follow:

$$u_{1xx} - u_{1tt} = \alpha_1 A\omega_0^2 \sin(\omega_0 x)\sin(\omega_0 t), \tag{2.440}$$

$$u_1 = 0 \quad \text{for} \quad x = 0, \tag{2.441}$$

$$u_{1x} + 2u_1 = A_1 \sin(\omega_0 t)[\ln\ (A^2\sin^2\omega_0) + \ln\ (\sin^2\omega_0 t)] \quad \text{for} \quad x = 1, \tag{2.442}$$

where $A_1 = -A\sin\omega_0$.

The particular solution to Equation (2.440), which satisfies the BC (2.441), has the following form:

$$u_1^{(1)} = -0.5\alpha_1 A\omega_0 x\ \cos(\omega_0 x)\sin(\omega_0 t).$$

Let us choose α_1 in a way to remove secular term occurring on the r.h.s. of Equation (2.442):

$$\alpha_1 = 2R_1/(\omega_0(6+\omega_0^2)),$$

where $R_1 = \ln\ (0.25eA^2\ \sin^2\omega_0)$.

Nonsecular harmonics that appeared on the r.h.s. of BC (2.442) yield the following solution:

$$u_1^{(2)} = 4A_1 \sum_{k=2}^{\infty} T_k \sin(\omega_0 kx)\sin(\omega_0 kt)\frac{1}{k^2-1}, \tag{2.443}$$

where $T_k = 1/[k\omega_0\cos(k\omega_0) + 2\sin(k\omega_0)]$.

A complete solution to the first order approximation has the following form:

$$u_1 = u_1^{(1)} + u_1^{(2)}.$$

In order to construct the second order approximation the r.h.s. of the BC (2.439) is rewritten into the form

$$u_{2x} + 2u_2 = -2u_1 - 0.5u_1 \ln\ u_0^2 \quad \text{for} \quad x = 1. \tag{2.444}$$

Let us separate secular harmonics on the r.h.s. of the BC (2.444):

$$u_{2x} + 2u_2 = -R_2 \sin\omega_0 t + \varphi(t) \quad \text{for} \quad x = 1, \tag{2.445}$$

where $R_2 = A\omega_0 \sin\omega_0[\alpha_1(\cot\omega_0 + 0.5(\omega_0 - \cot\omega_0)R_1) + R_3]$, $R_3 = 4\sum_{k=3,5,\ldots} T_k$, $T_k = k^2/[(k^2-1)(k\omega_0 + 2\tan(k\omega_0))^2]$, and by $\varphi(t)$ the sum of nonsecular harmonics is denoted.

We construct a particular solution to Equation (2.444) satisfying the BC (2.438) regarding the secular part of the r.h.s.:

$$u_2^{(1)} = -0.5A\sin(\omega_0 t)x\omega_0[\alpha_2\cos(\omega_0 x) + \alpha_1^2\sin(\omega_0 x)]. \tag{2.446}$$

Constant α_2 is chosen from a condition of lack of secular terms on the r.h.s. of the BC (2.442):

$$\alpha_2 = \frac{\alpha_1^2(\tan\omega_0 + \omega_0) + \alpha_1(2 - R_1) + \tan\omega_0(\alpha_1\omega_0 + 2R_3)}{\omega_0\tan\omega_0 - 3}.$$

Therefore, the following frequency of vibrations is obtained with an accuracy up to the third order terms with respect to δ:

$$\omega \cong \omega_0\sqrt{1 + \alpha_1\delta + \alpha_2\delta^2}. \tag{2.447}$$

If the PA is applied then

$$\omega \cong \omega_0\sqrt{\frac{\alpha_1 + (\alpha_1^2 - \alpha_2)\delta}{\alpha_1 - \alpha_2\delta}}. \tag{2.448}$$

Assuming $\delta = 1$, formulas (2.447), (2.448) yield the solution to our problem governed by (2.424)–(2.426).

It should be emphasized that the procedure so far illustrated and applied loses its benefits with an increase in frequency values. However, it is not difficult to guess that in the latter cases one may use asymptotics regarding an inverted frequency value. Then, in the zero order approximation we get the linear BVP (2.431), (2.432) and

$$u_{0x}(0,t) = 0,$$

whereas nonlinear terms of BCs are taken into account with the help of known perturbation methods.

Let us consider one more application of the δ perturbation method, using it for the following ODE:

$$\ddot{x} + x^{\frac{1}{2n+1}} = 0 \tag{2.449}$$

with attached initial conditions

$$x(0) = 0, \qquad x(0) = A \tag{2.450}$$

for $n \to \infty$.

Although a solution to Equation (2.449) can be found directly through special functions, this approach is not convenient for direct applications. In the limit $n \to \infty$ we get

$$\ddot{x}_0 + \mathrm{sgn}(x_0) = 0, \quad \mathrm{sgn}(x) = \begin{cases} +1 & \text{for} \quad x > 0, \\ -1 & \text{for} \quad x < 0. \end{cases} \tag{2.451}$$

Table 2.6 Comparison of analytical and numerical results

T	$n = 2$	$n = 3$	$n = 4$	$n = 5$	$n = 10$	$n = 100$
Runge-Kutta	1.2198	1.1651	1.1320	1.1099	1.0598	1.0068
(2.457)	1.3057	1.2077	1.1575	1.1269	1.0644	1.0065
$\Delta_{(2.457)}$, %	7.04	3.66	2.25	1.53	0.434	0.0047
(2.458)	-	-	-	1.2537	1.0793	1.0066
$\Delta_{(2.458)}$, %	-	-	-	12.9	1.85	0.0160
(2.459)	-	-	1.2231	1.1612	1.0710	1.0066
$\Delta_{(2.459)}$, %	-	-	8.05	4.62	1.06	0.0103

Analytical solution to Equation (2.451) can be sought in the form of a Fourier series or saw-tooth functions proposed by Pilipchuk [67]. One can also use a fitting method.

Let us introduce the parameter $\delta = (2n+1)^{-1}$, $\delta << 1$. We consider only values $x > 0$, since for $x < 0$ we get a symmetric solution. Applying the series

$$x^{\delta} = 1 + \delta \ln\ |x| + O(\delta^2), \tag{2.452}$$

a solution to the Cauchy problem (2.449), (2.450) is sought in the following form

$$x = x_0 + \delta x_1 + O(\delta^2). \tag{2.453}$$

Substituting Ansatz (2.453) into Equation (2.449), (2.450) and taking into account relations (2.452) yield the following equation:

$$\ddot{x}_1 = -\ln\ |x_0|. \tag{2.454}$$

We consider a solution to Equation (2.454) in the interval of the 1/4 period part. The zero order solution of Equation (2.451) has the following form:

$$x(t) = \begin{cases} -\frac{t}{2}(t-2A) & \text{for} \quad 0 \le t \le 2A, \\ \frac{t^2}{2} - 3At + 4A^2 & \text{for} \quad 2A \le t \le 4A, \end{cases}$$

$$x(t+nt) = x(t), \quad T = 4A.$$

First order approximation is

$$\ddot{x} = -\ln\left(tA - \frac{t^2}{2}\right), \tag{2.455}$$

$$x_1(0) = 0, \quad x(0) = 0. \tag{2.456}$$

Integrating Equation (2.455) gives

$$x_1(t) = 2At - 3t^2 - t^2 \ln\ 2 - 4A(A-t)\ln\ |-2A| + t^2 \ln\ |2A - t| +$$
$$t^2 \ln\ t + 4A^2 \ln\ |-2A+t| - 4At \ln\ |-2A+t|.$$

and finally we get

$$x \approx x_0 + \delta x_1. \tag{2.457}$$

The PA yields

$$x \approx \frac{x_0^2}{x_0 - \delta x_1}. \tag{2.458}$$

The exponential approximation follows

$$x \approx x_0 \exp\,(\delta x_1/x_0). \tag{2.459}$$

The results obtained by the obtained formulas have been compared with those obtained through Runge-Kutta method (Table 2.6). Analysis of the results allows us to conclude that best results are given by the formula (2.457). Relative error decreases fast with increase of n. Besides, the exponential approximation give better results than the PA.

2.7.5 Boundary Conditions and First Order Approximation Theory

Theories devoted to improvement in accuracy regarding beams, plates and shells have been widely applied [40]. It should be noted that all mentioned theories have been constructed on the basis of certain introduced phenomenological approaches. As has been shown in the series of works [38], the direct use of the improved equations without the associated modification of the BCs may result in errors. Therefore, this section is aimed at presentation of relatively simple examples, showing how modifications of the BCs may improve the solution approximation accuracy.

To begin with we study vibration of a stretched beam governed by the equation

$$\rho F W_{tt} - T W_{xx} + EI W_{xxxx} = 0, \tag{2.460}$$

where T is the stretching force.

The following BCs are applied:

$$W = W_{xx} = 0 \quad \text{for} \quad x = 0, L. \tag{2.461}$$

Let us rescale BVP (2.516), (2.517) taking into account $\xi = x/L$, $\varepsilon^2 = EI/(TL^2)$:

$$W_{\tau\tau} - W_{\xi\xi} + \varepsilon^2 W_{\xi\xi\xi\xi} = 0, \tag{2.462}$$

$$W = W_{\xi\xi} = 0 \quad \text{for} \quad \xi = 0, 1. \tag{2.463}$$

For $\varepsilon = 0$ the BVP (2.516), (2.517) yields the string model

$$W_{\tau\tau} - W_{\xi\xi} = 0, \tag{2.464}$$

$$W = 0 \quad \text{for} \quad \xi = 0, 1. \tag{2.465}$$

We note the second order derivative regarding the longitudinal coordinate, which essentially simplifies further analysis [48]. In what follows we show how to keep the second order of the approximating equation with a simultaneous improvement of the approximation accuracy. Namely, for the differential operator $-\partial^2/\partial\xi^2 + \varepsilon^2\partial^4/\partial\xi^4$ the following PA is applied:

$$\frac{-\partial^2/\partial\xi^2}{(1+\varepsilon\partial^2/\partial\xi^2)}.$$

Then, Equation (2.518) with accuracy of ε^2 can be recast to the form

$$\left(1+\varepsilon^2\frac{\partial^2}{\partial\xi^2}\right)W_{\tau\tau} - W_{\xi\xi} = 0. \tag{2.466}$$

BCs for Equation (2.466) have the form (2.465). If the BVP (2.464), (2.465) approximates eigenvalues of the studied problem with accuracy ε^2, then BVP (2.466), (2.465) does the same with accuracy of ε^4, whereas the order of the governing equation with respect to the spatial coordinate does not change.

Observe that Equation (2.466) may be obtained by the differential approach. Namely, if the operator $1+\varepsilon^2\partial^2/\partial\xi^2$ acts on Equation (2.518), then all terms of the accuracy order higher than two are neglected.

Analogously, one may consider vibrations of a plate with the small bending stiffness governed by the equation

$$W_{\tau\tau} - \nabla^2 W + \varepsilon^2\nabla^4 W = 0, \tag{2.467}$$

$$W = \nabla^2 W = 0 \quad \text{for} \quad \xi, \eta = 0, 1, \tag{2.468}$$

where $\nabla^2 = \frac{\partial^2}{\partial\xi^2} + \frac{\partial^2}{\partial\eta^2}$.

Furthermore, one gets the membrane model from Equation (2.467) for $\varepsilon = 0$:

$$W_{\tau\tau} - \nabla^2 W = 0, \tag{2.469}$$

$$W = 0 \quad \text{for} \quad \xi, \eta = 0, 1. \tag{2.470}$$

The accuracy improved model takes the form

$$(1+\varepsilon^2\nabla^2)W_{\tau\tau} - \nabla^2 W = 0 \tag{2.471}$$

with BCs (2.470).

In what follows we consider the problem of formulation of the BCs for the Equation (2.466). In the case of a simple support the approach so far works excellently.

Let us study the beam clamped on its ends:

$$W = W_x = 0 \quad \text{for} \quad x = 0, L.$$

In this case Equation (2.464) does not allow satisfying all BCs, and hence the following state (boundary layer) should be attached:

$$\varepsilon^2 W_{n0xx} - W_{n0} = 0, \tag{2.472}$$

$$\varepsilon W_{n0x} = -W_{0x} \quad \text{for} \quad x = 0, L. \tag{2.473}$$

The solution to Equation (2.472) has the following form:

$$W_{n0} = \lfloor C_1 \exp(-\varepsilon^{-1}x) + C_2 \exp(\varepsilon^{-1}(x-\pi))\rfloor \sin\omega t. \tag{2.474}$$

We assume that the beam possesses enough length that the mutual influence of the boundary layers governed by formula (2.474) can be neglected.

Solution (2.474) suffers for the incorrectness of ε order, which has occurred in the BCs with respect to w:

$$W_0 + \varepsilon W_{n0} \neq 0 \quad \text{for} \quad x = 0, L.$$

Consequently, application of the improved Equation (2.466) is not allowed in this case. However, taking into account solution (2.474) in BCs (2.473), one may reduce the BVP in a way that the obtained BCs consisting of Equation (2.466) and BCs

$$W_0 - \varepsilon W_{0x} \quad \text{for} \quad x = 0, L. \tag{2.475}$$

yield a solution with the accuracy of ε^2.

The examples illustrated and discussed so far can also be applied in the case of nonlinear problems. For instance, consider the following nonlinear elastic formulation:

$$\varepsilon^2 W_{xxxx} - W_{xx} + W^3 + W_{tt} = 0.$$

In this case, the zeroth order nonlinear equation is of a second order regarding the spatial coordinate with the following form:

$$-W_{0xx} + W^3 + W_{0tt} = 0. \tag{2.476}$$

The accuracy improved equation, suitable for the case of simple support, has the form:

$$-W_{xx} + \left(1 + \varepsilon^2 \frac{\partial^2}{\partial x^2}\right)(W^3 + W_{tt}) = 0.$$

In the case of clamping one can use Equation (2.476) with BCs (2.475).

2.8 Links between the Adomian and Homotopy Perturbation Approaches

Recently the development of various practical approaches of analytical approximate integrations of nonlinear differential equations are mainly matched with the application of the solution continuation with respect to a parameter within two main modifications: the Adomian decomposition method (ADM) and HPM. ADM in spite of a lack of any links with either small or large parameters [1], [3], [25], practically coincides with the special form perturbation method (see [16]). On the other hand, the method of HPM initially uses the artificially introduced parameter and takes into account a special function, called a homotopic mapping. Today's variant of this method influenced theory of analytical solutions of nonlinear DEs considerably. However, the authors of the papers devoted to those problems usually do not study links of the homotopic functional properties and the help obtained with its approximations (existence, domain of applicability, stability, convergence, etc.). Although for the ADM the convergence properties have been widely studied via various methods [1], [3], [25], [26], [46] the HPM has been less investigated. This problem has been analyzed for instance in [64], where only algebraic equations have been studied. However, numerous references illustrate the efficiency of the method devoted to solve numerous model and real problems.

If the known functions occurred in an ODE can be presented in the form of the generalized series regarding the unknown variable, the function and its derivative being sought, then the approximate solution obtained either by the ADM or the HPM has a polynomial form with the power of the integration variable. The ADM variant described so far is called the modified ADM approach [46]. In some cases solutions obtained via those different approaches coincide, but not always. Therefore, a problem of relation between approximations obtained via the ADM and HPM appears. It should be emphasized that in the case of algebraic equations this question has been rigorously solved in [64], where the coincidence of the results has been shown for the suitable chosen parameters. In what follows we show how two methods can be unified through the novel synthetic approach.

In the beginning we study the model examples of ODEs, where a solution obtained via either the ADM or HPM do not coincide, and then we study the qualitative behavior of the solutions.

Let us take a nonhomogenous linear singularly ODE perturbated with the following initial conditions:

$$\varepsilon z' + z = 1, \quad z(0), \quad 0 < \varepsilon << 1. \tag{2.477}$$

Exact solution of this problem $z = 1 - \exp(-x/\varepsilon)$ while developing into the Maclaurin series regarding the independent variable is presented in the following form

$$z = \frac{x}{\varepsilon} - \frac{x^2}{2\varepsilon^2} + \cdots + (-1)^{n+1}\frac{1}{n!}\left(\frac{x}{\varepsilon}\right)^{n} + \ldots . \tag{2.478}$$

Equations (2.477) and (2.478) show that either the exact solution or its representation in the power of the series have the logarithmic singularity at point $\varepsilon = 0$, where the series is divergent. Therefore, accuracy of the solution approximation through the Maclaurin series part is nonhomogenous in the neighborhood of the singularity.

Let us introduce parameter ε_1 in the following way:

$$z' = \frac{1}{\varepsilon} - \varepsilon_1\frac{z}{\varepsilon}. \tag{2.479}$$

This idea of the parameter introduction reduces the problem to a sequence of the ADM approximations [16]. Let us present z in the form of the series with respect to ε_1, and let us substitute this series into Equation (2.479), and then let us compare to zero the terms standing by the same power of this parameter. We obtain:

$\varepsilon_1^0\colon z_0' = \frac{1}{\varepsilon}, z_0(0) = 0 \Rightarrow z_0 = \frac{x}{\varepsilon},$

$\varepsilon_1^1\colon z_1' = -\frac{z_0}{\varepsilon} = -\frac{x}{\varepsilon^2}, z_1(0) = 0 \Rightarrow z_1 = -\frac{x^2}{2\varepsilon^2} = (-1)^1\frac{1}{2!}\left(\frac{x}{\varepsilon}\right)^2,$

$\varepsilon_1^2\colon z_2' = -\frac{z_1}{\varepsilon} = \frac{x^2}{2\varepsilon^2}, z_2(0) = 0 \Rightarrow z_2 = \frac{x^3}{6\varepsilon^3} = (-1)^2\frac{1}{3!}\left(\frac{x}{\varepsilon}\right)^3,$

$\ldots$

$\varepsilon_1^n\colon z_n' = -\frac{z_{n-1}}{\varepsilon} = (-1)^n\frac{1}{(n-1)!}\left(\frac{x}{\varepsilon}\right)^n\left(\frac{1}{\varepsilon}\right),$

$z_n(0) = 0 \Rightarrow z_n = (-1)^n\frac{1}{n!}\left(\frac{x}{\varepsilon}\right)^{n+1},$

$\ldots$

$$z = \frac{x}{\varepsilon} - \frac{x^2}{2\varepsilon^2}\varepsilon_1 + \frac{x^3}{6\varepsilon^3}\varepsilon_1^2 \cdots + (-1)^n\frac{1}{n!}\left(\frac{x}{\varepsilon}\right)^{n+1}\varepsilon_1^n + \ldots .$$

For $\varepsilon_1 = 1$ we get the approximate solution in the form overlapping with series (2.478), which is an agreement with the results reported in reference [16].

In order to remove the nonhomogeneity in the neighborhood of zero point the PA is applied. Since in the approximation two variable quantities appear (the introduced parameter and independent variable), the PA is constructed according to the following three steps:

(i) with respect to ε_1 for $x = const$, $x \neq 0$;
(ii) with respect to x for $\varepsilon_1 = const$, $\varepsilon_1 \neq 0$;
(iii) two-dimensional form with respect to both ε_1 and x, simultaneously.

In what follows we apply the method of V. Vavillov [86] while the 2D PA is constructed. The method allows to get uniquely defined coefficients of the approximation, as well as it guarantees that the 2D PA possesses a whole spectrum of properties in the sense of the Montessus de Ballore theorem [18].

If the considerations are restricted to the first order approximation, then for all studied cases and for $\varepsilon = 1$ we obtain approximation z_{ε_1} regarding ε_1 in the following form:

$$z_{\varepsilon_1} = \frac{x}{\varepsilon}\left(1 - \frac{3x}{6\varepsilon + 2x}\right),$$

whereas the coinciding approximations of z_x regarding x as well as the two-dimensional z_{2D} have the following forms:

$$z_x = z_{2D} = \frac{2x}{2\varepsilon + x}.$$

It is clear that the first PA, contrary to the second and third, does not remove the singularity at the point $\varepsilon = 0$.

If all terms of Equation (2.477) are shifted into its l.h.s., and if then we add a higher order derivative to both sides of the obtained equation and we introduce ε_1 in the following form:

$$z' = \varepsilon_1(1 - z + (1 - \varepsilon)z') \tag{2.480}$$

one gets the following sequence of approximations regarding to the HPM [43]:

$$\begin{aligned}
&\varepsilon_1^0 : z_0' = 0,\ z_0(0) \Rightarrow z_0 = 0;\\
&\varepsilon_1^1 : z_1' = 1 - z_0 + (1-\varepsilon)z_0' = 1,\ z_1(0) = 0 \Rightarrow z_1 = x;\\
&\varepsilon_1^2 : z_2' = -z_1 + (1-\varepsilon)z_1' = -x + (1-\varepsilon),\ z_2(0) = 0\\
&\Rightarrow z_2 = -\tfrac{x}{2!} + (1-\varepsilon)x;\\
&\varepsilon_1^3 : z_3' = -z_2 + (1-\varepsilon)z_2' = \tfrac{x^2}{2} - (1-\varepsilon)x + (1-\varepsilon)(-x + (1-\varepsilon)),\\
&z_3(0) = 0 \Rightarrow z_3 = \tfrac{x^3}{3!} - (1-\varepsilon)x^2 + (1-\varepsilon)^2 x;\\
&\varepsilon_1^4 : z_4' = -z_3 + (1-\varepsilon)z_3' = -\tfrac{x^3}{6} + (1-\varepsilon)x^2 - (1-\varepsilon)^2 x + (1-\varepsilon)(\tfrac{x^2}{2} - 2(1-\varepsilon)x +\\
&\quad (1-\varepsilon)^2),\ z_4(0) = 0 \Rightarrow z_4 = -\tfrac{x^4}{4!} + (1-\varepsilon)\tfrac{x^3}{2} - 3(1-\varepsilon)^2\tfrac{x^2}{2} + (1-\varepsilon)^3 x.
\end{aligned}$$

Therefore, the following approximation of the HPM is obtained:

$$z = x\varepsilon_1 + \left(-\frac{x^2}{2!} + (1-\varepsilon)x\right)\varepsilon_1^2 + \left(\frac{x^3}{3!} - (1-\varepsilon)x^2 + (1-\varepsilon)^2 x\right)\varepsilon_1^3 + \dots .$$

Equivalently, for $\varepsilon_1 = 1$ we get

$$\begin{aligned} z =&(1 + (1-\varepsilon) + (1-\varepsilon)^2 + (1-\varepsilon)^3 + \dots)x + \\ &\left(-\frac{1}{2!} - (1-\varepsilon) - 3(1-\varepsilon)^2\frac{1}{2} + \dots\right)x^2 + \\ &\left(\frac{1}{3!} + (1-\varepsilon)\frac{1}{2} + \dots\right)x^3 - \frac{x^4}{4!} + \dots, \end{aligned} \tag{2.481}$$

which coincides with series (2.478) taking into account the development of its coefficients into the series regarding ε in the neighborhood of $\varepsilon = 1$.

In what follows we construct the PA in the way analogous to that of the ADM construction. For $\varepsilon = 1$ we get approximations regarding ε_1 in the following form:

$$z_{\varepsilon_1} = z_{2D} = \frac{2x}{2\varepsilon + x},$$

whereas the approximation regarding x is

$$z_x = \frac{2(2-\varepsilon)^2 x}{2(2-\varepsilon) + x}.$$

Parameter ε_1 is introduced in the following form:

$$z' = \varepsilon_1 \frac{1-z}{\varepsilon}, \tag{2.482}$$

which yields the novel system of subsequent approximations.

We present z in the series form with respect to powers of ε_1, substitute this series into Equation (2.482), and then after the comparison of terms standing by the same parameter into zero we get

$$\begin{aligned} &\varepsilon_1^0 : z_0' = 0,\ z_0(0) = 0 \Rightarrow z_0 = 0; \\ &\varepsilon_1^1 : z_1' = \tfrac{1-z_0}{\varepsilon} = \tfrac{1}{\varepsilon},\ z_1(0) = 0 \Rightarrow z_1 = \tfrac{x}{\varepsilon} = (-1)^2\tfrac{1}{1!}\left(\tfrac{x}{\varepsilon}\right)^1; \\ &\varepsilon_1^2 : z_2' = -\tfrac{z_1}{\varepsilon} = -\tfrac{x}{\varepsilon^2},\ z_2(0) = 0 \Rightarrow z_2 = -\tfrac{x^2}{2\varepsilon^2} = (-1)^3\tfrac{1}{2!}\left(\tfrac{x}{\varepsilon}\right)^2; \\ &\dots \\ &\varepsilon_1^n : z_n' = -\tfrac{z_{n-1}}{\varepsilon} = (-1)^{n+1}\tfrac{1}{(n-1)}\left(\tfrac{x}{\varepsilon}\right)^{n-1}\left(\tfrac{1}{\varepsilon}\right),\ z_n(0) = 0 \\ &\Rightarrow z_n = (-1)^{n+1}\tfrac{1}{n!}\left(\tfrac{x}{\varepsilon}\right)^n; \\ &\dots \\ &z = \tfrac{x}{\varepsilon}\varepsilon_1 - \tfrac{1}{2!}\left(\tfrac{x}{\varepsilon}\right)^2\varepsilon_1^2 + \tfrac{1}{3!}\left(\tfrac{x}{\varepsilon}\right)^3 + \dots. \end{aligned}$$

Observe that for $\varepsilon_1 = 1$ we obtain the solution being sought in the form coinciding with series (2.478).

PA for $\varepsilon_1 = 1$ in this case yields

$$z_{\varepsilon_1} = z_x = z_{2D} = \frac{2x}{2\varepsilon + x}.$$

As the second example we consider the nonhomogenous linear singular perturbated ODE with variable coefficients with the following BC:

$$\varepsilon z' + xz = x, \quad z(0) = 2, \quad 0 < \varepsilon << 1. \tag{2.483}$$

Exact solution of Equation (2.483) $z = 1 + \exp\left(-\frac{x^2}{2\varepsilon}\right)$ can be developed into the Maclaurin series:

$$z = 2 - \frac{x^2}{2\varepsilon} + \frac{x^4}{8\varepsilon^2} \cdots + (-1)^n \frac{1}{n!}\left(\frac{x^2}{2\varepsilon}\right)^n + \ldots. \tag{2.484}$$

As in the first case in order to proceed with the ADM, we introduce the parameter ε_1

$$z' = \frac{x}{\varepsilon} - \varepsilon_1 \frac{xz}{\varepsilon}. \tag{2.485}$$

Let us present z in the series form with powers of ε_1, and let us introduce this series into Equation (2.485). After splitting with respect to ε_1 we get

$$\varepsilon_1^0 : z_0' = \tfrac{x}{\varepsilon},\ z_0(0) = 2 \Rightarrow z_0 = 2 + \tfrac{x^2}{2\varepsilon};$$
$$\varepsilon_1^1 : z_1' = -\tfrac{xz_0}{\varepsilon} = -\tfrac{x}{\varepsilon}\left(2 + \tfrac{x^2}{2\varepsilon}\right),\ z_1(0) = 0 \Rightarrow z_1 = -\tfrac{x^2}{\varepsilon} - \tfrac{x^4}{8\varepsilon^2};$$
$$\varepsilon_1^2 : z_2' = -\tfrac{xz_1}{\varepsilon} = \tfrac{x}{\varepsilon}\left(\tfrac{x^2}{\varepsilon} + \tfrac{x^4}{8\varepsilon^2}\right),\ z_2(0) = 0 \Rightarrow z_2 = \tfrac{x^4}{4\varepsilon^2} + \tfrac{x^6}{48\varepsilon^3};$$
$$\ldots$$
$$z = 2 + \tfrac{x^2}{2\varepsilon} + \left(-\tfrac{x^2}{\varepsilon} - \tfrac{x^4}{8\varepsilon^2}\right)\varepsilon_1 + \left(\tfrac{x^4}{4\varepsilon^2} + \tfrac{x^6}{48\varepsilon^3}\right)\varepsilon_1^2 - \ldots.$$

For $\varepsilon_1 = 1$ the approximate solution is

$$z = 2 + \frac{x^2}{2\varepsilon} - \frac{x^2}{\varepsilon} - \frac{x^4}{8\varepsilon^2} + \frac{x^4}{4\varepsilon^2} + \frac{x^6}{48\varepsilon^3} - \ldots.$$

Observe that it coincides with series (2.484) up to the summation term with power $2n$ regarding x, which is in agreement with the result presented in [16].

Diagonal PA of the first order with respect to ε_1 and for $\varepsilon_1 = 1$ has the following form:

$$z_{\varepsilon_1} = 2 + \frac{x^2}{2\varepsilon} - x^2 \frac{(8\varepsilon + x^2)^2}{384\varepsilon^3 + 32x^2\varepsilon^2 + 8x^4},$$

whereas the expecting approximation z_x and z_{2D} do not exist.

Since the solution practically depends on x^2, let us apply the following change $\tilde{x} = x^2$ during construction of the PA. Approximation regarding ε_1 will not change in this case, whereas approximation with respect to x and the 2D form are as follows

$$z_x = z_{2D} = \frac{8\varepsilon}{4\varepsilon + x^2}.$$

In order to carry out the HPM, we shift all terms of Equation (2.483) into its l.h.s. We add a higher order derivative to both sides of the obtained equation, and we introduce ε_1 in the following way:

$$z' = \varepsilon_1(x - xz + (1 - \varepsilon)z'); \tag{2.486}$$

$$\varepsilon_1^0 : z_0' = 0,\ z_0(0) = 2 \Rightarrow z_0 = 2;$$
$$\varepsilon_1^1 : z_1' = x - xz_0 + (1-\varepsilon)z_0' = x - 2,\ z_1(0) = 0 \Rightarrow z_1 = -\tfrac{x^2}{2};$$
$$\varepsilon_1^2 : z_2' = -xz_1 + (1-\varepsilon)z_1' = \tfrac{x^3}{2} - (1-\varepsilon)x,\ z_2(0) = 0$$
$$\Rightarrow z_2 = \tfrac{x^4}{8} - (1-\varepsilon)\tfrac{x^2}{2};$$
$$\ldots$$
$$z = 2 - \tfrac{x^2}{2}\varepsilon_1 + \left(\tfrac{x^4}{8} - (1-\varepsilon)\tfrac{x^2}{2}\right)\varepsilon_1^2 + \ldots.$$

For $\varepsilon_1 = 1$ we get approximation of the HPM in the following form:

$$z = 2 - (1 + (1-\varepsilon) + \ldots)\frac{x^2}{2} + \left(\frac{1}{2} + \ldots\right)\frac{x^2}{4} + \ldots, \tag{2.487}$$

which corresponds to series (2.484) taking into account the development of its coefficients into the series regarding ε in the vicinity of $\varepsilon = 1$.

Diagonal PA of both first order regarding ε_1 and the 2D approximation have the following forms

$$z_{\varepsilon_1} = z_{2D} = \frac{8\varepsilon}{4\varepsilon + x^2}$$

and

$$z_x = 2 - x^2\frac{8 - 4\varepsilon + \varepsilon^2}{8 - 4\varepsilon + x^2}.$$

Solving Equation (2.483) with respect to the higher derivative and introducing the parameter ε_1 in the following way:

$$z' = \varepsilon_1\frac{x - xz}{\varepsilon}, \tag{2.488}$$

we get

$$\varepsilon_1^0 : z_0' = 0,\ z_0(0) = 2 \Rightarrow z_0 = 2;$$
$$\varepsilon_1^1 : z_1' = \tfrac{x - xz_0}{\varepsilon} = -\tfrac{x}{\varepsilon},\ z_1(0) = 0 \Rightarrow z_1 = -\tfrac{x^2}{2\varepsilon} = (-1)^1\tfrac{1}{1!}\left(\tfrac{x^2}{2\varepsilon}\right)^1;$$
$$\varepsilon_1^2 : z_2' = -\tfrac{xz_1}{\varepsilon} = \tfrac{x^3}{2\varepsilon^2},\ z_2(0) = 0 \Rightarrow z_2 = \tfrac{x^4}{8\varepsilon^2} = (-1)^2\tfrac{1}{2!}\left(\tfrac{x^2}{2\varepsilon}\right)^2;$$
$$\ldots$$
$$\varepsilon_1^n : z_n' = -\tfrac{xz_{n-1}}{\varepsilon} = (-1)^n\tfrac{1}{(n-1)!}\left(\tfrac{x^2}{2\varepsilon}\right)^{n-1},\ z_n(0) = 0$$
$$\Rightarrow z_n = (-1)^n\tfrac{1}{n!}\left(\tfrac{x^2}{2\varepsilon}\right)^n;$$
$$\ldots$$
$$z = 2 - \tfrac{x^2}{2\varepsilon}\varepsilon_1 + \tfrac{x^4}{8\varepsilon^2}\varepsilon_1^2 + \ldots.$$

For $\varepsilon_1 = 1$ we obtain the approximate solution being sought in the form coinciding with series (2.484) without a need for summation with respect to x and ε.

Diagonal PA of the first order for $\varepsilon_1 = 1$ has the following form:

$$z_{\varepsilon_1} = z_x = z_{2D} = \frac{8\varepsilon}{4\varepsilon + x^2}.$$

Now we study nonhomogenous singularly Riccati ODE [68], with the initial condition of the form:

$$\varepsilon z' = z^2 + x, \quad z(1) = 1, \quad 0 < \varepsilon << 1. \tag{2.489}$$

It is known that a general solution can be obtained using the special functions, and it has the following form:

$$z = -\varepsilon\frac{w_x'}{w}, \quad w = \sqrt{x}\left(C_1 J_{\frac{1}{3}}\left(\frac{2}{3\varepsilon}x^{\frac{3}{2}}\right) + C_2 Y_{\frac{1}{3}}\left(\frac{2}{3\varepsilon}x^{\frac{3}{2}}\right)\right),$$

where J_α, Y_α are the Bessel functions of the order α ([2], chapter 9), and C_1, C_2 are arbitrary constants.

It is known that for $\alpha = \frac{1}{3}$ the solution cannot be obtained in the quadratures of elementary functions, as well as the determination of the constants do not belong to a trivial problem.

General solution can be presented in the following form:

$$z = -\varepsilon\left(\frac{1}{2\sqrt{x}} + \frac{C\frac{d}{dx}J_{\frac{1}{3}}\left(\frac{2}{3\varepsilon}x^{\frac{3}{2}}\right) + \frac{d}{dx}Y_{\frac{1}{3}}\left(\frac{2}{3\varepsilon}x^{\frac{3}{2}}\right)}{CJ_{\frac{1}{3}}\left(\frac{2}{3\varepsilon}x^{\frac{3}{2}}\right) + Y_{\frac{1}{3}}\left(\frac{2}{3\varepsilon}x^{\frac{3}{2}}\right)}\right), \quad C = \frac{C_1}{C_2},$$

and

$$C = -\frac{\frac{2+\varepsilon}{2\varepsilon}Y_{\frac{1}{3}}\left(\frac{2}{3\varepsilon}\right) + \frac{d}{dx}Y_{\frac{1}{3}}\left(\frac{2}{3\varepsilon}\right)}{\frac{2+\varepsilon}{2\varepsilon}J_{\frac{1}{3}}\left(\frac{2}{3\varepsilon}\right) + \frac{d}{dx}J_{\frac{1}{3}}\left(\frac{2}{3\varepsilon}\right)}.$$

Since the illustrated methods are particularly suitable to be applied having the initial condition in zero, we change the variable and we obtain the following problem equivalent to (2.489):

$$\varepsilon z' = -z^2 + x - 1, \quad z(0) = 1, \quad 0 < \varepsilon << 1. \tag{2.490}$$

To realize the ADM we solve Equation (2.490) with respect to the derivative and we introduce the parameter ε_1 in the following way:

$$z' = -\varepsilon_1\frac{z^2}{\varepsilon} + \frac{x-1}{\varepsilon}, \quad z(0) = 1. \tag{2.491}$$

We present z in the series regarding ε_1, substitute this series into Equation (2.491), and compare to zero terms standing by the same powers of ε.

We obtain

$$\varepsilon_1^0 : z_0' = \frac{x-1}{\varepsilon}, z_0(0) = 1 \Rightarrow z_0 = \frac{x^2}{2\varepsilon} - \frac{x}{\varepsilon} + 1;$$

$$\varepsilon_1^1 : z_1' = -\frac{x^4}{4\varepsilon^3} + \frac{x^3}{\varepsilon^3} - \frac{2x^2}{\varepsilon^3} + \frac{2x}{\varepsilon^2} - \frac{1}{\varepsilon}, z_1(0) = 0$$

$$\Rightarrow z_1 = \frac{1}{60\varepsilon^3}(-3x^5 + 15x^4 - 40x^3 + 60\varepsilon x^2 - 60\varepsilon^2 x);$$

$$\varepsilon_1^2 : z_2' = -\frac{1}{60\varepsilon^5}(-3x^5 + 15x^4 - 40x^3 + 60\varepsilon x^2 - 60\varepsilon^2 x)(x^2 - 2x + 2\varepsilon),$$

$z_2(0) = 0 \Rightarrow z_2 = -\frac{1}{60\varepsilon^5}\left(-\frac{3x^8}{8} + 3x^7 - \frac{(3\varepsilon+55)^2x^6}{6} + (18\varepsilon + 16)x^5 - \right.$

$\left.(15\varepsilon^2 + 50\varepsilon)x^4 + 80\varepsilon^2x^3 - 60\varepsilon^3x^2\right);$

...

$z = 1 - \frac{x}{\varepsilon} + \frac{x^2}{2\varepsilon} + \frac{1}{60\varepsilon^3}(-30x^5 + 15x^4 - 40x^3 + 60\varepsilon x^2 - 60\varepsilon^2 x)\varepsilon_1 +$

$\frac{1}{60\varepsilon^5}\left(\frac{3x^8}{8} + 3x^7 - \frac{(3\varepsilon+55)^2x^6}{6} + (18\varepsilon + 16)x^5 - \right.$

$\left.(15\varepsilon^2 + 50\varepsilon)x^4 + 80\varepsilon^2x^3 - 60\varepsilon^3x^2\right)\varepsilon_1^2 + \ldots.$

For $\varepsilon_1 = 1$ we obtain the approximation being sought in the following form:

$$z = 1 - \frac{2x}{\varepsilon} + \left(\frac{1}{2\varepsilon} + \frac{2}{\varepsilon^2}\right)x^2 + \frac{1}{60\varepsilon^3}(-30x^5 + 15x^4 - 40x^3) -$$

$$\frac{1}{60\varepsilon^5}\left(\frac{3x^8}{8} + 3x^7 - \frac{(3\varepsilon + 55)^2x^6}{6} + (18\varepsilon + 16)x^5 - \right.$$

$$\left.(15\varepsilon^2 + 50\varepsilon)x^4 + 80\varepsilon^2x^3\right).$$

Diagonal PA of the first order with respect to ε_1 and for $\varepsilon_1 = 1$ has the following form:

$$z_{\varepsilon_1} = 1 - \frac{x}{\varepsilon} + \frac{x^2}{2\varepsilon} + \frac{1}{60\varepsilon^3}(3x^5 + 15x^4 - 40x^3 + 60\varepsilon x^2 - 60\varepsilon^2 x)^2 \times$$

$$\left(3x^5 + 15x^4 - 40x^3 + 60\varepsilon x^2 - 60\varepsilon^2 x + \left(-\frac{3x^8}{8} + 3x^7 - \right.\right.$$

$$\left.\left.\frac{(3\varepsilon + 55)^2x^6}{6} + (18\varepsilon + 16)x^5 - (15\varepsilon^2 + 50\varepsilon)x^4 + 80\varepsilon^2x^3 - 60\varepsilon^3x^2\right)\right),$$

whereas the approximation z_x is

$$z_x = \frac{4\varepsilon + (\varepsilon - 6)x}{4\varepsilon + (\varepsilon + 2)x}$$

and the 2D approximation reads:

$$z_{2D} = \frac{2\varepsilon + (\varepsilon - 3)x}{2\varepsilon + (\varepsilon + 1)x}.$$

In order to proceed with the HPM, we introduce ε_1 in the following way:

$$z' = \varepsilon_1((1 - \varepsilon)z' - z^2 + x - 1); \tag{2.492}$$

$\varepsilon_1^0 : z_0' = 0,\ z_0(0) = 1 \Rightarrow z_0 = 1;$

$\varepsilon_1^1 : z_1' = x - 2,\ z_1(0) = 0 \Rightarrow z_1 = \frac{x^2}{2} - 2x;$

$\varepsilon_1^2 : z_2' = -x^2 + (5-\varepsilon)x - 2(1-\varepsilon),\ z_2(0) = 0 \Rightarrow$

$z_2 = -\frac{x^3}{3} + \frac{(5-\varepsilon)}{2}x^2 - 2(1-\varepsilon)x;$

...

$z = 1 + \left(\frac{x^2}{2} - 2x\right)\varepsilon_1 + \left(-\frac{x^3}{3} + \frac{(5-\varepsilon)}{2}x^2 - 2(1-\varepsilon)x\right)\varepsilon_1^2 + \dots.$

For $\varepsilon_1 = 1$ we obtain the following approximation of the HPM:

$$z = 1 - 2(2-\varepsilon)x + \frac{(6-\varepsilon)}{2}x^2, \qquad (2.493)$$

where the coefficients standing by x represent the series part regarding the development of ADM coefficients with respect to ε in the neighborhood of one.

Diagonal PA of the first order for $\varepsilon_1 = 1$ has the following form:

$$z_{\varepsilon_1} = 1 + \frac{3}{2}\frac{x(x-4)^2}{2x^2 + (3\varepsilon - 12)x - 12\varepsilon},$$

$$z_x = \frac{4(2-\varepsilon) + (-26 + 31\varepsilon - 8\varepsilon^2)x}{4(2-\varepsilon) + (6-\varepsilon)x},$$

$$z_{2D} = \frac{4\varepsilon - (\varepsilon - 4)x + 4\varepsilon(\varepsilon - 1)}{4\varepsilon - (\varepsilon - 4)x + 4x(2\varepsilon - 1)}.$$

Let us introduce the parameter ε_1 in the following way:

$$z' = \varepsilon_1 \frac{-z^2 + x - 1}{\varepsilon}, \qquad z(0) = 1. \qquad (2.494)$$

We obtain

$\varepsilon_1^0 : z_0' = 0,\ z_0(0) = 1 \Rightarrow z_0 = 1;$

$\varepsilon_1^1 : z_1' = \frac{x-2}{\varepsilon},\ z_1(0) = 0 \Rightarrow z_1 = \frac{x^2}{2\varepsilon} - \frac{2x}{\varepsilon};$

$\varepsilon_1^2 : z_2' = -\frac{x^2}{\varepsilon^2} + \frac{4x}{\varepsilon^2},\ z_2(0) = 0 \Rightarrow$

$z_2 = -\frac{x^3}{3\varepsilon^2} + \frac{2x^2}{\varepsilon^2};$

...

$z = 1 + \left(\frac{x^2}{2\varepsilon} - \frac{2x}{\varepsilon}\right)\varepsilon_1 + \left(-\frac{x^3}{3\varepsilon^2} + \frac{2x^2}{\varepsilon^2}\right)\varepsilon_1^2 + \dots.$

For $\varepsilon_1 = 1$ the solution being sought has the following form:

$$z = 1 - \frac{2x}{\varepsilon} + \left(\frac{1}{2\varepsilon} + \frac{2}{\varepsilon^2}\right)x^2 - \frac{x^3}{3\varepsilon^2},$$

which fully coincides with the first terms of the ADM formula.
Diagonal PA of the first order for $\varepsilon_1 = 1$ is

$$z_{\varepsilon_1} = 1 + \frac{3}{2}\frac{x(x-4)^2}{2x^2 + (3\varepsilon - 12)x - 12\varepsilon},$$

$$z_x = z_{2D} = \frac{4\varepsilon + (\varepsilon - 4)x}{4\varepsilon + (\varepsilon + 4)x}.$$

It is seen that z_{ε_1} overlaps with analogous approximation of the HPM.
Therefore we have illustrated that:

- governing equations of the ADM are solved regarding higher derivative with respect to the independent variable, whereas the HPM allows to get the similar result only for $\varepsilon = 1$;
- coefficients of the ADM approximation yield the exact solution after the summation of the corresponding approximations, whereas the HPM, after both the same summation procedure and development of the exact solution coefficients versus the real small parameter ε, yields the exact solution in the neighborhood of $\varepsilon = 1$;
- in the given examples both approximations coincide in a limit approaching the real solution in the domain of its holomorphicity with respect to both the independent variable and small parameter;
- proposed method allows us to get values of coefficients standing by powers of the independent variable without their summation with respect to higher order approximations.

Let us formulate the problem more rigorously. Namely, let us formally introduce the ADM and HPM for the system of ODEs in the asymptotic terminology language.

It is recognized [90], that an ODE or system of ODEs can be recast to the so called normal form of the first order regarding unknown functions $\{u_i = u_i(\xi)\}_{i=1}^n$ in the same interval $\omega : \xi \in]0, 1[$ of the following form:

$$Lu_i + R_i(\xi, u_1, \dots, u_n) + N_i(\xi, u_1, \dots, u_n) = g_i(\xi), \quad L = \frac{d}{d\xi}, \quad i = \overline{1, n}, \tag{2.495}$$

with the BCs on the boundary $\partial\omega : \xi = 0 \cup 1$

$$G_j(u_1, \dots, u_n)|_{\partial\Omega} = 0, \quad j = \overline{1, n}, \tag{2.496}$$

where: L is the operator of the derivative with respect to the independent variable ξ; R_i is the linear differential operator with respect to the sought function; N_i and G_j are nonlinear differential operators regarding the function being sought $g_i = g_i(\xi)$.

We also assume the point $\xi_0 = 0$ belonging to a closure of Ω, whereas R_i, N_i and G_j are the holomorphic functions regarding $\{u_i\}_{i=1}^n$. Owing to the ADM, a solution can be presented in the following form:

$$u_i = \sum_{j=0}^{\infty} u_{ij}^A, \qquad i = \overline{1, n}. \tag{2.497}$$

Component of a solution can be found from the equations

$$Lu_{ij}^{A} = A_{ij}, \qquad i = \overline{1,n}, \qquad j = \overline{0,\infty}. \tag{2.498}$$

where A_{ij} are the Adomian polynomials [3] defined through the following formulas:

$$A_{i0} = g_i, \quad A_{ij} = -\frac{1}{j!}\frac{\partial^j}{\partial \lambda^j}\left(N_i\left(\sum_{m=0}^{j} u_{im}^{A}\lambda^m\right) + R_i\left(\sum_{m=0}^{j} u_{im}^{A}\lambda^m\right)\right)\Bigg|_{\lambda=0}, \tag{2.499}$$
$$i = \overline{1,n}, \quad j = \overline{1,\infty}.$$

As has been shown in [16], the ADM is equivalent to both development of the governing equation and its solution with respect to powers of the artificial parameter λ, which is introduced in the following way:

$$u_i = \sum_{j=0}^{\infty} u_{ij}^{A}\lambda^j, \quad Lu_i + \lambda(R_i(u_1,\dots,u_n) + N_i(u_1,\dots,u_n)) = g_i, \quad i = \overline{1,n}$$

for $\lambda = 1$.

In the case of the HPM, the governing equation (in general) cannot be solved regarding a higher derivative, therefore the system of ODEs can be cast to the following form:

$$\begin{aligned} &Lu_1 + R_1(u_1,\dots,u_n) + N_1(u_1,\dots,u_n) + F(Lu_1,u_1,\dots,u_n) = g_1, \\ &Lu_i + R_i(u_1,\dots,u_n) + N_i(u_1,\dots,u_n) = g_i, \quad i = \overline{2,n}, \end{aligned} \tag{2.500}$$

where F is the nonlinear differential operator, and ε is introduced in the following way:

$$\begin{aligned} &u_i = \sum_{j=0}^{\infty} u_{ij}^{H}\varepsilon^j, \\ &(1-\varepsilon)(Lu_1 - Lu_1|_{\partial\Omega}) + \varepsilon(Lu_1 + R_1 + N_1 + F - g_1) = 0, \\ &(1-\varepsilon)(Lu_i - Lu_i|_{\partial\Omega}) + \varepsilon(Lu_i + R_i + N_i - g_i) = 0, \quad i = \overline{2,n}, \\ &G_j(u_1|_{\partial\Omega},\dots,u_n|_{\partial\Omega})|_{\partial\Omega} = 0, \quad j = \overline{1,n}, \end{aligned} \tag{2.501}$$

$u_1|_{\partial\Omega}$ are referred to as the "probe" functions satisfying the BCs [44].

System (2.501) can be rewritten in the following form:

$$Lu_i + \varepsilon(Lu_i|_{\partial\Omega} + R_i + N_i + F\delta_i^1 - g_i) = 0, \quad i = \overline{1,n}, \tag{2.502}$$

where δ_i^1 is the Cronecker symbol.

Assuming $\{u_i = u_i(\xi)\}_{i=1}^{n}$ and their derivatives as the independent arguments, let us substitute the operators R_i, N_i, F and G_j in the form of the generalized multidimensional Taylor series:

$$R_i + N_i = \sum_{j=1}^{n}\left(N_{ij}u_j + \frac{1}{2!}\sum_{p=1}^{n} N_{ijp}u_j u_p + \dots\right), \quad i = \overline{1,n}, \tag{2.503}$$

$$F = \left(F_0 L u_1 + \frac{1}{2!} \sum_{p=1}^{n} F_{0p} u_p L u_1 + \ \dots \ \right) +$$
$$\sum_{j=1}^{n} \left(F_j u_j + \frac{1}{2!} \sum_{p=1}^{n} F_{jp} u_j u_p + \ \dots \ \right),$$
$$G_j = \sum_{q=1}^{n} \left(G_{jq}(u_q - u_q|_{\partial\Omega}) + \frac{1}{2!} \sum_{p=1}^{n} G_{jqp}(u_q - u_q|_{\partial\Omega})(u_p - u_p|_{\partial\Omega}) + \ \dots \ \right),$$
$$j = \overline{1, n}. \tag{2.504}$$

Let us also develop the coefficients of the applied operators and functions g_i into the series with respect to powers of ξ:

$$N_{ij} = \sum_{r=0}^{\infty} N_{ij}^r \xi^r, \quad N_{ijp} = \sum_{r=0}^{\infty} N_{ijp}^r \xi^r, \dots \quad i, j, p = \overline{1, n},$$
$$F_j = \sum_{r=0}^{\infty} F_j^r \xi^r, \quad F_{jp} = \sum_{r=0}^{\infty} F_{jp}^r \xi^r, \dots \quad j, p = \overline{0, n}, \tag{2.505}$$
$$g_i = \sum_{j=0}^{\infty} g_{ij} \xi^i, \quad i = \overline{1, n}.$$

We substitute the series (2.503)–(2.505) into Equations (2.499):

$$A_{ij} = -\frac{1}{j!} \frac{\partial^j}{\partial \lambda^j} \left(\sum_{r=1}^{n} \left(N_{ir} u_r + \frac{1}{2!} \sum_{p=1}^{n} N_{irp} u_r u_p + \ \dots \ \right) \right) \Bigg|_{\lambda=0} =$$
$$-\frac{1}{j!} \frac{\partial^j}{\partial \lambda^j} \left(\sum_{r=1}^{n} \left(N_{ir} \sum_{k=0}^{j} u_{rk}^A \lambda^k + \frac{1}{2!} \sum_{p=1}^{n} N_{irp} \sum_{k=0}^{j} u_{rk}^A \lambda^k \sum_{q=0}^{j} u_{pq}^A \lambda^q + \ \dots \ \right) \right) \Bigg|_{\lambda=0} =$$
$$-\sum_{r=1}^{n} \left(N_{ir} u_{rj}^A + \sum_{p=1}^{n} N_{irp} \sum_{k=0}^{j} u_{rk}^A u_{p(j-k)}^A + \ \dots \ \right). \tag{2.506}$$

The following sequence of the ADM problems is obtained:

$$\varepsilon^0 : \quad L u_{i0}^A = g_i, \quad G_j(u_{10}^A, \dots, u_{n0}^A)|_{\partial\Omega} = 0, \quad j = \overline{1, n},$$
$$\varepsilon^0 : \quad L u_{i1}^A = -\sum_{r=1}^{n} \left(N_{ir} u_{r0}^A + \sum_{p=1}^{n} \sum_{m=0}^{1} N_{irp} u_{r0}^A u_{p0}^A + \ \dots \ \right),$$
$$u_{j1}^A|_{\partial\Omega} = 0, \quad j = \overline{1, n}, \tag{2.507}$$
$$\varepsilon^0 : \quad L u_{i2}^A = -\sum_{r=1}^{n} \left(N_{ir} u_{r1}^A + \sum_{p=1}^{n} N_{irp}(u_{r1}^A u_{p0}^A + u_{r0}^A u_{p1}^A) + \ \dots \ \right),$$
$$u_{j2}^A|_{\partial\Omega} = 0, \quad j = \overline{1, n}.$$

Let us substitute Ansatzes (2.500), (2.503)–(2.505) into Equations (2.502) and let us collect terms standing by the same powers of ε:

$$Lu_i + \varepsilon\left(Lu_i|_{\partial\Omega} + \delta_i^1 \left(F_0 Lu_1 + \frac{1}{2!}\sum_{p=1}^{n} F_{0p}u_p Lu_1 + \ldots \right) + \right.$$

$$\delta_i^1 \sum_{j=1}^{n}\left(F_j u_j + \frac{1}{2!}\sum_{p=1}^{n} F_{jp}u_j u_p + \ldots \right) + \sum_{j=1}^{n}\left(N_{ij}u_j + \right.$$

$$\left.\left.\frac{1}{2!}\sum_{p=1}^{n} N_{ijp}u_j u_p + \ldots \right) - g_i \right) = L\sum_{k=0}^{\infty} u_{ik}^H \varepsilon^k + \varepsilon\left(L\sum_{k=0}^{\infty} u_{ik}^H\varepsilon^k \right|_{\partial\Omega} +$$

$$\delta_i^1\left(F_0 L\sum_{k=0}^{\infty} u_{1k}^H\varepsilon^k + \frac{1}{2!}\sum_{p=1}^{n} F_{0p}\sum_{k=0}^{\infty} u_{1k}^H\varepsilon^k \sum_{q=0}^{\infty} u_{pq}^H\varepsilon^q + \ldots \right) +$$

$$\delta_i^1\sum_{j=1}^{n}\left(F_j\sum_{k=0}^{\infty} u_{jk}^H\varepsilon^k + \frac{1}{2!}\sum_{p=1}^{n} F_{jp}\sum_{k=0}^{\infty} u_{jk}^H\varepsilon^k \sum_{q=0}^{\infty} u_{pq}^H\varepsilon^q + \ldots \right) +$$

$$\left.\sum_{j=1}^{n}\left(N_{ij}\sum_{k=0}^{\infty} u_{jk}^H\varepsilon^k + \frac{1}{2!}\sum_{p=1}^{n} N_{ijp}\sum_{k=0}^{\infty} u_{jk}^H\varepsilon^k \sum_{q=0}^{\infty} u_{pq}^H\varepsilon^q + \ldots \right) - g_i \right) =$$

$$Lu_{i0}^H - g_i\varepsilon + \sum_{k=1}^{\infty}\varepsilon^k\left(Lu_{i(k-1)}^H\Big|_{\delta\Omega} + \delta_i^1\left(F_0 Lu_{1(k-1)}^H + \right.\right.$$

$$\left.\frac{1}{2!}\sum_{p=1}^{n} F_{0p}\sum_{q=0}^{k-1} u_{p(k-q-1)}^H Lu_{1q}^H + \ldots \right) + \delta_i^1\sum_{r=1}^{n}(F_r u_{r(k-1)}^H +$$

$$\left.\frac{1}{2!}\sum_{p=1}^{n} F_{rp}\sum_{q=0}^{k-1} u_{rq}^H u_{p(k-q-1)}^H + \ldots \right) + \sum_{r=1}^{n}(N_{ir}u_{r(k-1)}^H +$$

$$\left.\left.\frac{1}{2!}\sum_{p=1}^{n} N_{irp}\sum_{q=0}^{k-1} u_{rq}^H u_{p(k-q-1)}^H + \ldots \right)\right). \tag{2.508}$$

Comparing to zero the coefficients standing by the same power of ε, the following sequence of problems of the HPM is obtained:

$$\varepsilon^0: \quad Lu_{i0}^H = 0, \quad G_i(u_{10}^H, \ldots, u_{n0}^H)|_{\partial\Omega} = 0, \quad i = \overline{1,n},$$

$$\varepsilon^1: \quad Lu_{i1}^H - g_i + Lu_{i0}^H|_{\partial\Omega} + \delta_i^1\left(F_0 Lu_{10}^H + \frac{1}{2!}\sum_{p=1}^{n} F_{00}u_{p0}^H Lu_{10}^H + \ldots \right) +$$

$$\delta_i^1\sum_{r=1}^{n}\left(F_r u_{r0}^H + \frac{1}{2!}\sum_{p=1}^{n} F_{rp}u_{r0}^H u_{p0}^H + \ldots \right) +$$

$$\sum_{r=1}^{n}\left(N_{ir}u_{r0}^{H}+\frac{1}{2!}\sum_{p=1}^{n}N_{irp}u_{r0}^{H}u_{p0}^{H}+\ldots\right)=0,\quad u_{i1}^{H}|_{\partial\Omega}=0,\quad i=\overline{1,n},$$

$$\begin{aligned}
\varepsilon^2:\quad & Lu_{i2}^{H}+Lu_{i1}^{H}|_{\partial\Omega}+\delta_i^1\left(F_0Lu_{11}^{H}+\frac{1}{2!}\sum_{p=1}^{n}F_{00}\left(u_{p0}^{H}Lu_{11}^{H}+u_{p1}^{H}Lu_{10}^{H}\right)+\ldots\right)+\\
& \delta_i^1\sum_{r=1}^{n}\left(F_ru_{r1}^{H}+\frac{1}{2!}\sum_{p=1}^{n}F_{rp}\left(u_{r1}^{H}u_{p0}^{H}+u_{r0}^{H}u_{p1}^{H}\right)+\ldots\right)+\\
& \sum_{r=1}^{n}\left(N_{ir}u_{r1}^{H}+\frac{1}{2!}\sum_{p=1}^{n}N_{irp}\left(u_{r1}^{H}u_{p0}^{H}+u_{r0}^{H}u_{p1}^{H}\right)+\ldots\right)=0,\\
& u_{i2}^{H}|_{\partial\Omega}=0,\quad i=\overline{1,n}.
\end{aligned}\tag{2.509}$$

Substituting Ansatz (2.505) into Equations (2.507) and (2.509), the following sequence of the ADM approximations is obtained:

$$\varepsilon^0:\quad Lu_{i0}^{A}=\sum_{j=0}^{\infty}g_{ij}\xi^{i},\quad G_j(u_{10}^{A},\ldots,u_{n0}^{A})|_{\partial\Omega}=0,\quad j=\overline{1,n},$$

$$\Rightarrow\quad u_{i0}^{A}=u_i|_{\partial\Omega}+\sum_{j=0}^{\infty}\frac{g_{ij}}{(j+1)}\xi^{j+1},$$

$$\begin{aligned}
\varepsilon^1:\quad & Lu_{i1}^{A}=-\sum_{r=1}^{n}\left(\left[N_{ir}u_r|_{\partial\Omega}+\frac{1}{2!}\sum_{p=1}^{n}N_{irp}u_r|_{\partial\Omega}u_p|_{\partial\Omega}+\ldots\right]+\right.\\
& N_{ir}\sum_{j=0}^{\infty}\frac{g_{rj}}{(j+1)}\xi^{j+1}+\frac{1}{2!}\sum_{p=1}^{n}\left(N_{irp}\left(u_r|_{\partial\Omega}\sum_{j=0}^{\infty}\frac{g_{pj}}{(j+1)}\xi^{j+1}+\right.\right.\\
& \left.\left.u_p\Big|_{\partial\Omega}\sum_{j=0}^{\infty}\frac{g_{rj}}{(j+1)}\xi^{j+1}\right)+\ldots\right)+\left.\left(\frac{1}{2!}\sum_{p=1}^{n}N_{irp}\sum_{j=0}^{\infty}\sum_{q=0}^{\infty}\frac{g_{rj}g_{pq}}{(j+1)(q+1)}\xi^{j+q+2}+\right)\right),\\
& u_{j1}^{A}|_{\partial\Omega}=0,\quad j=\overline{1,n},\quad\Rightarrow\quad u_{i1}^{A}=-\sum_{r=1}^{n}\sum_{k=0}^{\infty}\left(\left(N_{ir}^{k}u_r|_{\partial\Omega}+\right.\right.\\
& \left.\frac{1}{2!}\sum_{p=1}^{n}N_{irp}^{k}u_r|_{\partial\Omega}u_p|_{\partial\Omega}+\ldots\right)\frac{\xi^{k+1}}{(k+1)}+\\
& N_{ir}^{k}\sum_{j=0}^{\infty}\frac{g_{rj}}{(j+1)(k+j+2)}\xi^{k+j+2}+\\
& \frac{1}{2!}\sum_{p=1}^{n}\left(N_{irp}^{k}\sum_{j=0}^{\infty}\frac{u_r|_{\partial\Omega}g_{pj}+u_p|_{\partial\Omega}g_{rj}}{(j+1)(k+j+2)}\xi^{k+j+2}+\ldots\right)+
\end{aligned}\tag{2.510}$$

$$\left(\frac{1}{2!}\sum_{p=1}^{n} N_{irp}^{k}\sum_{j=0}^{\infty}\sum_{q=0}^{\infty}\frac{g_{rj}g_{pq}}{(j+1)(q+1)(k+j+q+3)}\xi^{k+j+q+3}+\ldots\right)\Bigg)\Bigg),$$

$$j=\overline{1,n},$$

$$\varepsilon^2:\quad Lu_{i2}^A=-\sum_{r=1}^{n}\left(N_{ir}u_{r1}^A+\frac{1}{2!}\sum_{p=1}^{n}N_{irp}(u_{r1}^Au_{p0}^A+u_{r0}^Au_{p1}^A)+\ldots\right)\Bigg),$$

$$u_{j2}^A|_{\partial\Omega}=0,\quad j=\overline{1,n},$$

In the case of the HPM we get

$$\varepsilon^0:\quad Lu_{i0}^H=0,\quad G_i(u_{10}^H,\ldots,u_{n0}^H)|_{\partial\Omega}=0,\quad\Rightarrow\quad u_{i0}^H=u_i|_{\partial\Omega},\quad i=\overline{1,n},$$

$$\varepsilon^1:\quad Lu_{i1}^H-g_i+Lu_{i0}^H|_{\partial\Omega}+\delta_i^1\left(F_0Lu_{10}^H+\frac{1}{2!}\sum_{p=1}^{n}F_{00}u_{p0}^HLu_{10}^H+\ldots\right)+$$

$$\delta_i^1\sum_{r=1}^{n}\left(F_ru_{r0}^H+\frac{1}{2!}\sum_{p=1}^{n}F_{rp}u_{r0}^Hu_{p0}^H+\ldots\right)+\sum_{r=1}^{n}(N_{ir}u_{r0}^H+$$

$$\frac{1}{2!}\sum_{p=1}^{n}N_{irp}u_{r0}^Hu_{p0}^H+\ldots\Bigg)=0,\quad u_{i1}^H|_{\partial\Omega}=0\quad\Rightarrow\quad u_{i1}^H=\sum_{j=0}^{\infty}\frac{g_{ij}}{(j+1)}\xi^{j+1}-$$

$$\sum_{k=0}^{\infty}\frac{\xi^{k+1}}{(k+1)}\sum_{r=1}^{n}\Big((N_{ir}^k+\delta_i^1F_r^k)u_r|_{\partial\Omega}+ \tag{2.511}$$

$$\frac{1}{2!}\sum_{p=1}^{n}\left(N_{irp}^k+\delta_i^1F_{rp}^k\right)u_r|_{\partial\Omega}u_p|_{\partial\Omega}+\ldots\Bigg),\quad i=\overline{1,n},$$

$$\varepsilon^2:\quad Lu_{i2}^H+Lu_{i1}^H|_{\partial\Omega}+\delta_i^1\left(F_0Lu_{11}^H+\frac{1}{2!}\sum_{p=1}^{n}F_{00}\left(u_{p0}^HLu_{11}^H+u_{p1}^HLu_{10}^H\right)+\ldots\right)+$$

$$\sum_{r=1}^{n}\left(\left(N_{ir}+\delta_i^1F_r\right)u_{r1}^H+\frac{1}{2!}\sum_{p=1}^{n}(N_{irp}+\delta_i^1F_{rp})(u_{r1}^Hu_{p0}^H+u_{r0}^Hu_{p1}^H)+\ldots\right)=0,$$

$$u_{i2}^H|_{\partial\Omega}=0\quad\Rightarrow\quad u_{i2}^H=-\delta_i^1\sum_{k=0}^{\infty}\left(F_0^k+F_{00}^k\sum_{p=1}^{\infty}u_p\Big|_{\partial\Omega}\right)\times$$

$$\Bigg(\sum_{j=0}^{\infty}\frac{g_{1j}}{(j+1)}\xi^{k+j+1}-\sum_{q=0}^{\infty}\frac{\xi^{k+q+1}}{(q+1)}\sum_{l=1}^{n}\Big((N_{1l}^q+F_l^q)u_l|_{\partial\Omega}+$$

$$\frac{1}{2!}\sum_{p=1}^{n}(N_{1lp}^{q}+F_{lp}^{q})u_l|_{\partial\Omega}u_p|_{\partial\Omega}\Bigg)\Bigg)-$$

$$\sum_{s=0}^{\infty}\sum_{r=1}^{n}\Bigg(\left(N_{ir}^{s}+\delta_i^1F_r^s\right)\left(\sum_{j=0}^{\infty}\frac{g_{rj}}{(j+1)(s+j+2)}\xi^{s+j+2}-\right.$$

$$\sum_{t=0}^{\infty}\frac{\xi^{s+t+2}}{(t+1)(s+t+2)}\sum_{l=1}^{n}((N_{rl}^{t}+\delta_r^1F_l^t)u_l|_{\partial\Omega}+$$

$$\frac{1}{2!}\sum_{q=1}^{n}\left(N_{rlq}^{t}+\delta_r^1F_{lq}^{t}\right)u_l|_{\delta\Omega}u_q|_{\partial\Omega}+\ \dots\ \Bigg)\Bigg)+$$

$$\frac{1}{2!}\sum_{p=1}^{n}(N_{irp}^{s}+\delta_i^1F_{rp}^{s})\left(u_p\Big|_{\partial\Omega}\left(\sum_{j=0}^{\infty}\frac{g_{rj}}{(j+1)(s+j+2)}\xi^{s+j+2}+\right.\right.$$

$$\sum_{t=0}^{\infty}\frac{\xi^{s+t+2}}{(t+1)(s+t+2)}\sum_{l=1}^{n}((N_{rl}^{t}+\delta_r^1F_l^t)u_l|_{\partial\Omega}+$$

$$\frac{1}{2!}\sum_{q=1}^{n}\left(N_{rlq}^{t}+\delta_r^1F_{lq}^{t}\right)u_l|_{\delta\Omega}u_q|_{\partial\Omega}\Bigg)\Bigg)+$$

$$u_r|_{\partial\Omega}\left(\sum_{j=0}^{\infty}\frac{g_{pj}}{(j+1)(s+j+2)}\xi^{s+j+2}-\right.$$

$$\sum_{t=0}^{\infty}\frac{\xi^{s+t+2}}{(t+1)(s+t+2)}\sum_{l=1}^{n}((N_{pl}^{t}+\delta_p^1F_l^t)u_l|_{\partial\Omega}+$$

$$\frac{1}{2!}\sum_{q=1}^{n}\left(N_{plq}^{t}+\delta_p^1F_{lq}^{t}\right)u_l|_{\delta\Omega}u_q|_{\partial\Omega}\Bigg)\Bigg)\Bigg)+\ \dots\ \Bigg)+\ \dots,\qquad i=\overline{1,n},$$

... .

Summation of the coefficients standing by powers of ξ yields for the ADM:

$$u_i=\xi^0([u_i|_{\partial\Omega}]+[0]+[0]+\ \dots\ )+ \tag{2.512}$$

$$\xi^1\left([g_{i0}]+\left[-\left(\sum_{j=1}^{n}\left(N_{ij}^{0}+\frac{1}{2!}\sum_{p=1}^{n}N_{ijp}^{0}u_j|_{\partial\Omega}u_p|_{\partial\Omega}\right)+\ \dots\ \right)\right]+0+\ \dots\ \right)+$$

$$\xi^2\left(\left[\frac{g_{i1}}{2}\right]+\left[-\frac{1}{2!}\sum_{r=1}^{n}\left(\left(N_{ir}^{1}u_r|_{\partial\Omega}+\frac{1}{2!}\sum_{p=1}^{n}N_{irp}^{1}u_r|_{\partial\Omega}u_p|_{\partial\Omega}+\ \dots\ \right)-\right.\right.\right.$$

$$\left(N_{ir}^{0}\frac{g_{r0}}{2!}+\frac{1}{2!}\sum_{p=1}^{n}N_{irp}^{0}\left(\frac{g_{r0}u_p|_{\partial\Omega}+g_{p0}u_r|_{\partial\Omega}}{2!}+\right)+\ \dots\ \right)\Bigg]+\ \dots\ \Bigg)+\ \dots,$$

$$i=\overline{1,n},$$

and for the HPM

$$u_1=\xi^0([u_1|_{\partial\Omega}]+[0]+[0]+\ \dots\ )+$$

$$\xi^1\Bigg([0]+\Bigg[g_{10}-\sum_{r=1}^{n}((N_{1r}^{0}+F_r^0)u_r|_{\partial\Omega}+$$

$$\frac{1}{2!}\sum_{p=1}^{n}(N_{1rp}^{0}+F_{rp}^{0})u_r|_{\partial\Omega}u_p|_{\partial\Omega}+\ \dots\ \Bigg)\Bigg]+ \tag{2.513}$$

$$\Bigg[-\left(F_0^0+F_{00}^0\sum_{p=1}^{\infty}u_p|_{\partial\Omega}\right)\Bigg(g_{10}-\sum_{r=1}^{n}\left(\left(N_{1r}^{0}+F_r^0\right)u_r|_{\partial\Omega}\ +\right.$$

$$\frac{1}{2!}\sum_{p=1}^{n}(N_{1rp}^{0}+F_{rp}^{0})u_r|_{\partial\Omega}u_p|_{\partial\Omega}+\ \dots\ \Bigg)\Bigg)\Bigg]+\ \dots\ \Bigg)+$$

$$\xi^2\Bigg([0]+\Bigg[\frac{g_{i1}}{2!}-\frac{1}{2}\sum_{r=1}^{n}((N_{1r}^{1}+F_r^1)u_r|_{\partial\Omega}+$$

$$\frac{1}{2!}\sum_{p=1}^{n}(N_{1rp}^{1}+F_{rp}^{1})u_r|_{\partial\Omega}u_p|_{\partial\Omega}+\ \dots\ \Bigg)\Bigg]+$$

$$\Bigg[-\left(F_0^1+F_{00}^1\sum_{p=1}^{\infty}u_p|_{\partial\Omega}\right)\Bigg(g_{10}-\sum_{l=1}^{n}\left(\left(N_{1l}^{0}+F_l^0\right)u_l|_{\partial\Omega}\ +\right.$$

$$\frac{1}{2!}\sum_{p=1}^{n}(N_{1lp}^{0}+F_{lp}^{0})u_l|_{\partial\Omega}u_p|_{\partial\Omega}+\ \dots\ \Bigg)\Bigg)-$$

$$\left(F_0^0+F_{00}^0\sum_{p=1}^{\infty}u_p|_{\partial\Omega}\right)\Bigg(\frac{g_{11}}{2}-\frac{1}{2!}\sum_{l=1}^{n}\left(\left(N_{1l}^{1}+F_l^1\right)u_l|_{\partial\Omega}\ +\right.$$

$$\frac{1}{2!}\sum_{p=1}^{n}(N_{1lp}^{1}+F_{lp}^{1})u_l|_{\partial\Omega}u_p|_{\partial\Omega}+\ \dots\ \Bigg)\Bigg)+\sum_{r=1}^{n}\Bigg(\left(N_{1r}^{0}+F_r^0\right)\Bigg(\sum_{j=0}^{\infty}\frac{g_{rj}}{2}-$$

$$\frac{1}{2}\sum_{l=1}^{n}\Bigg(\left(N_{1l}^{0}+F_l^0\right)u_l|_{\partial\Omega}+\frac{1}{2!}\sum_{q=1}^{n}(N_{1lq}^{0}+F_{lq}^{0})u_l|_{\partial\Omega}u_q|_{\partial\Omega}+\ \dots\ \Bigg)\Bigg)+$$

$$\frac{1}{2!}\sum_{p=1}^{n}(N^0_{1rp}+F^0_{rp})\left(u_p|_{\partial\Omega}\left(\frac{g_{r0}}{2}-\frac{1}{2}\sum_{l=1}^{n}\left(\left(N^0_{rl}+F^0_{l}\right)u_l|_{\partial\Omega}+\right.\right.\right.$$

$$\left.\left.\left.\frac{1}{2!}\sum_{q=1}^{n}\left(N^0_{rlq}+F^0_{lq}\right)u_l|_{\partial\Omega}u_q|_{\partial\Omega}\right)\right)+u_r|_{\partial\Omega}\left(\frac{g_{p0}}{2}-\frac{1}{2}\sum_{l=1}^{n}\left(\left(N^0_{pl}+F^0_{l}\right)u_l|_{\partial\Omega}+\right.\right.$$

$$\left.\left.\left.\frac{1}{2!}\sum_{q=1}^{n}(N^0_{plq}+F^0_{lq})u_l|_{\partial\Omega}u_q|_{\partial\Omega}\right)\right)\right)+\ \dots\ \Bigg]\Bigg)+\ \dots,$$

$$u_i=\xi^0([u_i|_{\partial\Omega}]+[0]+[0]+\ \dots\ )+$$

$$\xi^1\left([0]+\left[g_{i0}-\sum_{r=1}^{n}\left(N^0_{ir}u_r|_{\partial\Omega}+\frac{1}{2!}\sum_{p=1}^{n}N^0_{irp}u_r|_{\partial\Omega}u_p|_{\partial\Omega}+\ \dots\right)\right]+[0]+\ \dots\ \right)+$$

$$\xi^2\left([0]+\left[\frac{g_{i1}}{2}-\frac{1}{2}\sum_{r=1}^{n}\left(N^1_{ir}u_r|_{\partial\Omega}+\frac{1}{2!}\sum_{p=1}^{n}N^1_{irp}u_r|_{\partial\Omega}u_p|_{\partial\Omega}+\ \dots\right)\right]+\right.$$

$$\left[-\sum_{r=1}^{n}\left(N^0_{ir}\left(\frac{g_{r0}}{2}-\frac{1}{2}\sum_{l=1}^{n}\left(N^0_{rl}u_l|_{\partial\Omega}+\frac{1}{2!}\sum_{q=1}^{n}N^0_{rlq}u_l|_{\partial\Omega}u_q|_{\partial\Omega}+\ \dots\right)\right)+\right.\right.$$

$$\frac{1}{2!}\sum_{p=1}^{n}N^0_{irp}\left(u_p|_{\partial\Omega}\left(\frac{g_{r0}}{2}-\frac{1}{2}\sum_{l=1}^{n}\left(N^0_{rl}u_l|_{\partial\Omega}+\frac{1}{2!}\sum_{q=1}^{n}N^0_{rlq}u_l|_{\partial\Omega}u_q|_{\partial\Omega}\right)\right)+\right.$$

$$\left.\left.\left.u_r|_{\partial\Omega}\left(\frac{g_{p0}}{2}-\frac{1}{2}\sum_{l=1}^{n}\left(N^0_{pl}u_l|_{\partial\Omega}+\frac{1}{2!}\sum_{q=1}^{n}N^0_{plq}u_l|_{\partial\Omega}u_q|_{\partial\Omega}\right)\right)\right)+\right)\right]+\ \dots\ \right)+$$

$$+\ \dots,\qquad i-\overline{2,n}.$$

Square brackets include expressions corresponding to subsequent approximations regarding powers of the parameter.

Analysis of relations between solutions yielded by the ADM and HPM is explicitly defined by the relation of the nonlinear operator of the studied ODEs in both normal and general forms. Although in general this approach seems to be difficult, but assuming that the considerations will be limited to the second order terms in the series (2.503), the normal form of the system (2.500) can be recast in the following way:

$$\begin{gathered}Lu_1+\sum_{j=1}^{n}\left(\frac{N_{1j}+F_j}{1+F_0}u_j+\right.\\ \left.\frac{1}{2}\sum_{p=1}^{n}\frac{(N_{1jp}+F_{jp})(1+F_0)-F_{0p}(N_{1j}+F_j)-F_{0j}(N_{1p}+F_p)}{2(1+F_0)^2}u_ju_p\right)=\frac{g_1}{1+F_0},\\ Lu_i+R_i(u_1,\dots,u_n)+N_i(u_1,\dots,u_n)=g_i,\quad i=\overline{2,n}.\end{gathered}\tag{2.514}$$

In other words, the application of the ADM to Equations (2.514) is equivalent to application of the HPM to Equations (2.500) taking into account terms of less than the second order in Equations (2.503).

There are also a few particular cases worthy to be studied from an application point of view. In what follows for a nonlinear equation having a real small parameter ε standing by a higher derivative $F_0 = F_0^0 = \varepsilon - 1 = const, F_j \equiv F_{jp} \equiv 0, j, p = \overline{1, n}$, formulas (2.513)–(2.514) yield:

$$Lu_1 + \sum_{j=1}^{n}\left(\frac{N_{1j}}{\varepsilon}u_j + \frac{1}{2}\sum_{p=1}^{n}\frac{N_{1jp}}{2\varepsilon}u_j u_p\right) = \frac{g_1}{\varepsilon}, \tag{2.515}$$

$$\begin{aligned}
u_1 = {} & \xi^0 u_1|_{\partial\Omega} + \xi^1\left(\left(g_{10} - \sum_{r=1}^{n}\left(N_{1r}^0 u_r|_{\partial\Omega} + \frac{1}{2!}\sum_{p=1}^{n}N_{1rp}^0 u_r|_{\partial\Omega}u_p|_{\partial\Omega} + \cdots\right) - \right.\right.\\
& (\varepsilon - 1)\left(g_{10} - \sum_{r=1}^{n}\left(N_{1r}^0 u_r|_{\partial\Omega} + \frac{1}{2!}\sum_{p=1}^{n}N_{1rp}^0 u_r|_{\partial\Omega}u_p|_{\partial\Omega} + \cdots\right)\right) + \cdots\Bigg) +\\
& \xi^2\left(\left(\frac{g_{i1}}{2!} - \frac{1}{2}\sum_{r=1}^{n}\left(N_{1r}^1 u_r|_{\partial\Omega} + \frac{1}{2!}\sum_{p=1}^{n}N_{1rp}^1 u_r|_{\partial\Omega}u_p|_{\partial\Omega} + \cdots\right) - \right.\right.\\
& (\varepsilon - 1)\left(\frac{g_{11}}{2} - \frac{1}{2!}\sum_{l=1}^{n}\left(N_{1l}^1 u_l|_{\partial\Omega} + \frac{1}{2!}\sum_{p=1}^{n}N_{1lp}^0 u_l|_{\partial\Omega}u_p|_{\partial\Omega} + \cdots\right)\right) - \\
& \sum_{r=1}^{n}\left(N_{1r}^0\left(\sum_{j=0}^{\infty}\frac{g_{rj}}{2} - \frac{1}{2!}\sum_{l=1}^{n}\left(N_{1l}^0 u_l|_{\partial\Omega} + \frac{1}{2!}\sum_{q=1}^{n}N_{1lq}^0 u_l|_{\partial\Omega}u_q|_{\partial\Omega} + \cdots\right)\right)\right) +\\
& \frac{1}{2!}\sum_{p=1}^{n}N_{1rp}^0\left(u_p|_{\partial\Omega}\left(\frac{g_{r0}}{2} - \frac{1}{2}\sum_{l=1}^{n}\left(N_{rl}^0 u_l|_{\partial\Omega} + \frac{1}{2!}\sum_{q=1}^{n}N_{rlq}^0 u_l|_{\partial\Omega}u_q|_{\partial\Omega}\right)\right) +\right.\\
& \left. u_r|_{\partial\Omega}\left(\frac{g_{p0}}{2} - \frac{1}{2}\sum_{l=1}^{n}\left(N_{pl}^0 u_l|_{\partial\Omega} + \frac{1}{2!}\sum_{q=1}^{n}N_{plq}^0 u_l|_{\partial\Omega}u_q|_{\partial\Omega}\right)\right)\right)\Bigg) + \cdots\Bigg) + \cdots,
\end{aligned}$$

In words, the coefficients standing by the powers of the independent variable in the HPM solution represent the development of the ADM coefficients with respect to the real small parameter in the vicinity of one.

Let us now apply PA. Relations (2.512)–(2.513) show that if the equation is solvable regarding its higher derivative, then the coefficients standing by the same powers of the variable ξ of both ADM and HPM tend to each other and overlap with increase of the approximation order. It has been shown in reference [16], if the ADM is convergent to the development of the real solution into the Taylor series in its domain of holomorphicity in the neighborhood of $\xi = 0$, then those properties will also be exhibited by a solution of the HPM for the solvable equation. This allows us to apply to the obtained approximations the meromorphic continuation in the PA form [18]. In the case of ADM this type of continuation procedure has been proposed by many researchers [16] and has been called the modified Adomian method supplemented by the PA (MHPM-Padé). One may also apply the PA in the case of the HPM through its modification method of the nonlinear terms into series regarding the independent variable and the functions being sought (MHPM-Padé).

It seems, however, that more perspectives are associated with application of two-dimensional PA in the form proposed by V. Vavilov [86]. The latter approach allows us to define the approximation coefficients and a set of coefficients of the two-dimensional Taylor series uniquely, which are further used in order to construct approximations, as well as which, this approach guarantees the PA optimality in the sense of the Montessus de Ballore theorem. This is matched with the input requirements of the 2D approximations satisfying their transformation into 1D approximation, for the case in which the second variable is equal zero [86]. Furthermore, in order to apply the continuation procedure, one requires applicability of this transition for the parameter equal one. This can be done via the construction described so far of the 2D of PA regarding the transformed parameter mapping one into zero.

The method proposed by us coincides with the HPM for $F \equiv 0$, and with the ADM for $g \equiv 0$, and hence generalizes both approaches. It should be emphasized that the method does not require introducing into the equations the probe functions satisfying the BCs. This is carried out in a natural way in the process of construction of the successive approximations and allows us to solve the problem even for complicated BCs [16].

In order to apply the so far described method, we introduce the parameter ε in the following way:

$$u_i = \sum_{j=0}^{\infty} u_{ij}^M \varepsilon^j,$$

$$Lu_i = \varepsilon(g_i - R_i(u_1, \dots, u_n) - N_i(u_1, \dots, u_n)), \quad i = \overline{1, n},$$

$$G_j(u_1|_{\partial\Omega}, \dots, u_n|_{\partial\Omega})|_{\partial\Omega} = 0, \quad j = \overline{1, n}.$$

Applying series (2.505) and equating to the zero coefficients standing by the same powers of ε one obtains:

$$\varepsilon^0: \qquad Lu_{i0}^M = 0, \quad G_i(u_{10}^M, \dots, u_{n0}^M)|_{\partial\Omega} = 0 \Rightarrow u_{i0}^M = u_i|_{\partial\Omega}, \quad i = \overline{1, n},$$

$$\varepsilon^1: \qquad Lu_{i1}^M - g_i + \sum_{r=1}^{n}\left(N_{ir}u_{r0}^M + \frac{1}{2!}\sum_{p=1}^{n} N_{irp}u_{r0}^M u_{p0}^M + \dots\right) = 0,$$

$$u_{i1}^H|_{\partial\Omega} = 0 \Rightarrow u_{i1}^M = \sum_{j=0}^{\infty}\frac{\xi^{j+1}}{(j+1)}\left(g_{ij} - \sum_{r=1}^{n}\left(N_{ir}^j u_r|_{\partial\Omega} + \right.\right.$$

$$\left.\left.\frac{1}{2!}\sum_{p=1}^{n} N_{irp}^j u_r|_{\partial\Omega} u_p|_{\partial\Omega} + \dots\right)\right), \quad i = \overline{1, n}, \tag{2.516}$$

$$\varepsilon^2: \qquad Lu_{i2}^M + \sum_{r=1}^{n}\left(N_{ir}u_{r1}^M + \frac{1}{2!}\sum_{p=1}^{n} N_{irp}\left(u_{r1}^M u_{p0}^M + u_{r0}^M u_{p1}^M\right) + \dots\right) = 0,$$

$$u_{i2}^M|_{\partial\Omega} = 0 \Rightarrow u_{i2}^M = -\sum_{s=0}^{\infty}\sum_{r=1}^{n}\left(N_{ir}^s\left(\sum_{j=0}^{\infty}\frac{\xi^{s+j+2}}{(j+1)(s+j+2)}(g_{rj} - \right.\right.$$

$$\sum_{l=1}^{n}\left(N_{rl}^{j}u_l|_{\partial\Omega}+\frac{1}{2!}\sum_{q=1}^{n}N_{rlq}^{j}u_l|_{\partial\Omega}u_q|_{\partial\Omega}+\ \dots\ \right)\Bigg)\Bigg)+$$

$$\frac{1}{2!}\sum_{p=1}^{n}N_{irp}^{s}\left(u_p|_{\partial\Omega}\left(\sum_{j=0}^{\infty}\frac{\xi^{s+j+2}}{(j+1)(s+j+2)}\right.\right.(g_{rj}-$$

$$\sum_{l=1}^{n}\left(N_{rl}^{j}u_l|_{\partial\Omega}+\frac{1}{2!}\sum_{q=1}^{n}N_{rlq}^{j}u_l|_{\partial\Omega}u_q|_{\partial\Omega}+\ \dots\ \right)\Bigg)\Bigg)+$$

$$u_r|_{\partial\Omega}\left(\sum_{j=0}^{\infty}\frac{\xi^{s+j+2}}{(j+1)(s+j+2)}\left(g_{pj}-\sum_{l=1}^{n}\left(N_{pl}^{j}u_l|_{\partial\Omega}+\right.\right.\right.$$

$$\frac{1}{2!}\sum_{q=1}^{n}N_{plq}^{j}u_l|_{\partial\Omega}u_q|_{\partial\Omega}\Bigg)\Bigg)\Bigg)+\ \dots\ \Bigg)\Bigg)+\ \dots,\quad i=\overline{1,n},$$

$$\dots.$$

Summing up coefficients by powers of the independent variables yields

$$u_i=\xi^0([u_i|_{\partial\Omega}]+[0]+[0]+\ \dots\ )+ \tag{2.517}$$

$$\xi^1\left([0]+\left[g_{i0}-\sum_{r=1}^{n}\left(N_{ir}^{0}u_r|_{\partial\Omega}+\frac{1}{2!}\sum_{p=1}^{n}N_{irp}^{0}u_r|_{\partial\Omega}u_p|_{\partial\Omega}+\ \dots\ \right)\right]+\right.$$

$$\left.[0]+\ \dots\ \right)+\xi^2\left([0]+\left[\frac{g_{i1}}{2}-\frac{1}{2}\sum_{r=1}^{n}(N_{ir}^{1}u_r|_{\partial\Omega}+\right.\right.$$

$$\left.\frac{1}{2!}\sum_{p=1}^{n}N_{irp}^{1}u_r|_{\partial\Omega}u_p|_{\partial\Omega}+\ \dots\ \right)\Bigg]+\left[-\sum_{r=1}^{n}\left(N_{ir}^{0}\left(\frac{g_{r0}}{2}-\frac{1}{2}\sum_{l=1}^{n}(N_{rl}^{0}u_l|_{\partial\Omega}+\right.\right.\right.$$

$$\frac{1}{2!}\sum_{q=1}^{n}N_{rlq}^{0}u_l|_{\partial\Omega}u_q|_{\partial\Omega}+\ \dots\ \Bigg)\Bigg)+\frac{1}{2!}\sum_{p=1}^{n}N_{irp}^{0}\left(u_p|_{\partial\Omega}\left(\frac{g_{r0}}{2}-\right.\right.$$

$$\frac{1}{2}\sum_{l=1}^{n}\left(N_{rl}^{0}u_l|_{\partial\Omega}+\frac{1}{2!}\sum_{q=1}^{n}N_{rlq}^{0}u_l|_{\partial\Omega}u_q|_{\partial\Omega}\right)\Bigg)+u_r|_{\partial\Omega}\left(\frac{g_{p0}}{2}-\right.$$

$$\frac{1}{2}\sum_{l=1}^{n}\left(N_{pl}^{0}u_l|_{\partial\Omega}+\frac{1}{2!}\sum_{q=1}^{n}N_{plq}^{0}u_l|_{\partial\Omega}u_q|_{\partial\Omega}\right)\Bigg)\Bigg)+\ \dots\ \Bigg)\Bigg]+\ \dots\ \Bigg)+\ \dots,$$

$$i=\overline{1,n}.$$

It has been shown in [16] that approximation yielded by the proposed method is equivalent to the development of the exact solution into the Taylor series in the neighborhood of zero and allows us to carry out the analytical continuation into its meromorphic space using PA. Analysis of the obtained approximation allows us to conclude that this approximation, in contrary to the ADM and HPM, yields coefficients standing by powers of the independent variable up to the order of the approximation order keeping the accuracy of development into a series of the sought functions. This guarantees a stability of computations with limitation of the approximation order regarding the independent variable.

Geometrically nonlinear static equations governing the behavior of thin-walled construction include products and squares of the sought functions and their derivatives [7]. In this case solution (2.461) takes the following form:

$$\begin{aligned}
u_i = {} & \xi^0 u_i|_{\partial\Omega} + \xi^1 \left(g_{i0} - \sum_{r=1}^{n} \left(N^0_{ir} u_r|_{\partial\Omega} + \frac{1}{2!} \sum_{p=1}^{n} N^0_{irp} u_r|_{\partial\Omega} u_p|_{\partial\Omega} \right)\right) + \\
& \xi^2 \left(\left(\frac{g_{i1}}{2} - \frac{1}{2} \sum_{r=1}^{n} \left(N^1_{ir} u_r|_{\partial\Omega} + \frac{1}{2!} \sum_{p=1}^{n} N^1_{irp} u_r|_{\partial\Omega} u_p|_{\partial\Omega} \right)\right) - \right. \\
& \sum_{r=1}^{n} \left(N^0_{ir} \left(\frac{g_{r0}}{2} - \frac{1}{2} \sum_{l=1}^{n} \left(N^0_{rl} u_l|_{\partial\Omega} + \frac{1}{2!} \sum_{q=1}^{n} N^0_{rlq} u_l|_{\partial\Omega} u_q|_{\partial\Omega} \right)\right) + \right. \\
& \frac{1}{2!} \sum_{p=1}^{n} N^0_{irp} \left(u_p|_{\partial\Omega} \left(\frac{g_{r0}}{2} - \frac{1}{2} \sum_{l=1}^{n} \left(N^0_{rl} u_l|_{\partial\Omega} + \frac{1}{2!} \sum_{q=1}^{n} N^0_{rlq} u_l|_{\partial\Omega} u_q|_{\partial\Omega} \right)\right) + \right. \\
& \left.\left.\left. u_r|_{\partial\Omega} \left(\frac{g_{p0}}{2} - \frac{1}{2} \sum_{l=1}^{n} \left(N^0_{pl} u_l|_{\partial\Omega} + \frac{1}{2!} \sum_{q=1}^{n} N^0_{plq} u_l|_{\partial\Omega} u_q|_{\partial\Omega} \right)\right)\right)\right)\right) + \ldots, \\
& i = \overline{1, n}.
\end{aligned} \tag{2.518}$$

Let us consider numerical aspects of the obtained approximations and their PA.

A typical behavior of the approximations in the case of the following initial value problem:

$$\varepsilon z' + z = 1, \qquad z(0) = 0.$$

ADM approximation describes the exact solution only in part of the interval compared with the real small parameter approach. In spite of that, the errors of solution obtained via the HPM are essentially of lower order than those introduced by the ADM. The latter one exhibits the boundary layer occurrence in the neighborhood of zero. On the contrary, the PA regarding the independent variable of the ADM approximation, when the proposed method is applied, gives reliable qualitative and quantitative results (Figure 2.67).

Analogous results are also given by analysis of initial value problem for the following equation with variable coefficients:

$$\varepsilon z' + xz = x, \qquad z(0) = 2,$$

and the obtained results are reported in Figure 2.68.

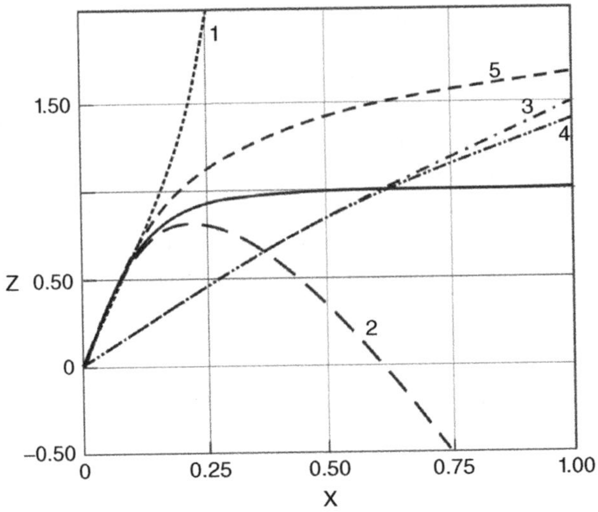

Figure 2.67 The exact solution (solid line) of Equation (2.477) when $\varepsilon = 0.1$ and approximate solutions (1−three terms ADM, 2−z_{ε_1} for ADM, 3−three terms HPM, 4−z_x for HPM, 5−2D Padé for MMPC, ADM and HPM)

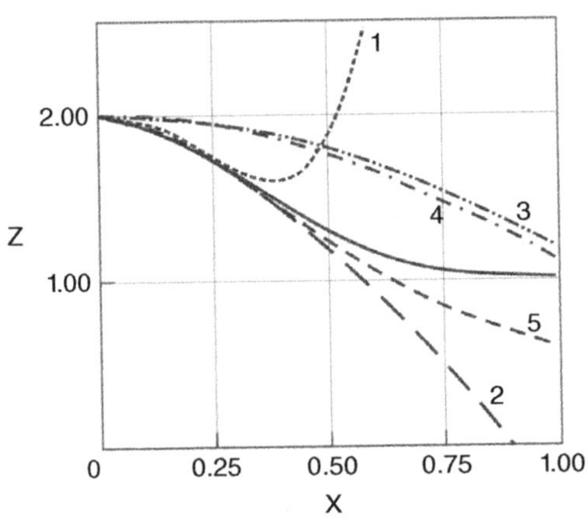

Figure 2.68 The exact solution (solid line) of Equation (2.483) when $\varepsilon = 0.2$ and approximate solutions (1 – three terms ADM, 2 – z_{ε_1} for ADM, 3 – three terms HPM, 4 – z_x for HPM, 5 – 2D Padé for MMPC, ADM and HPM)

Figure 2.69 presents graphs of approximation for the essentially nonlinear problem governed by the following initial value problem:

$$z' = -\varepsilon_1 \frac{z^2}{\varepsilon} + \frac{x-1}{\varepsilon}, \qquad z(0) = 1.$$

One can see that approximations of the HPM and MHPM-Padé describe relatively well the curve in average and rather badly in the case of the boundary layer. On the contrary, the ADM

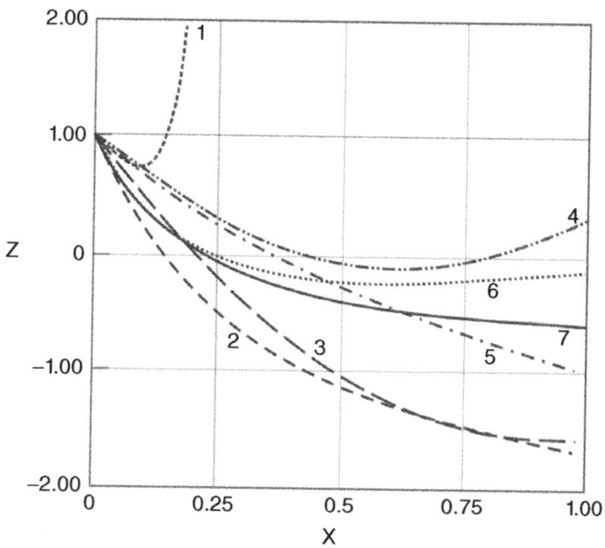

Figure 2.69 Approximate solutions of Equation (2.489) when $\varepsilon = 0.2$ (1 – three terms ADM, 2 – z_{ε_1} for ADM, 3 – z_x for ADM, 4 – three terms HPM, 5 – z_x for HPM, 6 – z_{ε_1} for HPM and MMPC, 7 – z_x and 2D Padé for MMPC)

and MADM-Padé approximations well coincide with the solution behavior in the vicinity of zero, and rather badly on its stationary part. However, the PA results obtained on a basis of the proposed method are in good agreement with the exact solution in the whole studied interval.

2.9 Conclusions

Let us finish our book with comments related to (i) advantages and (ii) disadvantages of the asymptotic methods.

(i.1.) Essential simplification of a solution, which in many cases can be found analytically.

(i.2.) In general, asymptotic methods are rather easily matched with computational approaches including numerical and variational ones. Namely, after a simplification of the studied BVP and after separation of its key features, one may apply a wide spectrum of the numerical methods. It should be emphasized that the asymptotic methods allow detecting main properties of the solution, and hence yield a proper recommendation regarding a choice of the approximating functions associated with the Bubnov-Galerkin, Rayleigh-Ritz, Trefftz, Kantonorovich, etc. approaches. Furthermore, sometimes a zeroth order asymptotic approximation can be directly applied in other iterational processes, such as for instance in the case of the Newton-Kantorovich method.

(i.3.) Since the asymptotic methods are tightly linked with the physical interpretation of a studied problem, they usually give the full picture of its behavior and allow for a deep understanding.

(i.4.) Asymptotic methods allow clarifying both mathematical and physical backgrounds of the engineering oriented procedures and approaches, as well as usually improving the accuracy, validity and reliability of the obtained solutions.

(i.5.) Asymptotic methods are universal ones, since in many cases they allow us to use one general approach to sometimes different, for a first glance, problems and exhibit their hidden universal property.

(ii.1.) In many cases the first approximating solution does not guarantee an expected accuracy.

(ii.2.) Construction of higher order approximations sometimes requires hard computational effort.

(ii.3.) Estimation of both the accuracy of asymptotic methods and intervals of their application does not belong to trivial tasks.

Finally, it seems that the further development of asymptotic methods is associated with the use of numerical-asymptotic approaches. Namely, a smooth solution part can be monitored numerically, whereas boundary layers can be constructed with the help of an asymptotic approach.

As for the HPM, one can conclude the following. HPM has often been applied recently [43], but it suffers for the lack of any rigorous treatment. We are aiming at an extension of the main concept of this method into effective solutions of complicated problems of mechanics. We have shown that matching of the HPM with PA allows to solve problems of natural, free and forced vibrations, as well as the SSS and stability of plates with mixed BCs in an analytical way.

In the case of nonlinear problems the proposed approach allows for the effective investigation of the infinite systems of nonlinear either algebraic or differential equations. Possible generalizations are associated with the solution to the mixed problems of the theory of elasticity (in the 2D-case, there are already promising results [16]) as well as mixed problems of the theory of shells. The proposed way of modification and validity extension also concerns the Ishlinskii-Leibenzon method applied in the theory of stability, numerous methods devoted to investigation of dynamic stability problems and the SSS of nonhomogenous constructions, as well as series of nonlinear problems, including those with nonlinear BCs.

References

[1] Abbaoui K., Cherruault Y.: Convergence of Adomian's method applied to differential equations. *Comp. Math. Appl.*, 28, 103–9 (1994).

[2] Abramowitz M., Stegun I.A.: *Handbook of Mathematical Functions, with Formulas, Graphs, and Mathematical Tables*. Dover Publications, New York, 1965.

[3] Adomian G.: A review of the decomposition method and some recent results for nonlinear equations. *Comp. Math. Appl.*, 21, 101–27 (1989).

[4] Aleksandrov V.M., Zelentsov V.B.: Dynamic problems on the bending of a rectangular plate with mixed fixing conditions on the outline. *J. Appl. Math. Mech.*, 43, 124–31 (1979).

[5] Ambartsumian S.A.: *Theory of the Anisotropic Shells*. World Scientific, Singapore, 1991.

[6] Andersen C.M., Geer J.F.: Power series expansion for the frequency and period of the limit cycle of the Van der Pol equation. *SIAM J. Appl. Math.*, 42(3), 678–93 (1982).

[7] Andrianov I.V., Awrejcewicz J., Manevitch L.I.: *Asymptotical Mechanics of Thin-Walled Structures: A Handbook*. Springer-Verlag, Berlin, 2004.

[8] Andrianov I.V., Danishevskyy V.: Asymptotic investigation of the nonlinear dynamic boundary value problem for rod. *Technische Mechanik*, 15(1), 53–5 (1995).

[9] Andrianov I.V., Danishevskyy V.: Asymptotic approach for nonlinear periodical vibrations of continuous structures. *J. Sound Vibr.*, 249(3), 465–81 (2002).

[10] Andrianov I.V., Ivankov A.O., Kurpa L.V.: Analytical and numerical investigations of stability of plates with mixed boundary conditions. *Izvestya VUZov. Civil Engineering and Architecture*, 8, 25–9 (1990) (in Russian).

[11] Andrianov I.V., Kholod E.G.: Intermediate asympotics in the nonlinear dynamics of shells. *Mech. Solids*, 28(2), 160–5 (1993).

[12] Andrianov I.V., Kholod E.G.: Non-linear free vibration of shallow cylindrical shell by Bolotin's asymptotic method. *J. Sound. Vibr.*, 160(1), 594–603 (1993).

[13] Andrianov I.V., Krizhevsky G.A.: Investigation of natural vibrations of circular and sector plates with consideration of geometric nonlinearity. *Mech. Solids*, 26(2), 143–8 (1991).

[14] Andrianov I.V., Krizhevsky G.A.: Free vibration analysis of rectangular plates with structural inhomogenity. *J. Sound Vibr.*, 162(2), 231–41 (1993).

[15] Andrianov I.V., Lesnichaya V.A., Manevitch L.I.: *Homogenization Methods in Statics and Dynamics of Ribbed Shells*. Nauka, Moscow, 1985 (in Russian).

[16] Andrianov I.V., Olevskii V.I., Tokarzewski S.: A modified Adomian's decomposition method. *J. Appl. Math. Mech.*, 62(2), 309–14 (1998).

[17] Andrianov I.V., Samoilenko O.G.: On the Ishlinskii–Leibenzon method in the theory of elastic stability. *Mech. Solids*, 36, 113–19 (2001).

[18] Baker G.A., Graves-Morris P.: *Padé Approximants*. Cambridge University Press, Cambridge, New York, 1996.

[19] Bartlett C.C.: The vibration and buckling of a circular plate clamped on part of its boundary and simply supported on the remainder. *Quart. J. Mech. Appl. Math.*, 16, 431–40 (1963).

[20] Birger I.L., Panovko Y.G. (Editors): *Strength, Stability, Vibrations*. vol. 3, Mashinostroyenie, Moscow, 1968 (in Russian)

[21] Bolotin V.V.: An asymptotic method for the study of the problem of eigenvalues of rectangular regions. *Problems of Continuum Mechanics*, SIAM, 56–8 (1961).

[22] Bolotin V.V.: *Random Vibration of Elastic Systems*. Martinus Nijhoff Publ., Boston, 1984.

[23] Chen G., Coleman M. P., Zhou J.: The equivalence between the wave propagation method and Bolotin's method in the asymptotic estimation of eigenfrequencies of a rectangular plate. *Wave Motion*, 16(3), 285–97 (1992).

[24] Chen G., Zhou J.: *Vibration and Damping in Distributed Systems*, vol. I, II. CRS, Boca Raton, Fl, 1993.

[25] Cherruault Y., Adomian G.: Decomposition methods: a new proof of convergence. *Math. Comp. Modeling*, 18, 103–6 (1993).

[26] Chrysos M., Sanchez F., Cherruault Y.: Improvement of convergence of Adomian's method using Padé approximants. *Kybernetes*, 31(6), 884–95 (2002).

[27] Collatz L.: *Eigenwertaufgaben mit Technischen Anwendungen*. Geest and Portig, Lepizig, 1963.

[28] Cox S., Zuazua E.: The rate at which energy decays in a string damped at one end. *Indiana Univ. Math. J.*, 44, 545–73 (1995).

[29] Darmawijoyo, Van Horssen W.T.: On boundary damped for a weakly nonlinear wave equation. *Nonlinear Dynamics*, 30(2), 179–91 (2002).

[30] Darmawijoyo, Van Horssen W.T., Clement: Ph. On a Rayleigh wave equation with boundary damping. *Nonlinear Dynamics*, 33(4), 399–429 (2003).

[31] Darmawijoyo, Van Horssen W.T.: On the weakly damped vibrations of a string attached to a spring-mass-dashpot system. *J. Vibr. Control*, 9(11), 1231–48 (2003).

[32] Dorodnitsyn A.A.: Using of small parameter method for numerical solution of mathematical physics equations. *Num. Meth. Cont. Mech. Probl. Solving*, 85–101 (1969) (in Russian).

[33] Eastep F.E., Hemmig F.G.: Natural frequency of circular plates with partially free, partially clamped edges. *J. Sound Vibr.*, 84, 359–70 (1982).

[34] Elishakoff I.: Bolotin's dynamic edge effect method. *The Shock and Vibr. Digest*, 8(1), 95–104 (1976).

[35] Elishakoff I., Sternberg A., Van Baten T.J.: Vibrations of multispan all-round clamped stiffened plates by modified dynamic edge effect method. *Comput. Meth. Appl. Mech. Eng.*, 105, 211–23 (1993).

[36] Fan S.C., Chueng Y.K.: Flexural free vibrations of rectangular plates with complex supported conditions. *J. Sound Vibr.*, 93, 81–94 (1984).

[37] Glikman B.G.: Free vibrations of a circular plate with mixed boundary conditions. *Mech. Solids*, 7(1), 115–20 (1972).

[38] Gol'denveiser A.L., Kaplunov J.D., Nolde E.V.: On Timoshenko-Reissner type theories of plates and shells. *Int. J. Solids Struct.*, 30, 675–94 (1993).
[39] Gontkevich V.S.: *Natural Oscillations of Plates and Shells*. Naukova Dumka, Kiev, 1964 (in Russian).
[40] Grigolyuk E.I., Selezov I.T.: *Nonclassical Theories of Oscillations of Beams, Plates and Shells*. VINITI, Moscow, 1973 (in Russian).
[41] Hamada M., Inoue Y., Hashimoto H.: Buckling of simply supported partially clamped rectangular plates uniformly compressed in one direction. *Bull. Japan Soc. Mech. Eng.*, 10(37), 35–40 (1967).
[42] Hamada M., Inoue Y., Mizushima I., Mifune T.: A new approach to bending problems of circular plates with mixed boundary conditions. *Bull. Japan Soc. Mech. Eng.*, 29(258), 4059–63 (1986).
[43] He Ji-Huan: A new approach to nonlinear partial differential equations. *Com. Non. Sc. Num. Sim.*, 2(4), 230–5 (1997).
[44] He Ji-Huan.: A coupling method of homotopy technique and a perturbation technique for non-linear problems. *Int. J. Nonlin. Mech.*, 35, 37–43 (2000).
[45] Hirano Y., Okazaki K.: Vibration of circular plate having partly clamped or partly simply supported edges. *Bull. Japan Soc. Mech. Eng.*, 19, 610–18 (1976).
[46] Hsu Jung-Chang, Lai Hsin-Yi, Chen C.K.: Free vibration of non-uniform Euler-Bernoulli beams with general elastically end constraints using Adomian modified decomposition method. *J. Sound Vibr.*, 318, 965–81 (2008).
[47] Ishlinskii A.Yu.: Considering the stability of equilibrium of elastic bodies from the standpoint of the mathematical theory of elasticity. *Ukrainian Math. J.*, 6(2), 140–6 (1954) (in Russian).
[48] Kaplunov J.D., Kossovich L.Ya., Nolde E.V.: *Dynamics of Thin Walled Elastic Bodies*. CA: Academic Press, San Diego, 1998.
[49] Keer L.M., Stahl B.: Eigenvalue problems of rectangular plates with mixed boundary conditions. *J. Appl. Mech.*, 39, 513–20 (1972).
[50] Keller J.B., Rubinow S.I.: Asymptotic solution of eigenvalue problems. *Ann. Phys.*, 9(1), 24–75 (1960).
[51] Korenev B.G.: *Questions of Calculation of Beams and Plates on the Elastic Foundation*. Strogizdat, Moscow, 1964 (in Russian).
[52] Kublanovskaya V.N.: Application of analytic continuation by means of change of variables in numerical analysis. *Trudy Matematicheskogo Instituta im. Steklov*, 53, 145–85 (1959) (in Russian).
[53] Kurpa L.V.: R-Functions method in problems of natural oscillations of the plates with mixed boundary conditions. *Gidroaeromekhanika i Teoriya Uprugosti*, 31, 8–84 (1983) (in Russian).
[54] Laura P.A.A., Gutierrez R.H.: Analysis of vibrating rectangular plates with nonuniform boundary conditions by using the differential quadrature method. *J. Sound Vibr.*, 173, 702–6 (1994).
[55] Leibenzon L.S.: About using of harmonical functions to question about stability of spherical and cylindrical shells. *Collected Papers*, 1, 50–86 (1951) (in Russian).
[56] Marchenko V.A., Khruslov E.Ya.: *Homogenization of Partial Differential Equations*. Birkhäuser, Boston, 2005.
[57] Mizusawa T., Kaijita T.: Vibration and buckling of rectangular plates with non-uniform elastic constraints in rotation. *Int. J. Solid Struct.*, 23, 45–55 (1986).
[58] Mizusawa T., Leonard J.W.: Vibration and buckling of plates with mixed boundary conditions. *Eng. Struct.*, 12, 285–90 (1990).
[59] Movchan A.B., Movchan N.V.: *Mathematical Modelling of Solids with Nonregular Boundaries*. CRC Press, Boca Raton, 1995.
[60] Morgül Ö., Rao B.P., Conrad F.: On the stabilization of a cable with a tip mass. *IEEE Trans. Autom. Control.*, 39(10), 2140–5 (1994).
[61] Narita Y.: Application of a series-type method to vibration of orthotropic rectangular plates with mixed boundary conditions. *J. Sound Vibr.*, 77, 345–55 (1981).
[62] Nayfeh A.H.: *Introduction to Perturbation Techniques*. John Wiley and Sons, Inc., New York, 1981.
[63] Nayfeh A.H.: *Perturbation Methods*. John Wiley and Sons, Inc., New York, 2000.
[64] Nayfeh A.H., Balachandran B.: *Applied Nonlinear Dynamics*. John Wiley and Sons, Inc., New York, 1995.
[65] Ota T., Hamada M.: Fundamental frequencies of simply supported but partially clamped square plates. *Bull. Japan Soc. Mech. Eng.*, 6(23), 397–403 (1963).
[66] Pal'tsev B.V.: The expansion of solutions of Dirichlet's problem and a mixed problem for a biharmonic equation in a series of solutions of reducing problems. *USSR Comp. Math. Math. Phys.*, 6(1), 59–72 (1966).
[67] Pilipchuk V.N.: *Nonlinear Dynamics: Between Linear and Impact Limits*. Springer, New York, 2010.

[68] Polyanin A.D., Zaitsev V.F.: *Handbook of Exact Solutions for Ordinary Differential Equations*. 2nd Edition, Chapman and Hall/CRC, Boca Raton, 2003.

[69] Pritulo M.F.: On the determination of uniformly accurate solutions of differential equations by the method of perturbation of coordinates. *J. Appl. Math. Mech.*, 26(3), 661–7 (1962).

[70] Rao G.V., Raju I.S., Murthy T.V.G.K.: Vibration of rectangular plates with mixed boundary conditions. *J. Sound Vibr.*, 30, 257–60 (1973).

[71] Shu C., Wang C.M.: Treatment of mixed and nonuniform boundary conditions in GDQ vibration analysis of rectangular plates. *Eng. Struct.*, 21, 125–34 (1999).

[72] Smirnov V.A.: *Calculation of Complex Shape Plates*. Stroizdat, Moscow, 1978 (in Russian).

[73] Striz A.G., Chen W., Bert C.W.: Static analysis of structures by the quadrature element method (QEM). *Int. J. Solid Struct.*, 31, 2807–18 (1994).

[74] Szemplinska–Stupnicka W.: *The Behaviour of Nonlinear Vibrating System*. Kluwer Academic Publishers, Dordrecht, 1990.

[75] Timoshenko S.P., Gere J.M.: *Theory of Elastic Stability*. McGraw-Hill, New York, 1961.

[76] Timoshenko S.P., Woinowsky-Krieger S.: *Theory of Plates and Shells*. 2nd ed., McGraw-Hill, New York, 1987.

[77] Timoshenko S.P., Young D.H., Weaver W. Jr.: *Vibration Problems in Engineering*. John Wiley and Sons, Inc., New York, 1974.

[78] Tseitlin A.I.: On the methods of dual integral equations and dual series and their application to problems of mechanics. *J. Appl. Math. Mech.*, 30, 322–33 (1966).

[79] Tseitlin A.I.: *Applied Methods for Solving of Boundary Value Problems of Structural Mechanics*. Stroisdat, Moscow, 1984 (in Russian).

[80] Tseitlin A.I., Glikman B.T.: Bending of elastic rectangular plate with mixed boundary conditions. *Nauchnye Trudy ZNIISK. Issledovaniya po Dinamike Sooruzhenii*, 17, 123–42 (1971) (in Russian).

[81] Tseitlin A.I., Glikman B.T.: Oscillations of rectangular plate with mixed boundary conditions. *Issledovanija po Teorii Sooruzhenii*, 19, 185–94 (1972) (in Russian).

[82] Tseitlin A.I., Glikman B.T.: Kolebanija pologoy cilindricheskoy obolochki pri smeshanyh kraevyh uslovyah. *Nauchnye Trudy ZNIISK. Issledovaniya po Dinamike Sooruzhenii*, 43, 4–9 (1975) (in Russian).

[83] Uflyand Y.S.: *Integral Transforms in the Theory of Elasticity Problems*. AN SSSR, Moscow, 1963 (in Russian).

[84] Vainberg D.V., Vainberg E.D.: *Calculations of Plates*. Budivel'nik, Kiev, 1970 (in Russian).

[85] Varvak P.M.: *Handbook of Theory of Elasticity*. Budivelnik, Kiev, 1971 (in Russian).

[86] Vavilov V.V., Tchobanou M.K., Tchobanou P.M.: Design of multidimensional recursive systems through Padé type rational approximation. *Non. Anal., Model. Control.*, 7(1), 105–25 (2002).

[87] Vyatchin A.V.: On the convergence of Padé approximants. *Moscow Univ. Math. Bull.*, 37(4), 1–4 (1982).

[88] Wang X., Bert C.W.: A new approach in applying differential quadrature to static and free vibration analysis of beams and plates. *J. Sound Vibr.*, 162, 566–72 (1992).

[89] Wang H., Elcrat A.R., Egbert R.I.: Modeling and boundary control of conductor galloping. *J. Sound Vibr.*, 161, 301–15 (1993).

[90] Wasov W.: *Asymptotic Expansions for Ordinary Differential Equations*. John Wiley and Sons, Inc., New York, 1965.

[91] Wei G.W., Zhao Y.B., Xiang Y.: The determination of natural frequencies of rectangular plates with mixed boundary conditions by discrete singular convolution. *Int. J. Mech. Sc.*, 43, 1731–46 (2001).

[92] Zelentsov V.B.: Method of homogeneous solutions in problems of bending of rectilinear plates with a mixed attachment condition along the contour. *Mech. Solids*, 15(5), 110–16 (1980).

[93] Zorski H.: Plates with discontinuous support. *Arch. Mech. Stos.*, X(3), 301–13 (1958).

[94] Zorski H.: A semi-infinite strip with discontinuous boundary conditions. *Arch. Mech. Stos.*, X(3), 371–98 (1958).

Index

Asymptotic Methods in the Theory of Plates with Mixed Boundary Conditions, First Edition.
Igor V. Andrianov, Jan Awrejcewicz, Vladislav V. Danishevs'kyy, Andrey O. Ivankov.

Printed and bound by CPI Group (UK) Ltd, Croydon, CR0 4YY
17/09/2026
14964300-0001